Ibn al-Haytham: The *Optics*, Books IV–V

The Warburg Institute gratefully acknowledges
the generous financial assistance of the
Foundation for Science, Technology and Civilisation (UK)
towards the publication of this volume.

Warburg Institute Studies and Texts
Edited by Charles Burnett and Jill Kraye

8

THE *OPTICS* OF IBN AL-HAYTHAM

BOOKS IV–V

On Reflection and Images Seen by Reflection

translated from the Arabic by

A. I. SABRA
Professor Emeritus of the History of Arabic Science,
Harvard University

prepared for publication by

J. P. HOGENDIJK
Professor of the History of Mathematics,
Utrecht University

First published 2023 by
The Warburg Institute
School of Advanced Study, University of London
Woburn Square, London WC1H 0AB

A CIP catalogue record for this book is available from The British Library.

ISBN 9781908590589

Cover image: Ibn al-Haytham, *Optics*, Book V, Chapter 2, Diagram 43; MS Köprülü Library, no. 952, fol. 98a.

Cover design for University of London Press by Nicky Borowiec.
Text set by Gavin Ward Design Associates in Times New Roman.

in appreciation of

Kamāl al-Dīn al-Fārisī (d. 718 AH / 1319 AD)

and

Muṣṭafā Naẓīf (d. 1971 AD)

whose invaluable labours eased my task

A. I. S.

Table of Contents

Detailed Table of Contents for Book IV of the *Optics*

Detailed Table of Contents for Book V of the Optics

Acknowledgements

Professor A. I. Sabra devoted many years of his life to editing and translating the seven Arabic books of the *Optics* by Ibn al-Haytham (c.354–432 AH / c.965–1041 AD). He published an Arabic edition of Books I–III (1983), an English translation with commentary of Books I–III (1989), and an Arabic edition of Books IV–V (2002). He had already started working on the English translation of Books IV–V in the 1980s, and his translation of six lemmata in Book V appeared in 1982.

When Sabra passed away in 2013, the manuscript of his English translation of the complete text of Books IV–V was in near-final form. The opportunity to make the finishing touches was entrusted to Jan Hogendijk, who checked the entire translation against the Arabic edition and, where necessary, the Arabic manuscripts. A few inconsistencies in the translation and use of symbols were eliminated, and the English text was reviewed throughout. Sabra presented the geometrical arguments in the second half of Book V in a very large number of extremely short paragraphs in a difficult layout, using mathematical symbols; these short sections have been combined into larger paragraphs. Further, the mathematical reasonings in Book V have been checked, and the remaining queries left by Sabra in his translation have been resolved.

To his translation of Books IV and V Sabra added a table of contents and a summary overview, but no detailed commentary. In the meantime, an excellent edition of the medieval Latin version of Books IV and V with English translation and commentary was published by Professor A. Mark Smith (2006). The Latin version is more concise than the Arabic, but the two versions are largely equivalent with regard to content. Smith's commentary can therefore be consulted to cast light on Sabra's English translation of the Arabic. To facilitate the comparison, a detailed concordance between the English translations of the Arabic and Latin versions has been added at the end of the volume, so that the reader can quickly identify where Smith's commentary relates to any given part of Sabra's translation.

Book V of the *Optics* includes forty-seven geometrical diagrams, which Sabra carefully reconstructed from the (defective) manuscript tradition in the 1980s. Sabra's reconstructions appear in this translation, together with some extra diagrams that he added for clarification. For researchers who want to study these extremely interesting diagrams further, a table of references to the sources has been added at the end the volume.

While the lion's share of the work of preparing the manuscript for publication is due to Jan Hogendijk, earlier versions of Sabra's texts have benefited from the advice of Elaheh Kheirandish and Saira Malik, to both of whom we are very grateful. We wish also to thank Adam Sabra for preparing a computer-readable version of his father's work, Joseph Spooner for his attentive editorial work, and Gavin Ward for the complicated typesetting, as well as Emma Gallon and Jamie Bowman of the University of London Press for guiding this book through to publication.

Charles Burnett
Warburg Institute

INTRODUCTION

1. Books IV and V of Ibn al-Haytham's Arabic *Optics*: *Kitāb al-Manāẓir*

The present volume contains an English translation of Books IV and V of Ibn al-Haytham's *Optics* by the editor of the critical edition, published in 2002, of the Arabic text of the same two Books.[1] It is now generally assumed that a Latin version of the Seven Books (*maqālāt*) that make up the Arabic *Optics* was not contemplated as late after the eleventh century AD as was previously conjectured, an assumption supported by the fact that the Arabic text has recently been shown to have arrived in Spain close to the end of the eleventh century.[2]

2. The Extant Arabic Manuscripts Forming the Basis for the Arabic Edition and English Translation of Books IV and V

The edition and English translation of Books IV and V are based on three manuscripts, all of them housed in Istanbul libraries, as follows.

MS F. Fatih Library, no. 3215, 332 fols, dated Jumādā II, 636 AH / 1239 AD, 24 × 18.5cm, 13 lines per page

> The decorated title-page displays the names of people who previously owned and/or read the volume, a feature also found in the extant volumes of what we have called the 'ʿAskarī Set', named for the copyist Aḥmad ibn Muḥammad ibn Jaʿfar, known as 'al-ʿAskarī' for his place of origin (al-ʿAskar in Iraq) and said to be a relation of Ibn al-Haytham's by marriage (*ṣihr al-muṣannif*). Al-ʿAskarī transcribed the seven books of *Kitāb al-Manāẓir* in 476 AH / 1083–4 AD, though the set now lacks Books IV and V. The text of Books IV–V in MS F is written by the hand of a single, unnamed scribe. The manuscript consists of two parts, Book IV, fols 1a–137b, and Book V, fols 138a–332b.

MS K. Köprülü Library, no. 952, 135 fols, 30 × 22cm, 31 lines per page

> The manuscript is mostly written in a single, Maghribī hand, but also evinces other Maghribī hands in a small number of places. It is the only work of Ibn al-Haytham's that I have seen written in a Maghribī style. The manuscript is not complete and only contains substantial parts of Books IV–VI and VII of *Kitāb al-Manāẓir*. The folios are not all correctly arranged. Max Krause, the German scholar who first published a complete list of Istanbul manuscripts of Ibn al-Haytham's works, in 1936, was of the opinion that the Köprülü copy of *Kitāb al-Manāẓir* was executed in the fourteenth century AD; Ramadan Šešen thinks that it goes back to the thirteenth century. It would be interesting to know exactly where it was executed.[3]

---

1. Sabra, *Kitāb al-Manāẓir (the* Optics*)*.
2. Hogendijk, 'Al-Mu'taman's Simplified Lemmas'.
3. Krause, *Stanbuler Handschriften*, p. 476. For Šešen, see Sabra, 'The Commentary'.

MS S. Ayasofya, no. 2448, 677 fols, 869 AH / 1464–5 AD, 22 × 12.5cm, 23 lines per page

The manuscript contains all seven books of *Kitāb al-Manāẓir* and was produced by the order of Sulṭān Muḥammad al-Fātiḥ (the Conqueror) 'for the use of those who thought and contemplated [*li-man naẓara wa-ta'ammala*]'. The whole text is written in a clear *ta'līq* style and was copied from the 'Askarī Set (complemented by MSS F4 and F5). MS S has no geometrical diagrams for any of the propositions in Book V, as is also the case with MS F, from which S was copied.[4] For this reason, we did not record the variant readings of S in the apparatus to the Arabic text of Books IV and V, as they faithfully reproduce those of F. Book IV in MS S occupies fols 315b–386a, and Book V fols 386b–508a.

MS F, which contains Books IV and V with continuous foliation throughout, warrants detailed comment. In the introduction to his *Al-Ḥasan ibn al-Haytham: Buḥūthuhu wa Kushūfuhu al-Baṣariyya*,[5] Naẓīf observes that the text is copied directly from the fourth and fifth volumes of the original 'Askarī Set (made some forty-three years after Ibn al-Haytham's death), and that damage (perhaps caused by damp) to the two 'Askarī volumes had necessitated a new copy that would match, he suggests, the exemplars in the number of lines per page. This complete text of the whole *Kitāb al-Manāẓir*, the 'Askarī Set with the replacement copies of Books IV and V, was sold to Yaḥyā ibn Muḥammad ibn al-Lubūdī, whose name appears in full on all the volumes, including MS F. We should add here that some folios (and torn parts thereof) at the end of Book IV have been rearranged incorrectly.

MS F might recommend itself to an editor of Books IV–V on account of its being copied directly from the 'Askarī Set, which from its extant parts is known to be generally clear and reliable. Unfortunately, the manuscript's copyist is distinguished by two qualities that quickly manifest themselves on many pages, namely ignorance of (or lack of experience in) the subject matter, and carelessness. It is not that we expect the copyist of a scientific and mathematical book to be familiar with the requisite technical vocabulary, though this is often the case with medieval copyists of Arabic manuscripts in scientific fields. But we do not see in MS F any evidence for experience of well-known vocabulary, such as one might reasonably expect; this is clear from the many examples recorded in the critical apparatus to the Arabic text of of Books IV–V, and from the obvious lack of attention to the shapes and dots that define the meanings of letters of the alphabet when used in reference to geometrical diagrams. There are frequent examples of confusion between the letters *bā'/tā'/thā'*; *bā'/dāl*; *bā'/fā'*; *dāl/jīm*; *hā'/dāl*; *dāl/zāy*; *'ayn/ghayn*; *fā'/qāf*; *nūn/rā'/zāy*; and *sīn/shin*. We also regularly find the *shape* of *ḥā'* for *both ḥā'* and *jīm* (except in two or three places in the whole text). The reader can easily guess how difficult it would be to follow a mathematical text, like that copied in F, in the absence of all diagrams and of manuscripts other than MS F. I, for one, confess that I would not have been able to reconstruct several important parts of Book V on the basis of MS F alone.

4. Sabra, *Kitāb al-Manāẓir (the* Optics*)*, pp. 41–42.
5. 2 vols, Cairo, 1942, p. 20 (abjad); reprinted with an introduction by Roshdi Rashed, Beirut, 2008.

3. Kamāl al-Dīn's *Tanqīḥ al-Manāẓir* and Its Use for Editing Books IV and V

The following is a list of the known manuscripts of the *Tanqīḥ*, in approximate date order.[6]

1. Istanbul: MS Ahmed III, no. 3340, 204 fols, 716 AH / 1316 AD; copied at Nishapur by Mas'ūd ibn Ḥakīm ibn 'Alī ibn al-Ḥusayn on 15 Sha'bān

2. Tehran: Sipahsalar, no. 551, 182 fols, 991 AH / 1583–4 AD[7]

3. Rampur: Raza Library, no. 9267M, 1052 AH / 1642 AD; wanting the Sequel (*Dhayl*) and the Appendices (*Lawāḥiq*) at the end of the *Tanqīḥ*

4. Rampur: Raza Library, no. 6444M, probably seventeenth century; contains only the text found on fols 1a–19b of manuscript no. 4 here

5. Jaipur: Library of the Maharaja Mansingh II Museum at Jaipur, MS no. 17, 150 fols, 1070 AH / 1659–60 AD[8]

6. Mashhad: Āstān Quds Riḍawī, no. 5480, 278 fols, 1073 AH / 1662 AD

7. Istanbul: Ayasofya, no. 2598, 317 fols, 18 Dhū-l-Ḥijja 1079 AH / 1669 AD[9]

8. Tehran: Majlis-i Shūrā-i Millī, no. 2451, 265 fols, 23 Rabī' II, 1096 AH / 1685 AD

9. Leiden: University Library, Or. 201, 318 fols, undated[10]

10. Jerusalem: Khālidiyya Library, undated[11]

11. Tehran: Sipahsalar, no. 552, 252 fols, 1097 AH / 1686 AD

12. Kuibyshev/Samara (Russia): Regional Library; reported by B. A. Rozenfeld, no date mentioned[12]

13. New York: Columbia University Library, Plimpton MS Or. 301, 310 fols, seventeenth century?

---

6. An autograph manuscript of Kamāl al-Dīn's *Tanqih al-Manāẓir*, dated 708 AH / 1309 AD, has been discovered recently. This manuscript is now in Qatar.

7. At the end of fol. 1a is written, apparently quoting MS Sipahsalar 554, the complete name of the author: 'Kamāl al-Dīn Ḥasan ibn 'Alī ibn Ḥasan Fārisī', who is often referred to as 'Kamāl al-Dīn al-Fārisī' for short, where the attributive 'al-Fārisī' is meant to distinguish him from others known by the same *laqab*, such as Kamāl al-Dīn ibn Yūnus of al-Mawṣil (d. 639 AH / 1241–2 AD), with whom Naṣīr al-Dīn al-Ṭūsī (d. 672 AH / 1274 AD) is said to have studied mathematics and astronomy.

8. King, 'A Handlist', p. 82.

9. Krause, *Stanbuler Handschriften*, p. 509.

10. Referred to as TL in this publication; see the Addendum to Book IV.

11. Sabra, *The Optics of Ibn al-Haytham. Books I–III*, II, p. lxxii n. 115.

12. Rozenfeld, 'A Medieval Physico-Mathematical Manuscript'.

14. Bankipore: General Library, no. 2455, 280 fols, seventeenth century?

15. Bankipore: General Library, no. 2456, 253 fols, eighteenth century

There is also the edition published in Hyderabad (India) by the Dāʾirat al-Maʿārif al-ʿUthmāniyya (Osmania Oriental Publications Bureau) in two volumes, 1347–1348 AH / 1928–1930 AD. It was apparently based on the Rampur manuscript (no. 3) and one of the two Bankipore manuscripts (more likely, no. 14) after the Leiden manuscript (no. 9) had been 'consulted' – a claim not supported by the publication itself.

It should be noted that the first manuscript here (no. 1) was produced in the life-time of Kamāl al-Dīn al-Fārisī, who died in 718 AH / 1319 AD, as noted by Muṣṭafā Ḥijāzīn in the Introduction to his edition of the 'first volume' of the *Tanqīḥ*, produced in collaboration with Maḥmūd Mukhtār.[13] Also to be noted are the dates derived by Muṣṭafā Mawāldī from Tehran MS Sipahsalar no. 554, which contains Kamāl al-Dīn's own summary of the *Tanqīḥ*, entitled *al-Baṣāʾir fī ʿilm al-manāẓir* and completed in 708 AH / 1309 AD. The copyist of the summary, al-Ḥusayn ibn al-Ḥasan Shahinshāh al-Si(u)mnānī, who completed the manuscript on 27 Dhū-l-Qaʿdah in the year 731 AH / 1331 AD, says that he transcribed the text from an exemplar written by 'the honoured hand of the author', al-Ḥasan ibn ʿAlī ibn al-Ḥasan al-Fārisī, and completed it 'in the months of the year 708'. Al-Si(u)mnānī then adds that the author died on 19 Dhū-l-Qaʿdah, in the lunar year 718 (12 January 1319), at Tabriz, at the age of fifty-three (lunar) years (he was thus born in 665 AH / 1266–7 AD). It may be that the Leiden manuscript (no. 9 above) is not much later than the Ahmed III manuscript (no. 1), but there is no indication as to its date. The earliest of the other manuscripts is MS Sipahsalar 551 (no. 2), which was produced in the tenth/sixteenth century, that is, 150 years later than the Ahmed III manuscript. The remaining manuscripts, with the exception of no. 12 (which is not dated) and the manuscript in the Khālidiyya Library (no. 10), are from the seventeenth or eighteenth century AD.

There are clear indications that readers of Ibn al-Haytham's *al-Manāẓir* in the Arab and Persian Middle East depended solely, for a long time, directly or indirectly on the *Tanqīḥ* rather than the original text. Here we might mention two rather late and significant mediators who had no access to Ibn al-Haytham's book, namely Mullā Fatḥallāh Shirwānī (d. 891 AH / 1486 AD), for his *Tadhnīb* or Appendix, forming part of his *Commentary* on Naṣīr al-Dīn al-Ṭūsī's *al-Tadhkira* (Library of Tehran University, MS no. 493); and Taqiyy al-Dīn ibn Maʿrūf (d. 993 AH / 1585 AD), author of *Nūr ḥadaqat al-ibṣār wa nūr ḥadīqat al-anẓār* (Oxford, Bodleian Library, MS Marsh 119, written by the author, dated 982 AH / 1574–5 AD).

I have not examined the manuscripts of the *Tanqīḥ* preserved at Rampur (nos. 3 and 4), Jaipur (5), Mashhad (6), Tehran (11), Kuibyshev/Samara (12), or Bankipore (14 and 15). I am of course not intending here to produce a critical edition of the *Tanqīḥ* itself, but in order to compare variant readings in Books IV and V, I have consulted five manuscripts of the *Tanqīḥ* (namely nos. 1, 2, 7, 8, and 9 above), as well as the Hyderabad edition; and I have almost always preferred to use MS Ahmed III (no. 1), referred to in the Arabic edition of Books IV and V by *Ṭāʾ* and in this publication as TA, not only

---

13. Cairo: Al-Majlis al-Aʿlā li-l-Thaqāfa and al-Hayʾa al-Miṣriyya al-ʿĀmma li-l-Kitāb, 1984, pp. 18–19.

because it is the oldest available copy of the *Tanqīḥ*, but because it contains the most convincing geometrical diagrams, as well as the most accurate text I have seen of these two Books.

It remains for us to highlight a useful and important feature of the *Tanqīḥ*, which we find in all manuscripts and in the Hyderabad edition, namely Kamāl al-Dīn's divisions of the text of *Kitāb al-Manāẓir* into inquiries (*mabāḥith*) and propositions (*maqāṣid*), which help the reader to follow the author's own arrangements. He explains the function of these labels as follows.

> Since [Ibn al-Haytham's] discourse in every chapter, while aiming to establish distinctive propositions, is unfolding [in Ibn al-Haytham's *Kitāb al-Manāẓir*] as a story in which beginning[s of explanations] merge with others [preceding them], I thought of identifying each question with a number distinguishing it from others, thus presenting them as separate inquiries and numbered propositions, as is usually the custom of mathematical books, while also indicating other [textual parts] of varied functions as 'Example', 'Result', 'Notice', and so on. And often a 'Notice' may be concerned with a proposition, or a 'Result' may be the 'Result' of a [certain] proposition or [of several] propositions.[14]

Since these divisions evince Kamāl al-Dīn's concern for presenting Ibn al-Haytham's explanations lucidly, I have decided to include these titles in the body of the text. In order to distinguish them from Ibn al-Haytham's original, I have always placed them between double angle-brackets: ⟪ ⟫. Single angle-brackets, ⟨ ⟩, enclose words in the edition and translation intended to restore Ibn al-Haytham's original.

We may mention here that Friedrich Risner, the first editor to publish what has generally been taken to be the medieval Latin 'translation' of Ibn al-Haytham's *Optics*, also introduced features sometimes comparable to those of Kamāl al-Dīn, though Risner's and Kamāl al-Dīn's texts differ widely from each other, as might be expected. In my 2002 edition of the Arabic Books IV and V, I wrote that it would still be interesting to compare the texts with each other, provided that the comparison were based entirely on a critical edition of the medieval Latin. This edition has since appeared: A. Mark Smith's *Alhacen on the Principles of Reflection* (2006).

There is a claim, previously noted in the Hyderabad text, to the effect that the single copy of the *Optics* used by Kamāl al-Dīn was written in Ibn al-Haytham's own hand. If true, the statement would grant the *Tanqīḥ* a special authority for anyone concerned with editing Ibn al-Haytham's book. The claim, however, seems to have been copied from the complete Bankipore manuscript (no. 14 above), which I have not been able to consult. The claim is also to be found in MS Tehran, Majlis-ī Shūrā-ī Millī (no. 8), added between lines 4 and 5 on fol. 3a and seemingly written by a different hand from that of rest of the copy; in MS Sipahsalar (no. 2), embedded in the text on fol. 2a, line 35; and also in MS Columbia University (no. 13), fol. 3a, line 20. The claim is however absent from MSS Ahmed III, Ayasofya, and Leiden, as well as the Khālidiyya manuscript in Jerusalem (nos. 1, 7, 9, and 10).[15]

---

14. *Tanqīḥ al-Manāẓir*, I, pp. 8–9.
15. For more details, see Sabra, *The Optics of Ibn al-Haytham. Books I–III*, II, p. lxxi n. 114.

4. The Geometrical Diagrams

Perhaps the single most important feature of both our edition of the Arabic text of Books IV and V and the English translation thereof are the geometrical diagrams; these allow the reader to follow the text of the *Optics* by visualizing the diagrams conceived by the author within the limits of his own methods of geometrical representation, something that is particularly important for understanding the complexities of Book V. We have already noted that Books IV and V in MSS Fatih, no. 3215 (F) and Ayasofya, no. 2448 (S) completely lack diagrams, and the same is true of the somewhat abbreviated Book IV in the *Tanqīḥ*. Moreover, the Köprülü manuscript (K) lacks no fewer than fourteen diagrams (those relating to the missing part of Book V in that manuscript; see below). As for the Latin text published by Risner, it contains about 17 diagrams for Book IV and more than 100 for Book V; these may have been felt necessary by the same person who felt compelled to complement the author's text with many explanatory additions, thereby producing a somewhat extended text and commentary.

It is fortunate, however, that Kamāl al-Dīn chose to follow a strict method for presenting Ibn al-Haytham's *Optics* in the *Tanqīḥ*. For although he did not always decide to reproduce Ibn al-Haytham's text, he was careful to remain close to the author's intentions and often retained whole long passages of the book exactly as he found them. In addition, Kamāl al-Dīn seems to have preserved for the most part the diagrams he found in the copy obtained for him by al-Shīrāzī, as may be surmised by comparing the diagrams in the *Tanqīḥ* with those in Ibn al-Haytham's surviving texts. We are therefore indebted in the present volume to Kamāl al-Dīn in our attempts – in which we hope to have been successful – to reconstruct the diagrams. We are also indebted to the copyist of Ahmed III MS 3340 (referred to as TA), which we have consulted repeatedly in preference to other manuscripts, as noted, because of its clearer text and better diagrams.

We should here note the following regarding the 24 extant diagrams for Book V in the truncated MS K: they are of course those that relate only to the 'propositions' and 'demonstrations' that have survived in this same manuscript. MS TA of the *Tanqīḥ* in the Ahmed III Library contains 47 diagrams that cover the whole text of Book V (with Diagram 47 also used, by the author, as Diagram 48), in addition to five diagrams that Kamāl al-Dīn added (without numbering them) in the body of some comments of his own on Ibn al-Haytham's text (MS TA, fols 98a, 109a, 114b, and 115b). All or most of the diagrams in the *Tanqīḥ* that refer to Ibn al-Haytham's text are represented, with various degrees of precision, in all the manuscripts we have consulted. Nevertheless, MS TA is distinguished by being the only one regularly to label the diagrams with letters according to the numerical values assigned to them in *abjad* order; and there are indications that the numbers in MS TA coincide with those in the now-lost manuscript from which TA was directly copied.

Finally, it is clear, from a comparison of MSS K and TA, that the diagrams in Book V in MS K are the same as the similarly numbered diagrams in MS TA; this numbering is that employed in our edition of the Arabic text and the English translation, namely diagrams 1–18, diagrams 40–43, and the two diagrams 46 and 47 – a total of 24 diagrams. This last observation alone is sufficient to demonstrate how much we need the *Tanqīḥ* in order to edit Book V of *Kitāb al-Manāẓir*, particularly for the important diagrams relating to the 'demonstrations', which constitute no less than half of the diagrams for Book V.

5. Editing the Arabic Text

We have therefore based our edition of the Arabic text on MSS F and K, while also carefully consulting MS TA of the *Tanqīḥ*, and indicated the folio numbers of Books IV and V in MS F in the margin of the Arabic text as well as in the English translation. The reader can further compare these numbers with the corresponding folio and line numbers in MS K by consulting the concordance tables in the edition of the Arabic text.

As was done in the edition of the Arabic Books I–III and their English translation, we have divided the text of Books IV and V into paragraphs, numbered continuously within every chapter. These paragraph numbers appear in square brackets in the translation, e.g., '[283]'. The same paragraph numbers have been used in the Summary Overview at the end.

6. Symbols Used in the English Translation and the Diagrams

The following punctuation marks have been used in the translation and the Summary Overview.

⟪ ⟫ Double angular brackets contain translations of words and expressions from the *Tanqīḥ*; these can be chapter headings, or passages used to restore the meaning of Ibn al-Haytham's original in places where the Arabic manuscripts of the *Optics* are defective.

⟨ ⟩ Single angular brackets contain text added to restore Ibn al-Haytham's original in passages where the Arabic manuscripts of the *Optics* are defective; these additions were often inspired by the *Tanqīḥ*.

[ ] Square brackets contain glosses added by the editor and translator.

(( )) Double parentheses contain the translations of passages that are found in MS K but not in MS F.

| A vertical bar in the translation indicates the beginning of a new folio in MS F (Istanbul, Fatih Library, no. 3215). The folio number is indicated in the margin, with a notation such as 'IV 33a' indicating the beginning of fol. 33 (recto) in that part of MS F containing the text of Book IV of the *Optics*, and 'V 239b' indicating the beginning of fol. 239 (verso) in that part of MS F containing the text of Book V of the *Optics*.

Using the above, it is easy to check the translation against the Arabic edition and the Arabic critical apparatus, in which the foliation of MS F is also indicated, using the *abjad* notation *F-dāl* for Book IV and *F-hāʾ* for Book V.

The following mathematical symbols have been used in the translation.

∠	angle sign
=	equals
:	expressing the Arabic 'the ratio of … to ...'
||	parallel to
>	greater than
<	less than
×	multiplication
AG^2	the square of line AG (whose side is line AG)
△	triangle
~	similar to
+	together with

The first part of Ibn al-Haytham's proof of Proposition ⟪V 2.IV.2⟫ is offered here as an example of how these symbols have been used. A literal translation of this part from the Arabic manuscripts would be as follows.

> [91] Furthermore, angle NGZ is equal to angle AGH, because it is equal to angle BGH. And angles EGN and EGH are right [angles], then by subtraction, angle AGE is equal to angle EGZ.
> [Then,] the ratio of AE to EZ is as the ratio of AG to GZ.
> We draw AH parallel to GZ. Then, angle AHG is equal to angle NGZ. And then angle AHG is equal to angle AGH. Then line AH is equal to line AG.
> And the ratio of GA to GZ is as the ratio of AE to EZ, then the ratio of AH to GZ is as the ratio of AE to EZ,
> and the ratio of AH to GZ is as the ratio of AN to NZ, then the ratio of AE to EZ is as the ratio of AN to NZ.

The English version of the same proof offered here is as follows.

> [91] [Figure V.5] Furthermore, ∠NGZ is equal to ∠AGH, because it is equal to ∠BGH; and angles EGN, EGH are right angles; therefore, by subtraction, ∠AGE = ∠EGZ.
> Therefore AE:EZ = AG:GZ.
> Draw AH || GZ. Then ∠AHG = ∠NGZ, therefore ∠AHG = ∠AGH; and thus line AH = line AG.
> But GA:GZ = AE:EZ, thus AH:GZ = AE:EZ.
> And AH:GZ = AN:NZ, therefore AE:EZ = AN:NZ.

Ibn al-Haytham and Kamāl al-Dīn labelled the points in the geometrical diagrams in Book V by means of letters of the Arabic alphabet, mostly following the Eastern Arabic *abjad* system. Points in diagrams are here labelled according to the following key.

alif	A	1
bāʾ	B	2
jīm	G	3
dāl	D	4
hāʾ	E	5
wāw	W	6
zāy	Z	7
ḥāʾ	H	8
ṭāʾ	T	9
yāʾ	Y	10
kāf	K	20
lām	L	30
mīm	M	40
nūn	N	50
sīn	S	60
ʿayn	O	70
fāʾ	F	80
ṣād	C	90
qāf	Q	100
rāʾ	R	200
shīn	J	300
tāʾ	t	400
thāʾ	U	500
khāʾ	X	600
dhāl		700 (not used)
ḍād	d	800
ẓāʾ	f	900
ghayn	g	1000

In figures with many labelled points, Ibn al-Haytham used two-letter combinations, which have been transcribed as follows: lām-alif = L_1, lām-bāʾ = L_2, shīn-alif = J'.

| *In the Name of God the Compassionate and Merciful* IV 1b

THE FOURTH BOOK OF THE *OPTICS*

OF AL-ḤASAN IBN AL-ḤASAN IBN AL-HAYTHAM

CHAPTERS OF THIS BOOK: THEY ARE FIVE

CHAPTER 1. PREFACE

[1] In the preceding three Books [of the *Optics*] we have shown how sight perceives visible objects | directly, and we have set out in detail all the properties perceptible by IV 2a
sight, and the different ways in which sight perceives all of what it perceives directly.

[2] But not all of what is perceptible by sight is perceived directly, nor do all of sight's acts of perception take place in straight lines. Rather, sight's perception of visible objects is of three kinds: direct perception, the manner of which we showed in the preceding Books [I–III]; perception by reflection from polished bodies; and perception by refraction from behind transparent bodies whose transparency differs from that of air. Sight perceives no visible object in a way other than [one of] these three; and in each one of these ways it perceives all the properties we set out in detail in Book II, in accordance with all the modes of vision [that were] also detailed in the last chapter of Book II. Now, sight perceives many
of the properties it perceives by reflection and many of those it perceives | by refraction IV 2b
correctly, but it may also err in regard to many of [the properties] it perceives by reflection or by refraction. In the present Book, we will show how sight perceives visible objects by reflection from polished bodies, and the manner in which this reflection takes place, as well as the position[s] of the lines along which the forms of visible objects are reflected; [moreover] we will advance all the notions that need to be advanced to clarify the manner of vision by reflection.

CHAPTER 2.
THAT THE FORMS OF VISIBLE OBJECTS ARE REFLECTED FROM POLISHED BODIES
⟨⟨In Six Propositions⟩⟩

⟨⟨IV 2.1⟩⟩

[1] It was shown in Book I that lights shine from every luminous body upon every facing body, whether the light in [the shining body] is essential or accidental. It was also shown [there] that form[s] of colour radiate from every coloured body onto every facing body when any kind of light exists in the coloured body; and that the form of the colour
IV 3a that radiates from the coloured body is always mixed | with the form of the light that shines from that body.[1] That being so, light will then shine from the light that exists in any luminous body facing a polished body upon the polished body; and if the luminous body is coloured, the form of the colour that exists in it will also shine upon the facing polished body, mixed with the form of the light shining [upon it]. Now, it is an [essential] property of polished bodies that the light that shines upon them is reflected from them, whether the shining light is a primary light that radiates from a body that is luminous from its essence, or a secondary light that radiates from a body that shines with accidental light.

[2] This state of affairs can be examined carefully through experimentation. In the case of strong lights, the matter is clear. For when an extremely polished body, such as a polished mirror, is placed in extremely strong light, if opposite the place where the light falls there is a wall or ceiling, on the side from which the light has come, then there will
IV 3b appear on that wall or ceiling | a strong light that was not there before the mirror was placed in the light. That light will cease when the mirror is removed and will return when the mirror is returned. These are the lights that we have shown in Book I always to be extended along straight lines.

[3] As for weak lights: they may be examined experimentally as we now describe. The experimenter should employ a chamber with an aperture in one of its walls that is exposed to the sky and through which light may enter. The aperture should be high above the ground, but not too high, so that the light entering through the aperture may reach the floor of the chamber; this way the experiment will be rendered more effective. Let the experimenter close the chamber's door and draw a thick curtain over it on the outside, so that only the light entering through the aperture may remain in the chamber. He should choose a time at which daylight, but not direct sunlight, may enter through the aperture. At the place where the light shows on the chamber's floor, let the experimenter place a white body [B_1], on which the light will be found. He should then bring near to this light a polished mirror facing the light; then bring close to the mirror from one side of it a[nother] white body [B_2] ⟨⟨on which only the light from the chamber's floor will be found⟩⟩; then

1. For an explanation of the technical meanings of 'form' in Ibn al-Haytham's *Optics*, see Sabra, *The Optics of Ibn al-Haytham. Books I–III*, II, pp. 68–70. Ibn al-Haytham discusses the forms of colours in Book I, Chapter 3.

incline the mirror | towards the body [B_2] that he has brought near it. Then, a certain light will shine on this body [B_2] that did not exist before the presence of the inclined mirror. Then, upon changing the position of the mirror, or removing it altogether from that place, this additional light will vanish, and there will remain on the body that was moved near to the mirror only the secondary light shining upon it from the chamber's floor, [this light now] being weaker than that which shone upon [B_2] when the mirror was inclined towards it. If the experimenter returns the mirror to its [former] position and he again inclines it as at first, the light will again shine on this body. IV 4a

[4] If [the experimenter] fixes the mirror in its inclined position and moves the body on which the light has appeared around the mirror on all sides, making sure that the distances of the body [from the mirror] remain equal to the body's first distance from the mirror, when the light [first] shone upon it, then no light will be found on the body in any of these locations, except for the weak light that shines from the floor. When the body is returned to its first place while the mirror retains its first, inclined position, that additional light will appear on the body [again]. If the experimenter examines | the position of the mirror IV 4b relative to the body on which the light appears, he will find the straight lines imagined between the mirror's surface and the light on the chamber's floor, or ⟨⟨some of them⟩⟩, to be inclined towards the mirror's surface; and find the straight lines between the mirror's surface and the light that shines on the moved body also to be inclined towards the mirror's surface; and find the inclination of these [latter] lines to be sensibly the same as that of the lines that lie between the mirror and the light shed on the chamber's floor, this [inclination] being the position proper to reflection from polished bodies.

[5] Thus from the appearance, in the mirror's presence, of light on the body [B_2] that was brought near to the mirror, and from the light's disappearance upon removing the mirror, it is clear that the cause of the additional light that appeared on the near body was the presence and the inclination of the mirror to that body. And from the relationship of the position of the mirror to it, and its position with respect to the light on the chamber's floor, and the light's disappearance when the mirror's position changed, thus losing the position proper to reflection while the mirror remained fixed in position, it is clear that the additional light | that appeared on the near body is but a light reflected from the mirror's IV 5a surface, and not a secondary light shining from the chamber's floor.[2] And since no light exists in the chamber other than what has entered through the aperture, it is manifest that the light reflected from the mirror is the secondary light that shone upon the mirror's surface from the light [found] on the chamber's floor, which is also a secondary light; and, therefore, the light that shone upon the mirror is a weak, not a strong light. In this manner it is shown that weak lights also are reflected from polished bodies.

⟨⟨IV 2.2⟩⟩

[6] A secondary light will also shine from the mirror upon a body brought near to it, just as it does from all bodies that shine with accidental light, but this secondary light

2. We have slightly emended the Arabic text using MS S (Ayasofya, no. 2448), fol. 317a, a manuscript that Sabra did not consider in his edition.

shining from the mirror will be very weak on account of its being mixed with the mirror's colour; and it will shine from the mirror upon the near body from all parts of the mirror. Thus, when the near body is moved around the mirror, a secondary light will invariably shine from the mirror upon it, whereas the additional, strong ⟨light⟩ will appear on the
IV 5b near body only when the latter occupies the place | proper to reflection, and no other. Therefore, the additional light that appears in the place [proper to] reflection is not one of the secondary lights.

[7] It is also possible to examine experimentally the secondary light that shines from a polished body while this light is being reflected from it; this can be done by using a mirror made of silver. For when the reflected light appears at the place proper to reflection, ⟪the experimenter should bring near to the mirror a third white body from a direction other than the direction of reflection⟫, then a secondary light will be found there, which will also be found [to be] weaker than the reflected light.

[8] To examine experimentally the reflection of light from polished bodies in any dark or poorly lit place, when opposite such a place there is a strong light, let a polished mirror be put in that place, facing the strong light, and let a white body be brought near to the mirror, and let the examination proceed as we have shown; the reflected light will then appear on the near body. Later on, we shall make clear the characteristics of the location proper to reflection and determine the position of the lines along which the light is reflected.

[9] From this experimental examination, it is clearly manifest that all the lights that
IV 6a shine on polished bodies are reflected from them to the locations proper to | reflection. And it was shown in Book I that reflected lights extend along straight lines. Therefore, all lights that shine on polished bodies are reflected from them to the locations proper to reflection along straight lines.

[10] Since it has been shown that all the lights that shine on polished bodies ⟨from⟩ coloured bodies that are illuminated by accidental lights are always accompanied by the forms of the colours of the coloured bodies, and that the forms of the colour and the light that emanate from an illuminated coloured body are always mixed together and never separate, then, from any coloured body illuminated by any light whatever, the forms of the light and colour that exist in that body will shine on any facing polished body, and will all be reflected from it in the direction proper to reflection. Further, if there is some body on the side where the light is reflected from the polished body, then the forms of the light and of the colour that are reflected from the polished body will both shine on that body if it occupies the location [proper to] reflection.

[11] An experimental examination can be made of this matter also, I mean the reflection
IV 6b of the forms of colours from | polished bodies. To do so, the experimenter should use the chamber we have described, namely that into which light enters through an aperture so as to reach the chamber's floor. He should wait for the light to enter through the aperture, then place a purple body where the light appears on the floor. The experimenter should then take a hollow object, such as a drinking cup or the like, in which he should insert a white cloth, and place it close to the lit spot. The [hollow] object should face the light sideways, so that the cloth within it is concealed from the adjacent light, thus receiving neither the secondary light nor the form of the colour; but let the orifice of the hollow object nevertheless be close to the lit spot on the floor. With the hollow object in this position, the lit form of the purple colour will not appear on the cloth inside the object.

Then let the experimenter bring the mirror near to the illuminated purple body from the side of the hollow object's orifice, but without introducing it into the light, so as to face the floor, and let him incline it by raising | the side of it next to the hollow object, so that the IV 7a
orifice is located where the light would be reflected from the mirror. In this situation, the form of the purple colour will appear on the cloth inside the hollow object.

[12] The experimenter should then examine the position of the mirror and ⟨then⟩ incline it in a direction other than that of reflection. The form that appeared on the cloth will vanish with the mirror's change of position, even though the mirror remains in its place, and despite the shining of the light and of the colour's form. Then let him incline the mirror's position as it was before. He will find that the form has returned to the cloth inside the hollow object. The form will be more manifest if a silver mirror is used in this experiment.

[13] This having been made clear, the experimenter should replace the purple body with one of a bright colour, such as bright green [*bāzil, rayḥānī*], red [*ṣaʿwī*], or purpure [*firfīrī*].[3] Then let him perform the experiment precisely in the manner we have described. He will find that the form of that colour will appear on the cloth inside the hollow object when the mirror takes the position in which the light is reflected to the object, | and it IV 7b
will vanish from ⟨the cloth⟩ upon changing that position. It is thus clearly manifest from this experiment that the form[s] of illuminated colours that shine on polished bodies are reflected from these bodies, and [that they] shine on facing bodies located in the position [proper to] reflection. And this will happen in the case of all coloured bodies that shine. From this it is clear that, from any coloured body that is illuminated with any [kind of] light, the form[s] of the light and colour that shine from it on a facing polished body will be reflected from the polished body and shine upon every body they may encounter in the place [proper to] reflection.

[14] As for [the reason] why the forms of colours do not appear at all times and in all conditions, even when the coloured illuminated body faces a polished body, while [other] bodies upon which these forms may shine are located in the place [proper to] reflection from the polished body: that arises from the causes that we showed in Book I, on account of which not all the forms of colours are visible on all facing objects.

⟨⟨IV 2.3⟩⟩

[15] Having shown that forms are reflected from polished bodies, I say that reflected forms are weaker than those from which | they have been reflected. This is clear in the IV 8a
case of light: for if in any place illuminated with accidental light a polished body is placed so that the light may be reflected from it, and if a [second] polished body is placed in the position [proper to] reflection so that the reflected light may appear on it, the reflected light on the latter body will always be found [to be] weaker than the light on the [first] polished body. A precise experimental examination of this state of affairs may be made [in the following way]: using the previously described chamber for this examination, let a white object be placed beside the polished body that occupies the illuminated spot on the chamber's floor, and let [another] white object like the first be placed in the position

3. On the terminology for colours, see Sabra, *The Optics of Ibn al-Haytham. Books I–III*, II, pp. 40–42 and 57–59.

[proper to] reflection. Then, when the reflected light found on the white object that has been set in the position [proper to] reflection is compared with the light that emanates from the other white object that was placed on the floor beside the polished body, the two [lights] will be found to be manifestly different from each other, and the reflected light will be found to be much weaker than that which appears on the floor. Now, this light that
IV 8b shows next to the polished body is of the same kind as that which appears on | the polished body. It is manifest from this experiment that the light reflected from the polished body is weaker than the light from which it has been reflected, and weaker than the light similar to that from which it has been reflected.

[16] It might be said that the reflected light becomes weaker only because it carries with it the colour of the polished body, and not because of its being reflected; and thus when the polished body is an iron mirror, the light reflected from it will be reduced as a result of its being mixed with the mirror's colour, because the colour of an iron mirror is dark. In reply to this, I say that the reflected light carries with it the colour of the polished body from which it is reflected, as has been said, but that the weakness and reduction it suffers is not simply due to its being mixed with the polished object's colour, but rather that the reduction caused by reflection is greater than that caused only by the colour of the polished body. For if a silver mirror is placed in the light and the experiment is performed in the manner we described, the light reflected from the mirror will be found [to be] weaker and lesser than the light that shines upon it, even though the reflected light
IV 9a is mixed with the colour of silver and is thereby made stronger | by the whiteness of silver. From this experiment, it is clearly manifest that reflection weakens the reflected light and reduces its strength.

⟨⟨IV 2.4⟩⟩

[17] The weakening of the forms of colours may be examined experimentally in the same manner as their reflection. For when the form of the colour is reflected from the polished body, and the reflected form appears on the cloth inside the hollow object previously mentioned [11], let another white cloth similar to the first be placed next to the coloured body that was set in the light and whose colour was reflected from the polished body; let the distance of this cloth from the coloured body that has been set in the light be the same as that of the polished body from this body; and let this body be placed beside the polished body. Then, when the form of the colour appears on this cloth, let it be compared with the form [of the colour] found inside the hollow object: the form on the cloth outside will be found stronger and more manifest than the form inside the hollow object. And [it will be] likewise when this state of affairs is examined using a silver mirror, namely, the form outside will always be found [to be] stronger and more manifest than the reflected
IV 9b form. It is clear | from this experiment that the forms of colours also are weakened and reduced upon being reflected from polished bodies, and that reflection weakens and reduces the forms of colours as it does lights.

⟨⟨IV 2.5⟩⟩

[18] The reflected forms of colours are reduced and altered by the colours of the polished bodies with which they are blended. And the effect of the colours of polished bodies on the reflected forms of colours is greater than their effect on the reflected forms of light, because the reflected forms of colours are very delicate and weaker than the forms of light, because when colours are mixed together, they are altered by being reduced, and because when mixed they darken and give rise to a colour different from each of them. Thus, the forms of colours reflected from polished bodies are weaker than they were before being reflected, and reflection itself weakens the reflected forms in addition to the alteration suffered by them.

[19] It might be said that forms, when reflected, are weakened only because they recede from their origin, | and that they weaken as they recede further, and that their weakening IV 10a
is thus not due to reflection. I say, in response to this statement, that the forms, as stated, do become weaker as they recede from their origin, and that distance weakens the forms, and yet that reflection has a weakening effect on the reflected forms of lights and colours that is due to reflection [itself]. This can be made clear as we [now] describe. To examine the weakening of reflected light, let the experimenter do so in the chamber in which the light enters through the aperture. Let him set the mirror at a point in the space directly between the aperture and the spot on the chamber's floor where the light falls. Let the mirror face the aperture, so as to receive the light on its surface, and make sure that not all the light [that was] reaching the floor is taken up by the mirror and that some of it is left around [the mirror], which can be effected by making the aperture wider than the mirror's surface. In the remainder of the light [that appears] on the floor, let him put a white body, and put another white body similar to it facing the mirror. Then let him incline the mirror so as to reflect the light to the white body facing the mirror. He should make sure that | the IV 10b
body to which the light is reflected has the same distance from the mirror's location as the body that has been placed on the floor from that location. Then, when the reflected light appears on the body facing the mirror, while maintaining the position described, let the experimenter compare the reflected light with the light on the chamber's floor. He will find that the reflected light is much weaker than the light on the floor and over the white body. Now, the distance from the origin of the light that, after reaching the mirror, was reflected to the body facing the mirror, is equal to the distance of the light that after reaching the location of the mirror has extended rectilinearly to the floor, since the mirror's location is equidistant from the reflected light and from the light on the floor; and thus the light that has been reflected in this manner and the light that has extended rectilinearly to the floor are equidistant from their origins and identical in all their conditions except for reflection; and yet the reflected light is found to be weaker. It is thus shown by this experiment that reflection itself weakens the reflected light.

[20] Similarly, when the forms of colours are experimentally examined | in the same IV 11a
manner, the reflected form will be found [to be] weaker than one that has extended rectilinearly. For when the reflected form of colour appears on the cloth inside the hollow object, let the experimenter take another white cloth and bring it near to the coloured body that has been placed in the light. Let their distance from each other be equal to the sum of the distance of the coloured body from the mirror's surface and the distance of the mirror from the cloth inside the hollow object. When the form of the colour appears on the cloth placed outside [the hollow object] facing the coloured body, let him compare the reflected form with the rectilinear form. He will find the reflected form weaker than the rectilinear form, despite their equal distances from the coloured body. The result will be the same if a silver mirror is used in the experiment, except that the form reflected from the silver will be more manifest. It is clear from this experiment that reflection weakens and reduces the forms of colours, just as it does the forms of lights.

⟪IV 2.6⟫

[21] I say also that the reflected form [of light] is stronger than the secondary form
IV 11b when they have | the same origin, or origins of equal [strength], and their distances from the origin [or origins] are equal. For example: let the light that shines on a polished body be reflected from it to some other body on which the reflected light will appear. Close to the polished body, and in a direction other than that of the reflection, let there be a body other than that to which the light is reflected, and let its distance from the polished body be equal to that of the body that receives the reflected light. Let the two bodies nonetheless be similar in colour and kind, and let there appear on the second body [placed] outside the position [proper to] reflection the secondary light that shines upon it from the polished body or from bodies surrounding it: the reflected light will be much stronger than the secondary light that appears on the body placed outside the position [proper to] reflection, for the secondary light extending from the polished body and its surroundings is a second secondary light.

[22] A precise experimental examination of this state of affairs can be made [as
IV 12a follows]: | employing the chamber in which the light enters through the aperture, let the experimenter narrow the aperture so as to narrow the light that is found on the chamber's floor; he should then place a silver mirror at the location of the light, making sure that the light does not exceed the mirror in magnitude. Let him bring a white body near to the mirror and set it in the direction of reflection, then move it back and forth until it receives the reflected light. He should then take another white body similar in kind to the first and bring it close to the mirror from a position other than that of reflection, and set it at a distance from the mirror equal to that of the first body: there will appear on [this second body] the secondary light that shines from the mirror on all sides around it. Comparing this light with the reflected light, the experimenter will find the reflected light to be much stronger than the secondary light, and with a manifest difference between them. Let the experimenter then widen the aperture, so that the light on the floor may become much wider than the magnitude of the mirror; then place a [third] white body in the additional illuminated area, beside the mirror and towards the [second] body positioned outside the place [proper] to reflection; and fix the [first] body, to which the light is reflected, in its own
IV 12b place: | he will find that the secondary light that appears on the body placed outside the place [proper] to reflection has grown in strength and increased because of the additional light, and yet he will find the reflected light to be stronger than the secondary light on the body placed outside the direction of reflection, despite the increase and greater strength of the latter. In this manner it is shown that reflected light is much stronger than secondary light when they have the same origin, or origins of equal [strength], and their distances from the origin [or origins] are equal.

[23] The forms of colours may also be examined in the way in which we have examined the reflection of light. If the form of the colour is examined by means of the hollow object, as previously shown, and the reflected form appears on the white cloth inside that object, and if the experimenter then brings another white body close to the mirror in a direction other than that of reflection and sets it at a distance from the mirror equal to that of the
IV 13a reflected form from the mirror, then there will appear on | that body a manifest colour owing to the mixture of the mirror's colour with the form of the colour that shines upon it. If, however, a silver mirror is used in this examination, or if beside the mirror the

experimenter places a white body on which the form of the colour appears, then this form will be equal [in strength] to the form that shines upon the mirror, and it will appear on the white body, but not on the mirror if the mirror is made of iron, because [the form of the colour] will be mixed with the mirror's colour. If, towards this white body on which the form of colour appears, the experimenter brings another white body and sets it at a distance from it equal to that between the reflected form inside the hollow object and the mirror, then contemplates this other white body, nothing of the form of the colour will appear on this body. If anything of the form appears on it, it will be extremely weak. It is seen from this experiment that the form that is reflected from the forms of colours is stronger than the secondary form that emanates from the primary form of colours.

⟨⟨Result of the chapter⟩⟩

[24] Thus it is shown from all that we have explained in this chapter that the forms of lights and colours are reflected from | polished bodies; that reflection weakens the reflected IV 13b forms; and that the direct form is stronger than the reflected form when they have the same origin, or origins of equal [strength], and their distances from the origin [or origins] are equal, ((and that the reflected form is stronger than the secondary form if they have the same origin, or origins of equal [strength], and they are equally distant from the origin [or origins])).

CHAPTER 3.
ON THE MANNER OF REFLECTION OF FORMS FROM POLISHED BODIES
⟨⟨Introduction and Nine Propositions⟩⟩

⟨⟨Introduction⟩⟩

[1] Polish is [the property] of the surface of a body when it is very smooth; smoothness of the surface of a body consists in the compactness of its parts, their [even] arrangement, and the narrowness of its pores; the surface of a body is very smooth when the pores in it are few and narrow; and perfect smoothness is the absence of pores, separations, and protrusions from the surface. And, in sum, the polish of a body's surface is the compactness of its parts when the pores in it are few and narrow; and perfect polish is the compactness of the parts of the surface in addition to the absence of pores from it. Now, polished bodies differ in shape and figure, and polished surfaces differ in figure while sharing the [property] of being polished; and all polished surfaces, | whatever their shapes IV 14a and figures, cause the light that shines upon them to be reflected from them on account of their polish. Thus, when all polished bodies are examined experimentally in the manner we have described, the light will be found to be reflected from them, whatever the shapes of their polished surfaces.

⟪IV 3.1⟫

[2] Now, all lights reflected from all polished surfaces are reflected in a single, special manner common to all reflected lights and subject to no variation; thus any light reflected from any point on a polished surface will do this in a straight line that lies in one and the same plane as the straight line along which the light has [previously] extended to that point and the perpendicular at that point to the plane tangent to the polished surface at that same point; and, moreover, the position of the line along which the light is reflected, relative to the said perpendicular, will be the same as the position of the [incident] line along which the light had extended to the reflection point, relative to that perpendicular. I mean that every line along which the light is reflected from a polished surface will form, IV 14b | with the perpendicular [drawn] at that point to the plane tangent to the polished surface at the same point, an angle equal to that formed between the first line along which the light has extended to that point and the same perpendicular; and the three lines will lie in the same plane, which is orthogonal to the plane tangent to the polished surface at the reflection point. If the line along which the light extends to the polished surface is perpendicular to the plane tangent to that surface at the point where it meets the polished surface, then the light will return on that same line. That is the manner of reflection, of all lights, from all polished bodies. If the polished surface is plane, then the plane tangent to it at the reflection point will coincide with the surface. If the polished surface is cylindrical, whether convex or concave, or conical, whether convex or concave, then the plane tangent to it at the reflection point will meet it in the straight line drawn from that point along the IV 15a length of | the cylindrical or conical surface. If, however, the polished surface, whether convex or concave, is such that a straight line cannot lie in it, then the plane tangent to it at the reflection point, and to which the perpendicular is drawn from the reflection point, will only meet the surface at a single point, which is the reflection point.

[Design of an instrument for examining the manner of reflection]

[The copper plate]

[3] The following is a description of how to examine this state of affairs experimentally in such a way as to achieve certainty; by 'this state of affairs' I mean the manner in which lights are reflected from polished bodies. Let [the experimenter] take a copper plate not less than twelve digits in length, and half of this in width; let it be somewhat thick to prevent it from bending or corrugating, and let it be as even as possible on both surfaces. Let [the experimenter] draw a line along the length of this plate close to one edge; bisect the line, and, with the dividing point as centre and with half the line [as radius], draw a semicircle. Let him establish from the centre a perpendicular to the diameter, which will cut across the width of the plate and bisect the curve of the semicircle. The experimenter should then divide one of the two [resulting] arcs into as many divisions as he likes, and IV 15b divide the other arc | into an equal number of divisions, each of which is equal to the one corresponding to it [on the first arc]. ⟪And let the divisions begin from the point at the middle.⟫

[4] [The experimenter] should then draw straight lines from the centre to the dividing points [on the arcs]. These lines should be permanently incised in the body of the plate.

About the same centre, let him draw another semicircle, smaller than the first by the width of one digit. Then let him mark off on the perpendicular, next to the centre, a segment also equal to the width of one digit, and establish at the dividing point a perpendicular that will be parallel to the circle's diameter. Let him extend this perpendicular along the length of the plate until its extremities meet the circumference of the first circle. He should then cut off the portion of the plate that remains between this line and the circle's diameter, and let the cutting terminate at the straight line drawn from the centre to the end of the first division on [one] arc. I mean the first of the divisions [of that arc] next to the middle of the plate. Then cut off also the remaining portion on the other side, [again] terminating the cutting at the straight line drawn on this side from the centre to the end of the first of the divisions on the other arc next to the middle of the plate.[4] There will result | a small IV 16a
triangle at the middle of [the edge of] the plate next to the centre. Let him then carefully cut away what is separated from the centre at the edge of the plate so that the point at the centre may form a vertex of the triangle so produced. He should then whet away the width of the plate next to the centre until the vertex of the triangle, i.e., the centre, is sharpened like a point; then cut away the redundant portion of the plate [above] the arc down to the circumference [of the larger circle].

[The ring-like wooden piece]

[5] Having finished with the plate in this manner, take a block of soft wood, such as juniper, cypress, or the like, and let it be four-sided, with a length exceeding that of the plate by two digits and with an equal breadth, and a height of six digits. Make its surfaces as plane as possible and, from a point marked in its middle as centre, draw a circle with a radius exceeding that of the larger circle on the [copper] plate by one fairly large digit, then draw about the same centre another circle equal to the smaller circle on | the plate. IV 16b
The distance between these two circles [on the wooden piece] will be equal to two digits. Then mark off on the circumference of the smaller circle an arc equal to the shorter arc on the plate; divide the arc into segments equal to those of the smaller circle on the plate; draw straight lines from the centre to the dividing points and extend them to meet the circumference of the larger circle. The lines should be drawn with an iron [implement] for making lasting impressions in the body of the piece of wood.

[6] Then mount this wooden block on a lathe at the centre of the circle and turn [the lathe] until the redundant portions surrounding the larger circle drop off all the way to the circumference of this circle. The middle of the wooden block should then be chipped out by being turned on the lathe again until the whole interior portion drops out from inside, when the turning ends, at the circumference of the smaller circle. To do so, attach this block to another block, [and] determine the point on the second block that is in line with the centre marked on the first block. Then mount the [combined blocks] on the lathe at this point and hollow out the first block all the way | to the circumference of the smaller circle. IV 17a
There will be left out of this block a ring two digits wide and six digits thick. The inner circumference [of this ring] will be equal to the smaller circle on the plate's surface; and on the surface of its width there will remain the ends of the lines that divided the arc into segments equal to those of the arc on the plate.

---

4. See the commentary and diagrams by Kamāl al-Dīn in the Addendum to Book IV, pp. 95–97

[7] Having finished this ring, draw on the exterior surface of its height straight lines extending from the extremities of the lines on its width and perpendicular to both surfaces of its width, i.e., the bases of the cylindrical surface. This can be done by laying a straight sharp-edged ruler on the cylindrical surface of the ring's height so that the ruler's [sharp] edge touches the extremity of the line on the ring's width, and [then] sliding the ruler over the cylindrical surface so that its [sharp] edge may coincide with the length of this surface, i.e., the ring's height, while the [short] edge of the ruler remains in touch with the circumference of the lower circle, i.e., the base of the cylindrical surface. Then the
IV 17b line drawn along the edge of the ruler will be | straight and perpendicular to both bases of the cylindrical surface, for it would not otherwise coincide with the upright cylindrical surface. A better way [of obtaining the same result] would be to make careful use of the extremity of the straight line on the ring's width as a centre and, with the help of a fine sharp compass centred on this extremity at the edge of the ring's height, draw about it an arc [⟪a semicircle⟫] on the cylindrical surface. Then bisect the arc and apply the ruler's edge to the dividing point and the line's extremity taken as centre, so that the ruler's edge will coincide with the cylindrical surface; then draw a line along the ruler's edge. In this manner a straight line will have been drawn on [the surface of] the ring's height.

[8] Again, on the concave surface of the ring,[5] and beginning with the extremities of the lines on the surface of the ring's width, draw lines corresponding to those on its exterior [convex surface]; these lines will then run along the interior of the height of the
IV 18a ring and will be perpendicular to the surfaces of both of its widths, i.e., | the two bases of the cylindrical surface. These lines can be drawn in the preceding manner. Then, having drawn from the extremities of each of the lines on the ring's width two straight lines perpendicular to the ring's bases, cut off the portion of the ring that is defined by the two divided [i.e., marked] arcs, and let the cutting at each end of this portion take place at the two opposed straight lines on the ring's height, one on the outside of the ring and the other inside.

[Using a wooden panel in the shape of a sector of a circle to draw two arcs on the concave surface of the ring's thickness]

[9] This portion of the ring having been cut off, draw on its interior surface an arc parallel to the arc formed by its edge, and let the distance between the [drawn] arc and the circumference of the base on which no divisions have been marked be equal to two digits. This can be done [as follows]: draw on a light panel of soft wood a circle equal to the small circle on the [copper] plate and equal to the limit of the interior surface of the ring, having made the surface of the panel as plane as possible. Then mark off on this circle an arc equal
IV 18b | to the shorter arc on the wooden piece, and remove what is left of the panel outside this arc all the way down to the curve of the arc, without interfering with its shape. Then cut off from the panel the sector defined by this arc.

[10] Then clearly mark off on the [rectilinear] edges of the height of the ring, at its concave surface, two line segments next to this undivided circle, each equal to two digits. Then slide the sector into the inside of the ring, apply its arc-shaped edge to the surface,

5. In relation to the ring, Sabra used 'concave' to denote the inner surface, and 'convex' to denote the outer surface.

setting the extremities of the arc at the points marked on the ring's height, so that the sector is perpendicular to the surface of the ring's ⟨height⟩ when its arc coincides with the concave surface. Then, with the help of an iron tool, [the experimenter] should draw an arc on the concave surface of the ring alongside the sector's edge. This [drawn] arc will be parallel to the circumference of the ring's base and at a distance from it equal to the [line] segment[s] marked off on the ring's height. Then, on the height of the ring, and next to the arc that has been drawn on the ring's surface, mark off, from the [lower] interval of two digits, two magnitudes [i.e., line segments] each equal to half the length of a full barley-seed. | Then apply the sector to these marks at right angles to the concave surface IV 19a
of the ring, and draw alongside its edge on this surface another arc parallel to the first [higher] arc, thereby producing on the surface of the ring's ⟨height⟩ two parallel arcs with a distance between them equal to half the length of a barley-seed. The second [lower] arc should then be engraved with a fine chisel, then should be deepened into a narrow groove not [quite] as wide as the thickness of the copper plate. Let the groove be one digit deep, thus reaching approximately the middle of the ring's body, and let its depth be approximately the same throughout its length.

[Mounting the plate into the ring]

[11] The copper plate should be mounted into this ring by inserting its arc-shaped edge into this groove with its marked surface facing the larger part of the ring's height and the divided base [i.e., the top] of the ring. Lightly tap the plate on either side of the triangle in the middle [of the plate], in order to force it gently and carefully into the groove. Let [the plate] be precisely positioned so that the shorter arc in its middle may reach the concave | IV 19b
surface of the ring and coincide with it. Make sure that the straight lines that divide the arc on the plate meet the extremities of the straight lines running along the [interior surface of] the ring's height. Thus, the arc on the plate will be parallel to the other arc on the [exterior] surface of the ring, which is parallel to the groove, and that other arc will be divided by the straight lines running along the [exterior surface of] the ring's height; and, further, the distances of the dividing points on the surface of the ring's ⟨height⟩ from the surface of the plate will be equal, each of them being equal to half a barley-seed, and all of these distances will be perpendicular to the plate's surface.

[Drilling the holes in the ring]

[12] Once the plate has been mounted into the ring and precisely positioned in the manner described, mark off, on the straight lines drawn on the interior and exterior [surfaces] of the ring, segments equal to the interval[s] between the dividing points on the interior arc on the ring's [inner] surface and the extremities of the straight lines on the top surface of the ring; and, at the dividing points on the exterior lines, mark manifest dots which together with the dots on the interior arc | will lie in a single plane parallel to IV 20a
the plate's surface. Then take a carpenter's drill with a head the size of a full barley-seed; apply the drill's ⟨head⟩ to ⟨one of the dividing points on⟩ the exterior surface of the ring's ⟨height⟩ from the outside, having mounted the ⟨ring⟩ on a firmly fixed and smooth body, and ensuring that nothing interferes with the ring's centre ⟨⟨namely, the triangle's vertex⟩⟩. Gently and carefully drill the ring's body to produce a straight hole that reaches the point

of intersection, on the interior of the ring, ⟨between the higher arc drawn on this interior and the line perpendicular to it⟩, i.e., the point opposite the one at which the drilling begins, and make sure that the hole [just] touches the plate's surface. Inserting a straight cylindrical rod into the hole and sliding it through the hole, it should rest on the plate and be wholly in touch with the plate's surface. Applying the same method for precisely positioning the hole, the ring should be drilled in this manner through all the dots marked on its exterior surface.

[Mounting the instrument on its base]

[13] This done, take a square board, also of soft wood, each side of which is not shorter than the chord of the base of the exterior surface of the ring; let the board be two
IV 20b fairly large digits thick, and let its surfaces be made as plane | as possible. Draw across the middle of the board a straight line ⟨⟨parallel to both edges⟩⟩; mark off from one end of this line a segment equal to the width of the [thickness of the] ring, i.e., the width of two digits; then mark off again from it a [further] segment equal to the radius of the smaller circle on the [copper] plate, which is equal to the edge of the ring's interior surface. With this second point as centre and with the distance to the first point [as radius], draw an arc of a circle, which will be equal to the smaller circle on the plate and the circle forming the [interior] circumference of the ring's base. Again, about the same centre, and with the distance to the line's extremity [as radius], draw another arc across the width of the board. Then cut off from the width of the board all that is left out outside this [second] arc. Then, on the line in the middle of the board, and from the part of it between the centre and the drawn arc[s], mark off a segment next to the centre equal to the width of one digit. The digit as a measuring unit should be a straight line precisely drawn on the
IV 21a copper plate, so that whenever | a magnitude has to be measured in digits, the legs of a compass should be placed at the extremities of this line, then the measuring can be made by means of the compass. At this dividing point, draw a straight line perpendicular to the middle line; extend this perpendicular on either side of the middle line, and mark off on this perpendicular, on both sides of the intersection [with the middle line], two segments each equal to two digits, thereby producing a total [line] of four digits.

[14] ⟨The experimenter⟩ should then draw from the extremities of these two segments two perpendiculars parallel to the middle line and extend them towards the remaining part of the board [i.e., away from the arcs]. Then mark off on these two perpendiculars two segments each equal to four digits and join their extremities with a straight line. There will result on the surface of the board a four-sided figure with equal sides. Engrave this square to a depth of one digit into the body of the board, make even the surface of this depression and ensure that it be as parallel as possible to the surface of the board; also make even the
IV 21b surrounding sides, and ensure that they are | as perpendicular as possible to the surface of the board and to the surface of the base of the depression.

[15] Having finished with this board, mount the ring upon it so that the circumference of the ring's interior surface may coincide with the shorter arc on the surface of the board; thus the circumference of the ring's exterior surface will coincide with the [curved] edge of the board, since the distance between the two arcs that have been drawn on the board's surface equals the depth of the ring. It should be secured, when the ring rests on the board, such that the distance of the surface of the [copper] plate from the board's surface is

everywhere the same, so that the two surfaces are parallel to each other; also make sure that the line in the middle of the plate is [directly] opposite the line in the middle of the board and parallel to it. With this precise position obtained, securely fasten the ring to the board and, this done, nail them together with fine nails, and do so gently so as not to disturb the ring's position. Having finished with all that, the instrument is now complete, and [with its] being completed, part of the depression in the board will lie beneath | the IV 22a
triangle at the middle of the [copper] plate's edge.

[The copper tube]

[16] ⟨The experimenter⟩ should then take a cylindrical copper tube of correct circularity and straightness, the diameter of whose exterior surface is equal to and not less than the diameter of each of the [drill-]holes in the ring; let the wall of the tube be of such width that the opening would accommodate a [surgeon's] probe, and let it be of uniform thickness, so that the tube may slide, when inserted, into the holes with a little difficulty and not easily. If the width of the tube exceeds that of the holes a little, then the holes should be carefully widened a little so that the tube may slide through each one of them with some effort and not with ease; this way the tube, once inserted, will be firmly fixed in position. It should be ensured that the tube, when inserted through each of the holes, will lie flat against the surface of the plate and be in touch with it throughout the length of the tube, which will be the case if the tube is of correct straightness and circularity. After insertion through [a given] hole, the surface of the tube should be in touch with that of the plate | IV 22b
along the straight line that has been drawn on the plate's surface from the centre to the hole's extremity. This condition will be fulfilled if, upon examining the tube's extremity when the tube lies flat against the plate's surface, it is found that the line on the plate's surface cuts the circumference of the tube's circle at the point where the perpendicular to the plate's surface appears to pass through the centre of the tube's opening. This position will be achieved only if the hole in the ring is straight and in a straight line with the line on the plate's surface.

[17] The [correct] shape and position of the tube having been precisely obtained when the tube slides through each one of the holes in the ring, [the experimenter] should then solder a thin copper ring around one extremity of the tube so that the latter may not move and shift too much upon being inserted through the hole. The length of this tube should be such that when it has been inserted through the hole and its extremity is in touch with the exterior surface of the [wooden] ring, the other end of it will reach the base of the triangle in the middle of [the edge of] the plate without reaching the centre or crossing the line [forming] the chord of the arc in the plate.

[Construction of the seven mirrors]

[18] Seven small mirrors of steel [*fūlādh*] should now be made ready: | one of them IV 23a
is plane; two are spherical in shape, of which one is convex and the other concave; two cylindrical, of which one is convex and the other concave; and two conical, of which one is convex and the other concave.

[19] The plane [mirror] should have a perfectly even surface; it should be circular, with a diameter equal to the width of three digits.

[20] The cylindrical convex [mirror] should have its convex surface perfectly sound and perfectly polished. It should be ⟨part⟩ of a cylinder the diameter of whose base is equal to six digits; the chords of the bases of its convex surface should each be equal to the width of three digits; its straight length should also be equal to three digits; and the two [straight] lines joining the opposed extremities of ⟨the peripheries⟩ of the mirror's two bases, namely the [lengthwise] edges of the [cut-off] mirror, should be straight.

[21] The cylindrical concave [mirror] should have its concave surface perfectly circular and perfectly polished. It should be part of a cylinder the diameter of whose base is equal to six digits. The chords of the bases of its concave surface should each be equal to three
IV 23b digits; | its straight length should also be equal to the width of three digits; and the two lines joining the opposed extremities of ⟨the peripheries⟩ of the mirror's two bases should be straight. Thus, if the diameter of the mirror's base is six digits, and the chord of the arc of the base is three digits, then the sagitta[6] of each of the arcs on both bases will be less than half a digit.

[22] The conical convex [mirror] should have its convex surface perfectly sound in regard to its conical shape, curvature, and straightness. It should be part of a cone the diameter of whose base does not exceed six digits; the chord of the base of its exterior, convex surface should equal the width of three digits; and its straight length should equal the width of four and a half digits.

[23] ((The conical concave [mirror] should have its concave surface perfectly sound in regard to its conical shape, ⟨curvature, and straightness⟩. It should be part of a cone the diameter of whose base does not exceed six digits; the chord of the base of its ⟨interior⟩, concave surface should equal the width of three digits, and thus the sagitta of its base will be less than half a digit; and its rectilinear length should be equal to four digits and a half.))

[24] The spherical convex [mirror] should be perfectly sound in its sphericity. It should be part of a sphere whose diameter does not exceed six digits, and it should be perfectly polished; the diameter of its circle should equal the width of three digits; and so its sagitta[7] will be less than half a digit.

[25] The spherical concave [mirror] should be perfectly sound in regard to sphericity,
IV 24a and it should be perfectly | polished; it should be part of a sphere whose diameter does not exceed six digits; the diameter of its circle should equal the width of three digits; and hence its sagitta will be less than half a digit.

[26] Let each of these mirrors be thick enough to keep its shape.

[The wooden (holding) posts]

[27] With these mirrors having been readied, the [experimenter] should obtain [a number of] wooden posts, each being six digits long, four digits wide, and three digits thick, with their surfaces being perfectly rectangular. Their surfaces, including their [top and bottom] ends, should be made perfectly even and parallel to each other. Each of these posts should then be mounted into the depression in the board previously described, with the four-digit width of the post being fitted into the four-digit width of the depression, and

6. That is, the line segment joining the midpoint of the arc to the midpoint to its chord. The Arabic word that Sabra translates as sagitta is *sahm*, 'arrow'.

7. That is, perpendicular distance to the mirror from the centre of the circular plane bounding the mirror.

with the six-digit length [of the post] being orthogonal to the surface of the depression. The posts should slide easily into the depression and should not fall short of it in width, and it should be ensured that the surfaces of their ends [i.e., bases] coincide with the surface of the depression.

[28] Once the post is fitted into the depression in this manner, | let its [front] surface IV 24b
come into contact with the centre of the plate. Mark a point on the surface of the post where it touches the centre; and, from this point of contact on each one of the posts, draw a straight line parallel to the edges of the post's length. On every one of the lines thus drawn on the surfaces of the posts, mark off from the longer segment next to the given point, a small line equal to the distance between the arc in the middle of the plate and the arc lying above it on the concave surface of the ring and passing through the centres of the [drilled] holes, namely half a barley seed. The point of division on each of the lines drawn on [the surfaces of] the posts will thus bisect the line in the middle of the post: for the distance of the dividing point from the surface of the board equals the distance between [on the one hand] the arc lying on the [concave] surface of the ring and passing through the centres of the holes, and [on the other] the ring's base, which in this situation coincides with the board's surface, this distance being equal to two digits; and thus the distance of the dividing point on the post's surface from the surface of the board, when the post stands
| upright on the board, will be equal to the width of two digits. Now, the part of this post IV 25a
that is sunk into the depression is equal to one digit, i.e., the depth of the depression; and so the distance of the dividing point from the end [i.e., base] of the post is equal to three digits; but the length of the post being six digits, the dividing point will bisect the straight [lengthwise] line in the middle of the surface of the post.

[29] From the division-point on each of the lines [marked] on the posts, draw a perpendicular to the line in the middle of the post and extend it across the width of the post. Bisect both segments of the line extending along the length of the post that is bisected by the perpendicular; and from the bisection points draw across the post's width perpendiculars to the line along the length of the post. The surface of the post will thus be divided by [all] these lines into four equal divisions, each equal to one digit and a half, and so the two divisions in the middle will [together] equal three digits.

[Mounting the mirrors on the posts]

[30] With this division completed in all | of the posts, each of them should now be IV 25b
carved in the middle, ((where the mirrors should be embedded)).

[31] For the plane mirror: its polished surface should be level with the [front] surface of the post; what is left at the ends of the straight line extending along the middle of the post should be continuous with the diameter imagined at the middle of the [circular] mirror. This can be achieved precisely by applying a ⟨sharp-edged⟩ ruler to the mirror [at] the line ((imagined [vertically] along the middle)).

[32] For the cylindrical convex mirror: carve out on both sides of the middle of the post a circular cavity in which the mirror should be fitted. Adjust the mirror so that the straight line imagined to extend through its middle will be in the [plane of the] surface of the post and in a straight line with the line drawn at the middle of the post. This can be done precisely by bisecting the arcs of the bases of the mirror and making sure, when mounting the mirror, that the bisection points touch the extremities of the remaining segments of the

line already drawn at the middle of the post, and by applying a sharp-edged ruler to these two points so that the ruler's edge coincides with the cylindrical surface and with the two
IV 26a straight line [segments] at both ends of the post. Then firmly affix | the mirror to the post in this position.

[33] For the cylindrical concave mirror: carve out in the middle of the surface of the post a cylindrical cavity in which the mirror should be embedded. Make sure that the cavity occupies the middle of the post; that the four endpoints of [the mirror's] bases lie in [the plane of] the post's surface; that the two lines drawn across the width of the post form two chords of the mirror's bases; that the extremity of each of the two remaining segments of the line that was drawn along the middle of the post bisects the arc of one base or the other of the mirror; and thus the line extending along the middle of the mirror's surface, joining the points bisecting the arcs of the mirror's bases, will be parallel to the straight line drawn lengthwise along [the middle of] the post's surface. Then firmly affix the mirror to the cavity.

[34] For the conical convex mirror: carve out for it a conical cavity in the post. Begin the carving from [both] end[s] of the post, with the vertex of the cone being at the extremity of the line in the middle of the post's surface, and with the base of the cone being at the transverse line which is four and a half digits distant from this extremity. Fasten the mirror
IV 26b to this carved-out area, | making sure that the point at the vertex of the mirror coincides with the extremity of the line that was [drawn] in the middle of the post's surface; that the middle of the cone's base will coincide with the extremity of the remaining [segment of the] line drawn in the middle of the post; and that the middle [point] on the cone's base lies at the extremity of what is left of the line drawn in the middle of the post; so that the straight line extending from the vertex of the conical surface to the mid[point] on its base will lie in [the plane of] the surface of the post and in a straight line with the line [segment(s)] left out of the line drawn in the middle of the post.

[35] For the conical concave mirror: carve out for it a conical cavity in the middle of the post. The cavity should extend from [one] end of the post to the line whose distance from [this] end is four and a half digits. Make sure that the carving is so deep that the line extending from the vertex of the cavity to the mid[point] on the arc of its base will be parallel to the line that was drawn in the middle of the post. Then the mirror should be embedded in this cavity, so that its vertex will be in [the plane of] the surface of the post's end; with the distance of this vertex from the front surface of the post being equal to the distance between the extremity of the remaining line [segment] at the middle of the
IV 27a post and the mid[point] on the arc of the base of the conical surface, namely half | a digit. Thus, the straight line imagined to extend in the middle of the cone from the vertex to the mid[point] of the base will be parallel to the straight line that was drawn in the middle of the post's surface.

[36] For the spherical convex mirror: carve out for it a spherical shape in the middle of the post. This can be done by taking the mid[point] on the post as centre and describing about it a circle with a three-digit diameter, thus equal to the diameter of the mirror's base. This circle should then be engraved [starting] from the circumference, thus producing the [hollow] segment of a sphere into which the mirror would be inserted. The limit [i.e., top] of the mirror's convexity should be at the middle of the post's surface. To do this precisely, apply to the surface of the post a sharp-edged ruler of the same length as the post, and let the edge of the ruler be very sharp. Bisect the ruler's [sharp] edge and mark

the bisection point. Slide the ruler over the surface of the post until its [sharp] edge falls upon the remaining ends of the line that was drawn in the middle of the post's surface, while touching the convexity of the mirror at the bisection point on the sharp-edged | ruler. IV 27b
With this position precisely obtained, the mirror should be firmly affixed to the post.

[37] For the spherical concave mirror: draw for it in the middle of the post's surface a circle with a diameter equal to that of the mirror's base [i.e., three digits]. Hollow out this circle [starting] at the middle in the shape of a sphere with the same concavity as the mirror, and embed the mirror in it. Make sure that the round edge of the mirror's concave surface lies on the surface of the post, with a diameter of the mirror's circle being in a straight line with the remaining ends of the line [that was drawn] in the middle of the post's surface. The centre of the mirror's [circular] base should be at the middle of post's surface. This position can be precisely secured [as follows]: bisect the length of the sharp-edged ruler and mark the bisection point; mark off on either side of this point two equal line [segments] each equal to half the diameter of the circle ⟨⟨forming the base⟩⟩ of the mirror; mark the division[s] by two points. Apply the edge of the ruler to the surface of the post in which the mirror has been embedded, so that this ⟨sharp⟩ edge may fall upon the remaining line [segments] on the surface of the post, while the two points [marked] on the sharp ruler will meet the circumference of the mirror's circle at | two opposed points. Then IV 28a
firmly affix the mirror to the post in this precise position.

[38] Mark off with a dot on the line in the ⟨middle⟩ of the [copper] plate, next to the centre of the plate, a segment equal to the sagitta of the mirror's concavity. This sagitta can be determined [as follows]: place the sharp-edged ruler upon the surface of the post as previously described, so that the ruler's edge may fall upon the two line [segments] on the post, with the two points [marked] on the edge coinciding with the extremities of the diameter of the mirror's circle. In this situation, and without shifting the ruler sideways, [the experimenter] should lift the ruler a little so as to apply a fine needle to its edge at the point in the middle of the edge, which point is now the centre of the mirror's circle. He should hold the needle perpendicularly to the ruler's edge and [advance it] into the mirror's cavity until the tip of the needle meets the concave surface of the mirror. Then, while [the experimenter] holds the ruler and the needle [in place], someone else should carefully mark with a fine pen the point at which the needle touches the ruler; then the needle should be removed. The distance between | the tip [of the needle] and the point IV 28b
marked on it will then be equal to the sagitta of the mirror's concavity. Then mark off an equal segment on the line ((extending through the middle of the triangle [at the plate's edge] and passing through the triangle's vertex, i.e., the centre of the plate.

[39] Having marked off this line [segment], the post should be mounted into the depression in the board, then moved forward until it comes to a stop when its concave surface reaches the centre of the plate. The [experimenter] should then apply the edge of the ruler to the surface of the post that is orthogonal to the surface of the [copper] plate, then lower it until it meets the surface of the plate; then mark the point at which it meets the line in)) the middle of the plate. This marking point will fall short of the first point [marked] on this line, namely the point whose distance from the centre of the plate equals the sagitta of the mirror's concavity. The reason is that the sagitta is perpendicular to the surface of the post at its midpoint, which is [now] higher than the surface of the plate by half [the length] of the barley seed; thus the centre of the plate will meet the concave surface of the mirror at a point other [i.e., lower] than the extremity of the sagitta; and, therefore, the part of the line on the plate that will penetrate the concavity of the mirror will be shorter than the sagitta of that concavity. And for this reason, the sharp-edged ruler will meet the line on the plate at a point before the first point, if the mirror's concavity is perfectly spherical, though the distance between the two points will be very small. With these two points marked on the line on the plate, let a point be marked on the concave surface of the mirror at that same point at which the centre of the plate meets the surface of the mirror.

[40] [The experimenter] should then remove the post and drill through the point
IV 29a marked in the mirror's concavity a small ⟨⟨conically shaped⟩⟩ hole equal in depth to | the remaining segment between the two points assumed on the line [drawn] on the plate. This done, the post should again be mounted into the depression, then moved to let the plate's centre penetrate the concavity of the mirror; the ⟨post⟩ should then be gently adjusted in this position until this centre of the plate enters the hole and reaches its end. In this situation, let the post be firmly settled in the depression, and apply the ruler's edge to the surface of the upright post, making sure that the edge reaches the surface of the plate and falls upon the first point that was marked on this line [upon the plate] at a distance from the plate's centre equal to the sagitta of the mirror. If the sharp edge of the ruler fails to meet that point, remove the post carrying the mirror, drill the hole precisely and accurately adjust the position of the post when mounted in the depression, so that: the centre of the plate will enter the hole in the mirror's concavity; the first point marked on the line on the plate will lie in [the plane of] the surface of the post; and with this position exactly secured, the centre of the plate, and the extremity of the sagitta of the mirror, will together
IV 29b lie in a plane | parallel to the surface of the post; the line imagined between the centre of the plate and the extremity of the sagitta of the mirror's concavity will be at right angles to the surface of the plate; the extremity of the mirror's sagitta will be in the middle of the plane parallel to the surface of the post; the sagitta will lie in the plane passing through the centres of the holes [in the ring], because it has the same distance as these centres are from the plate's surface; and this sagitta will be at right angles to the plane parallel to the surface of the post.

[Experimental examinations of the manner of reflection of lights]

[Examination of the reflection of light from plane mirrors]

[41] To examine the manner of reflection of lights from all [the types of] mirrors described, let the experimenter [first] mount the post with the plane mirror in the depression at the middle of the board, and let the centre of the plate touch the surface of the mirror. Place an object in the gap left in the depression in front of the post to stabilise the position of the post and prevent it from shifting. If a small gap also remains in the depression behind the post, press a chip of wood into it to fix the post in place and prevent it from
IV 30a changing position when the instrument | is moved, turned over, lifted, or put down. Once the post is fixed in this manner, with the centre of the plate in contact touch with the mirror, all the holes in the [ring] should be stopped except for one of the inclined holes. The holes should be stopped on the interior [concave surface of the ring] using small bits of paper lightly attached [to the surface]. Press the tip of one finger on [each] of the holes themselves, causing the round edge of the hole to make a visible impression on the paper. Using a fine pen, draw a circle round that visible impression.

[42] Having done that, the experimenter should place the instrument in the sun's light, with the [convex surface of the ring] facing the sun and with the open hole in it facing the body of the sun. He should tilt the instrument so that the light of the sun may enter through the hole and appear on the surface of the mirror, whereupon he should contemplate the interior [concave] surface of the [ring], namely that with the stopped holes. He will find the light reflected from the mirror's surface upon the [interior concave] surface of the [ring], and [also] find the light reflected upon the hole corresponding to the open hole, namely the one that is | as distant as the open hole is from the line drawn through the centre IV 30b
of the plate and bisecting the arc on the plate. The experiment will be clearer and more reliable if conducted in a chamber into which sunlight enters through a small opening.

[43] When the reflected light shows up, the experimenter should stop this hole on the interior [surface] of the ring with a piece of paper, open the hole on which the reflected light appeared, turn this opened hole towards the sun, and examine the light. He will find the light reflected onto the hole that was formerly open. He should then insert the copper tube into one of these two holes until the extremity of the tube reaches the surface of the [ring]. If the tube stays firm then it should be left alone; but if somewhat loose, it should be stabilised by being attached to the surface of the plate with a piece of wax. Make sure that the line drawn on the plate below the tube is in touch with the tube and parallel to the axis of the tube, a position realised when the perpendicular ⟨⟨to the plate⟩⟩ drawn from the extremity of the line sensibly passes through the centre of the hole – as mentioned | earlier. IV 31a

[44] With the tube precisely positioned, the experimenter should now use it to examine the light by facing the end of the tube towards the sun's body, so that the sun's light may pass through it and appear on the mirror. Contemplating the [concave] surface of the [ring] as the light appears on the mirror, he will find the light reflected on the hole corresponding to the one holding the tube. The experimenter should then take a piece of wax, roll it into a thread, wind it around the end of the tube [at the ring's convex surface] and attach it to the tube so as to narrow the tube's opening into a small aperture. Make sure that the resulting aperture is at the middle of the tube's opening. Do the same thing with the other end of the tube, so that the tube now has, at both ends, two small, opposed apertures at the middle of their openings. [The experimenter] should now return the instrument to its [original] position and examine the reflected light. He will find the reflected light to be small, as it shows up at the centre of the hole corresponding to the hole [reached by] the tube.

[45] Now, the two opposed apertures at the centre[s] of the tube's ends are on the straight axis of the tube; the light that is in them is a light that extends | along this axis; IV 31b
and the light that is reflected from the extremity of the tube's axis to the centre of the corresponding hole has been shown to extend along the straight line drawn from the reflection point to the centre of the corresponding hole, it having been shown that reflected lights only extend along straight lines. This experiment therefore shows that the light that extends along the axis of the tube will, upon reaching the surface of the mirror, be reflected along the straight line drawn from the extremity of the tube's axis to the centre of the hole corresponding to the hole [through which] the tube [passes].

[46] The experimenter should now remove the wax, take the tube out of its hole, stop the hole on the interior [surface of the ring] with a bit of paper, open the hole on which the reflected light appeared [earlier], insert the tube into this hole in the manner previously determined, set the instrument to face the sun as he did first, so that the sun's light may pass through the tube and appear on the mirror, then contemplate the interior surface of the

IV 32a [ring]. He will find the light reflected upon the hole | in which the tube was first inserted, namely that equally distant from the line in the middle of the plate as the one now fitted with the tube. If he then encircles the end of the tube with wax, as he did before, he will find the light reflected at the centre of the hole on which it appears.

[47] With that made clear, let him take the tube out of its position, stop the hole it is in with the piece of paper, open a hole in the [ring] other than these two, and examine the light passing through it in the absence of the tube. He will find the passing light reflected upon the hole corresponding to it. Upon closing this hole and opening the one corresponding to it, and examining the passing light, he will find the light to be reflected upon the first hole; he should then insert the tube in each of these two holes and examine the light passing through the tube. He will find the light to be reflected upon the hole corresponding to the one holding the tube, and, upon narrowing the opening of the tube with wax at both ends and examining the light passing through it, he will find the light to be reflected upon the
IV 32b centre of the hole corresponding to the one | with the tube.

[48] The experimenter should then similarly examine the light in each one of the holes in the [ring]. He will find the light always to be reflected upon the hole that corresponds to the opened hole. When the tube is mounted, he will find the light to be reflected upon the centre of the hole corresponding to that with the tube, and this light will be ((narrower than the first light; and upon narrowing the tube's opening he will find the light [also] reflected upon the centre of the corresponding hole, while being)) extremely small [in area].

[49] The experimenter should then close all holes except the middle one, and insert in it a straight cylindrical rod whose width matches the hole's opening and whose end has been sharpened in the shape of a cone; slide the rod through the hole until it meets the surface of the mirror at a point, which should then be marked; then take the rod out and face the instrument towards the sun until the light passes through the middle hole and appears on the mirror's surface. This light, when it appears, will be round, and wider than the hole, because lights diverge in the shape of a cone as they come out of holes, and widen further as they recede from the hole, the width of the visible light that comes out
IV 33a of the hole being dependent on | the distance between the end of the hole and the place where the light appears, and dependent on the length of the hole, when it is fairly oblong. The experimenter should make sure that the point marked on the mirror's surface lies at the middle of the light when it appears. He should then mark a point on the circumference of the light, and, with the first point as centre, draw a circle with the distance of the second point [as radius]. He should then [re-]examine the light, making sure that the circumference of light on the surface of the mirror coincides with the circumference of the circle that has been drawn on it.

[50] With this position precisely reached, let him erase the point [marked] at the middle of the mirror and contemplate the middle hole from the interior of the [ring]. He will find that a circular light surrounds the hole on all sides, and that the light will be equal in width. The [particular] appearance of this light is due to the fact that the light on the mirror is wider than the hole, because of the divergence of the light as it comes out of the hole, and because reflected light also diverges and widens.

[51] Having precisely determined [the features of] the reflected light in this manner, the experimenter should now narrow the middle hole on both sides [of the ring] with wax,
IV 33b ensuring that the | remaining [apertures] at both openings are opposite each other, then examine the reflected light. He should make sure in this examination that the light that has

passed through the opposite holes and appeared on the mirror is found at the middle of the circle drawn on it, and that the circumference of the light is uniformly distant from that of the circle. Contemplating the middle hole from the interior of the instrument, he will find: that the light that formerly appeared around the hole has vanished and that the light is now narrowed; and the more he narrows the opening at both ends, the narrower this light will be, until a limit is reached when no light appears around the hole.

[52] It is clear from this examination that the light that has extended along the axis of the hole is reflected along the axis itself, and not along any other line. For the light that appeared around the hole is a light reflected from the light existing around the axis when the hole is wide, and not from the light that extends along the axis. For if the light extending along the axis were reflected along a line other than the axis, then, upon narrowing the hole while the light [still] extends along the axis, the [reflected] light would appear at | a point IV 34a
on the hole's circumference. But since no reflected light appears [there] while the light extending along the axis passes through to the mirror's surface, it is clear that the light extending to the mirror's surface along the axis is not reflected along any line other than the axis. Further, since it is an [essential] property of light to be reflected from a polished surface upon meeting it, then the light that has extended along the axis will be reflected from the mirror's surface; and if the reflection does not take place along a line other than the axis, then the light will be reflected along the axis itself.

[53] The experimenter should now tilt the post with the mirror backwards, having removed what had jammed it in place, while leaving it in the depression and only slightly inclined; he should then fix the post in the tilted position and examine the light. He will find that the light passing through the middle hole along its axis [now appears] on | the IV 34b
[interior] surface of the [ring] above the hole, on the vertical line drawn along the height of the ring through the centre of the hole. Upon increasing the inclination of the post, the experimenter will find the reflected light raised higher and farther from the hole; and upon decreasing the inclination, the reflected light will be less displaced from the hole. But the light will always be found on the line drawn perpendicularly to the plate and passing through the centre of the hole. And as the inclination of the post is decreased, the reflected light will approach the hole; and upon restoring the post to its first position the reflected light will return to the hole itself.

[54] From this experiment it is clear that when the light passing through this [middle] hole meets the mirror, while the post stands in the upright position, it will be reflected from the mirror to the hole itself; and if the light meets the mirror while the post is inclined, it will be reflected to a spot other than that of the hole; and the light reflected upon this hole will always be on the line passing through the centre of the hole [in the direction] perpendicular to the surface of the plate.

[Examination of the reflection of light from the other types of mirror]

[55] All these notions having been established, the experimenter should then remove the post with | the plane mirror, replace it with one of the other posts that have been IV 35a
fitted with their mirrors, and examine the sun's light through all the holes, with the tube [inserted as before]. He will find the light passing through each of the holes to be reflected upon the corresponding hole, as was the case with the plane mirror. Upon narrowing the opening of the tube, the reflected light will be found at the centre of the hole on which the

reflected light shows. Examining the middle hole, the light will be found reflected upon the middle hole itself, where it will appear. If this hole is narrowed at both ends, as before, the reflected light will narrow and disappear from around the hole. If the post is inclined [backwards], the light reflected on the [concave] surface of the instrument will be found to have risen above the hole on the vertical line [drawn] along the height of the ring. Upon increasing the inclination of the post, the reflected light will move farther from the hole. When the inclination is decreased, the reflected light will approach the hole; and upon restoring the post to its first position, the reflected light will return to the hole itself.

[56] The experimenter should examine the mirrors one by one in all the ways we have IV 35b determined. He will find that the light | will be reflected from each of the mirrors and from each of the holes in the instrument, and will appear on the hole corresponding to the one through with the light enters; all the features we have noted in the case of the plane mirror will hold when the experiments are conducted with the help of the narrow tube.

[57] After the experimenter has mounted the post fitted with the cylindrical [convex] mirror, and examined the light by means of it in the preceding ways, he should loosen the post out of the depression, incline it to one side with the corner of the post touching the bottom of the depression, and bring the mirror into contact with the centre of the plate; he should make sure that this centre lies on the line extending along the length of the mirror continuously with the line in the middle of the post's surface, and that this surface is perpendicular to the plate's surface. Thus, the line extending along the mirror's length will be inclined to the plate's surface. He should stabilise and firmly fix the post in this position by [pressing] some wax on its sides at the bottom, then examine the light. He will find that the light coming out of the hole is reflected to the hole corresponding to it, and that the IV 36a middle of the light appears on the centre of the hole. Examining each | of the holes in this position of the mirror, he will find the reflection to be uniform and unchanged for every hole examined, and that the light passing through the hole is reflected to the corresponding hole and appears at the centre of the hole.

[58] When the experimenter mounts the post that has been fitted with the concave mirror[s], he should move it [forward] in the depression until the centre of the plate enters the concavity. In the case of the cylindrical concave and the conical concave mirror the centre of the plate should penetrate the concavity of each until it meets the mirror's surface; then the post should be settled in the depression and the light examined in the preceding fashion. In the case of the spherical concave mirror, the plate's centre should be made to penetrate the concavity of the mirror until this centre enters the small aperture [already drilled] in it, while the post stands at right angles with the [base-]board. The post should then be stabilized and the light examined. As for the spherical convex mirror, it should be mounted in the depression with its surface close to the plate's centre but not in contact with it; the sharp-edged ruler should be applied to the face of the vertical post carrying the mirror, then the ruler should be lowered until its edge meets the centre of the plate. In this IV 36b situation, | the plate's centre will be on the surface of the vertical post. With this position precisely obtained, the post should be settled in the depression and the light examined. When the experimenter examines the cylindrical concave mirror, he should also incline it as he did with the spherical mirror, then examine it.

[59] To examine the middle hole by means of the cylindrical and conical mirrors, the experimenter should determine the point on the mirror's surface opposite the centre of the hole, using [for this purpose] the straight rod as shown earlier. This point will then be on

the straight line extending along the middle of the length of the mirror and the middle of the post's surface. Let him mark this point; mark off from it, and on the straight line imagined to extend along the middle of the length of the mirror, a segment equal to the radius of the circle drawn in the middle of the plane mirror and surrounding the light that appears there; mark this point and erase the first point. Then mount the post in the instrument and examine the light that passes through the middle hole. Move the instrument until the top of the circumference of the light lies on the higher point marked on the mirror's surface while the two sides of the light are | equally distant from the lower point facing the centre IV 37a
of the hole. This equality can be determined by means of a compass. Mark [the distances of the sides] with two points. With the position of the light thus precisely determined, the light passing along the axis of the tube will meet the mirror at the point that was previously marked on the mirror's surface.

[60] In this precise position, let the experimenter contemplate the middle hole on the interior [concave surface] of the instrument. He will find it surrounded with the reflected light, also that the vertical line [drawn] along the height of the ring through the centre of this hole bisects this light. [The experimenter] should then complete the examination as before.

[61] To examine the middle hole by means of the two spherical mirrors, let the experimenter draw on each of them a circle equal to the one surrounding the light on the plane mirror. The centre [of this circle] should be the point opposite the centre of the hole, which point can be determined by means of the rod as before. He should then examine the light passing through the middle hole in the case of these [two mirrors]. He will find the reflected light surrounding the middle hole, and that the vertical straight line along the height of the ring bisects this light. It is in this manner that the sun's light may be examined. | IV 37b

[Examination of accidental lights]

[62] Accidental light may be examined experimentally as we [now] describe. The experimenter should use a chamber whose door faces a white wall on which sunlight can shine. Let the wall be close to the door. Let him drill in the same door an opening as large as a dirham. When the sun shines on the facing wall he should enter the chamber, close the door, and draw over it from inside a heavy curtain to prevent the light from entering the chamber. He should then cut a hole in the curtain opposite the opening in the door. The experimenter should then open one of the holes in the instrument, bring the instrument in contact with the door, place the open hole [of the instrument] over the hole in the door, and gently move it until the secondary light [from the outside wall] appears on the mirror's surface; he should then contemplate the [interior] surface of the instrument. He will find on it a reflected light that appears on the hole corresponding to the opened hole. Having observed this, he should examine the light by means of the tube, going through all the preceding experiments with all the holes in the instrument and all the mirrors. He will find the case with accidental light to be the same as that of [direct] sunlight with regard to all the preceding notions, the only difference being that the reflected sunlight will be stronger than | the reflected secondary light. IV 38a

[63] The experimenter should then make an opening in the door large enough to accommodate the instrument, cut a similar hole in the curtain, then open two holes on one half of the ring, and close the remaining holes. Then he should mount the instrument in the

opening so that the side with the two opened holes faces the wall on which the sun shines. (Let the wall be wide enough to face [all] the holes in the instrument.) This done, let him close all remaining gaps in the opening surrounding the instrument so that no light may enter [through them] into the chamber. Then move the instrument until the light passes through both opened holes and reaches the mirror. The passage of the light through these holes should be tested by facing them both from inside the instrument with a thin white object, such as paper and the like; bring this object close to the holes and contemplate it while gently moving the instrument until the light from the holes appears on it. When the light thus appears, [the experimenter] should remove the object and contemplate the interior [surface] of the instrument. He will find that two reflected lights [appear] on the
IV 38b other half of the ring, and these will be [located] on the two holes | corresponding to the two opened holes.

[64] If [the experimenter] closes those two holes, then opens two others, and examines the light as before, he will find the light to be reflected to the two holes corresponding to those others. If he opens all the holes in one half of the ring and contemplates the light, he will find on the other half as many reflected lights as those holes. If he obstructs one of the holes through which the light enters, the reflected light will vanish from the hole corresponding to the obstructed one. Upon removing the obstruction, the reflected light will return to its [former] place. If the experimenter narrows both ends of each one of the holes through which the light enters, he will find the reflected light at the centres of the corresponding holes. [And he will find] that the lights coming out from all the holes in the instrument all meet at a single place on the surface of the mirror, namely the middle of the post, and that they are reflected from that single place to all places on which the reflected lights appear. It is thus seen from this experiment that the lights extending along different
IV 39a lines to a single place on the polished body are reflected from that | place to different places; and that each of the different lights is reflected from that place along a line whose inclination to the polished surface is similar to the inclination of the line along which that light has extended to that place.

[65] If the experimenter examines the light when daylight, instead of direct sunlight, shines upon the wall facing the door, he will find the reflected light to be as previously described. And if he examines the light with all the holes and instruments [i.e., mirrors], he will find it as he did when examining sunlight, with no difference between the two [kinds of] light, except that the secondary light that extends from the sun's light is stronger than the secondary light that extends from daylight. It is in these ways that accidental light may be examined.

[Conclusions from the preceding experiments; the concept of least light]

[66] Having shown how to examine experimentally the reflection of lights, and what the experiments reveal, we now say, based on them: that light is reflected from a point on a polished surface only in the plane orthogonal to the plane tangent to the polished
IV 39b surface at the point of reflection; that the light extending along the perpendicular | to the plane tangent to the polished surface is reflected only along the perpendicular itself; and that the light extending along a line inclined to the polished surface is reflected only along the straight line that is drawn from the point at which that line meets the surface, and that contains, together with the line along which the light has extended to that point,

two equal angles made with the perpendicular to the plane tangent to the polished surface at that point.

[67] For if the light passing through one of the holes of the instrument is reflected to the hole corresponding to it, then it is clear that the position of the line along which the light is reflected, relative to the mirror's surface, is the same as [*musāwī*] the position of the line along which the light has extended to that surface. [That is so] because the position of these two holes relative to the middle of the mirror, where the light emerging from the hole appears, i.e., the point of reflection, is similar, and the position of the straight lines extending from the reflection point to each pair of these holes is the same, and [because] light only extends along straight lines and is reflected only along straight lines.

[68] Further, when | [the experimenter] closes the hole through which the light enters, IV 40a
and opens the corresponding hole on which the reflected light appears, then examines the light by means of the second hole, he will find the light reflected upon the hole that was open at first. Now, the light emerging from each of these two holes has reached one and the same point on the surface of the mirror; and the straight lines extending from that point to one of the holes are the same unchanged lines; and the straight lines that extend from that same point to the other hole are also the same, unchanged lines; and the straight lines along which the light is reflected in the first examination are the same as those along which the first light extends in the second examination; and the lines along which the first light extends in the first examination are those along which the light is reflected in the second examination. Thus, it is seen from this experiment that the light is reflected only along the lines whose position relative to the polished surface is the same as the position of the lines along which the light has extended to the point of reflection.

[69] Moreover, when the hole is made narrower by the tube, the reflected light also narrows, and this light is also found | on the same hole corresponding to the one [fitted] IV 40b
with the tube. When the tube is again narrowed with wax, so that only the light passing along the axis of the tube remains, the reflected light is found on the centre of the hole corresponding to the hole with the tube. But the position of the straight line extending from the reflection point to the centre of the corresponding hole, relative to the plane tangent to the mirror's surface, is the same as the position of the tube's axis along which the light extends to the reflection point. This experiment thus makes it clearer that the light is reflected only along the line whose position relative to the plane tangent to the polished surface is the same as the position of the line along which the light has extended to the reflection point.

[70] Again, when the light is examined by means of each of the holes inclined to [i.e., on either side of] the mid[point] on the instrument, the reflected light will be found as previously described. If the examination is made by means of the middle hole, the reflected light will be found on the middle hole itself.

[71] It is thus clear from examining all the holes: that the light[-ray] perpendicular to the plane tangent to the polished surface is reflected along the perpendicular itself | along IV 41a
which the light has extended to the polished surface; that the light inclined to the polished surface is reflected on inclined lines, and that the inclination of the lines along which the first light extends to the polished surface is always equal to that of the lines along which that light is reflected; and that when the inclination of the first light to the polished surface increases, so will the inclination of the reflected light to that surface; for when the hole by means of which the light is examined gets farther from the middle hole, so will the

reflected light; and the distances of the first [i.e., incident] light and of the reflected light from the middle hole will always be equal.

[72] Thus, by considering the equality of the inclination of the reflected light with that of the first [incident] light from which it has been reflected, and by the uniformity of this property of light at various inclinations, and from considering the reflection of the light along the perpendicular itself along which it has [first] extended, it is clear that when light meets a polished surface, it is reflected only along those lines whose position relative to the plane tangent to the polished surface is identical to that of the lines along which it has extended to the polished surface. Further, considering the fact that this property holds for
IV 41b reflected light | when the hole is narrowed by the tube and when narrowed [further] with wax, and [from] the uniformity of that [property] for all narrowed holes, it becomes clear that the reflection of light, whether plenty or little, behaves in one and the same manner. And considering the fact that this property holds for sunlight and accidental light, it is clear that this property belongs to all lights. If moonlight and the firelight are experimentally examined by means of the instrument described, it will be found that their light is also reflected in the preceding manner.

⟪IV 3.2⟫

[The rules of reflection apply to the least light]

[73] It is clear from all this that this property is a natural property, inseparable from all lights, whether they are little or plenty, essential or accidental, strong or weak. This being the case, this property is inseparable from the least light, as long as it keeps the form of light. And this also being the case, the least light that extends along the axis of the tube previously described can be reflected from the surface of each one of the mirrors described only along the line whose position relative to the plane tangent to the mirror's surface is the same as that of the tube's axis from that surface. Thus, if the light that shows when the
IV 42a examination is made with | the narrowed tube is the least light, then it is sensibly apparent that the least light is reflected along the line whose position at the point of reflection is the same as that of the line along which the light has extended to that point: for the line that extends from the extremity of the axis of the tube at the mirror's surface to the centre of the hole corresponding to the one with the tube, has the same position with regard to the plane tangent to the mirror's surface as that of the tube's axis to this surface. And if the least light cannot be perceived by sense, or discerned, or subjected to experimentation, then it is shown by reasoning that the [essential] property of that light, namely, the least light than which there can be no subtler light, is the same as the property of the wider light. And if the light that sense perceives, whether it is little or plenty, is [found to be] inseparable from the property described earlier and to be uniformly [exhibiting] the same manner [of behaviour], then the least light, even if not perceptible by sense, is also inseparable from the property described earlier, as long as it keeps the form of light. Thus, it is seen from
IV 42b the earlier experimental examinations, and from this | [latter] reasoning, that all lights, whether little or plenty, strong or weak, essential or accidental, are inseparable from the property we have described, and that the least light that extends along the axis of the tube previously described is reflected along the corresponding line, whose position at the point of reflection is the same as that of the axis at that point.

[Precise determination of the line of reflection for each one of the seven mirrors]

[74] Having shown the preceding, we will now determine precisely the position of the line along which the light is reflected. I say [that] the face of the post, being set up vertically in the instrument during the experiment, is orthogonal to the surface of the [base-]board, for it is orthogonal to the surface of the depression, which is parallel to the board's surface; and since the latter surface is parallel to the surface of the [copper] plate, the face of the post is orthogonal to the plate's surface if the latter is imagined to extend so as to cut the post. Now, the width of the post, i.e., the intersection between its face and the board's surface, will cut the line in that surface at right angles, if the line is imagined to be extended; and the line on the surface of the board is parallel to the line in the plate that issues from the centre [of the plate] and bisects the arc in the plate; and the intersection between the surface of | the board and the face of the post, if the former is imagined to IV 43a
extend over the depression, is parallel to the intersection between the face of the post and the plate's surface. Therefore, the intersection between the face of the post and the plate's surface will contain two right angles with that line in the middle of the plate. Again, the straight line that extends lengthwise in the middle of the face of the post is perpendicular to the width of the post that lies on the surface of the depression; and this width is the intersection between the face of the post and the surface of the depression. Therefore, the line along the middle of the post is perpendicular to the surface of the depression. But the surface of the depression is parallel to the board's surface, and the latter is parallel to the plate's surface. Therefore, the line along the middle of the post is perpendicular to the plate's surface, and, therefore, the line along the middle of the post, which meets the centre of the plate, if imagined to be extended, will contain a right angle with the line drawn from the plate's centre to the mid[point] on the plate['s arc]. It is also perpendicular to the intersection between the face of the post and the surface of the plate. | This line in IV 43b
the middle of the plate is, therefore, perpendicular to the face of the post and to the surface of the plane mirror mounted on the post.

[75] Now, it was shown earlier that the plane through the centres of the holes [in the ring] is parallel to the plate's surface. The distance of this plane from the plate's surface is half [the length of] a barley seed, and the diameter of each of the holes and of the exterior circumference of the tube is equal to one barley seed. Thus, the plane through the centres of the holes cuts the middle of [the opening of] the tube when the latter is fitted into one of the holes; the axis of the tube will thus [also] lie, during the experiment, in the plane through the centres of the holes. Again, the straight line extending in the middle of the plate and tangent to the tube's surface is parallel to the axis of the tube, for every straight line in a cylindrical surface is parallel to the axis of this surface; and the point reached by the tube's axis on the surface of the mirror lies in the plane through the centres of the holes; and the line between this point and the centre of the plate is perpendicular to the surface of the plate, because it is continuous with the line along the middle of the post, | and is IV 44a
thus equal to all the perpendiculars from the centres of the holes to the surface of the plate. Therefore, the straight line that joins the point on the mirror's surface at which the axis of the tube ends, with the centre of the hole corresponding to the one fitted with the tube, is, during the experiment, parallel to the line extending at the middle of the plate from the centre of the plate to the extremity of the diameter of the hole corresponding to the one with the tube. This is so because the vertical lines along the height of the ring, which

passed through the centres of the holes and past the extremities of the lines in the plate, are perpendicular to the surface of the plate, and thus they are equal to one another and to all perpendiculars drawn from the plane through the centres of the holes to the surface of the plate. Therefore, the angle contained between the axis of the tube and the line drawn from its endpoint to the centre of the hole corresponding to the one with the tube equals the angle contained by the two lines on the surface of the plate, of which one touches the tube's surface and the other extends from the centre to the extremity ⟨of the diameter⟩ of the hole corresponding to the one with the tube.

IV 44b **[76]** If we now imagine a line | stretching from the extremity of the tube's axis, [which extremity is located] on the mirror's surface, to the centre of the middle hole, it will lie in the plane through the centres of the holes; it will further be the axis of the middle hole; and it will be parallel to the line extending from the centre of the plate and bisecting the arc in it, since the centre of the middle hole is as distant from [the surface of] the plate as the extremity of the tube's axis on the mirror's surface is from the centre of the plate. But the line in the middle of the plate bisects the angle contained by the two lines previously mentioned, namely the one tangent to the tube's surface and the one extending from the centre to the hole corresponding to the one with the tube. Thus, the line imagined to extend from the extremity of the tube's axis located on the mirror's surface to the centre of the middle hole will bisect the angle contained by the [tube's] axis and the line drawn from its extremity to the centre of the hole corresponding to the one with the tube, since this [imagined] line is parallel to the line drawn from the centre of the plate so as to
IV 45a bisect the arc in the plate; and this middle line is perpendicular to the face of the post, | because the line parallel to it, which is drawn from the centre of the plate, has been shown to be perpendicular to the face of the post. Therefore, the axis of the tube and the line corresponding to it, i.e., the line extending from its extremity to the centre of the hole corresponding to the one with the tube, together contain two equal angles made with the perpendicular drawn from their meeting-point to the centre of the middle hole. Moreover, these three lines are in a single plane, i.e., the plane through the centres of the holes, which is orthogonal to the face of the post; and the face of the post, if imagined to be extended, is tangent to the surface of the mirror at the endpoint of the tube's axis. Therefore, the line drawn from the endpoint of the tube's axis at the mirror's surface to the centre of the corresponding hole, is the line whose position with respect to the reflection point is precisely the same as that of the axis of the tube with respect to that point.

[77] Now, it has been shown by experiment that the light that extends along the tube's axis is reflected from the surface of the mirror along the straight line drawn from the
IV 45b endpoint of that axis to the centre of the hole corresponding | to the one with the tube. It has also been shown by reasoning that the least light extending along the tube's axis is reflected along the straight line whose position with respect to the reflection point is the same as that of the axis of the tube to that point. Therefore, the least light extending along the tube's axis is reflected on the straight line that extends from the endpoint of that axis to the centre of the hole corresponding to the one with the tube. It has been further shown that this line lies, together with the axis, in the plane orthogonal to the plane tangent to the mirror's surface; and that it contains, together with the axis, equal angles with the perpendicular to the face of the post at the point of reflection. Therefore, the least light that extends along the tube's axis is reflected from the surface of the plane mirror in the plane orthogonal to the plane tangent to the mirror's surface, [and this reflection takes place]

along the line which contains, together with the axis of the tube, two equal angles with the perpendicular | to the tangent plane through the point of reflection. As for the light that IV 46a extends along the perpendicular to the face of the post, namely the line drawn from the centre of the middle hole in the plane through the centres of the hole, it has been shown to be reflected along this line itself.

[78] It was shown, moreover, that when the post is inclined [backwards] the reflected light rises above the hole [on which it appears]; that the reflected light moves farther from or closer to the hole as the inclination increases or decreases; and that in all these cases the light is [found] on the straight line that stands vertically along the height of the ring, passing through the centre of the middle hole.

[79] It is thus clear from this experiment that the light extending along the axis of the middle hole, when the latter is inclined to the polished surface, will be reflected in the plane containing [both] this axis and the line perpendicular to it, passing through the centre of the hole. Now, this plane remains orthogonal to the face of the post and to the surface of the mirror in all experiments with the middle hole. For when the post is perpendicular to the surface of the board, then | the axis of the middle hole will be perpendicular to the face IV 46b of the post, and every plane through the axis will be perpendicular to the face of the post; and when the post is inclined backwards in the position indicated in the experiment, then the width of it that cuts the surface of the board will cut at right angles the line that extends along the middle of the board. Thus, the axis of the middle hole, being parallel to this line, will cut at right angles the intersection between the face of the post and the plane through the centres of the holes; furthermore, that line in the middle of the [copper] plate, being parallel to the axis of the middle hole, will perpendicularly cut the intersection between the face of the post and the surface of the plate. Therefore, the plane through these two parallel lines will be orthogonal to the face of the post while the latter is inclined. But the vertical line extending along the height of the ring through the centre of the middle hole lies in the plane of these two parallel lines, because it meets both of them. Therefore, the lights reflected to this line, | when the post is inclined in the position we have indicated, IV 47a will always be in the plane orthogonal to the face of the post.

[80] Thus it is clear from this experiment, together with the first one, that the light is reflected from the plane mirror only in the plane orthogonal to the surface of the mirror; that when the line along which the light extends to the surface of the mirror is inclined to it, then the light will be reflected along a similarly inclined line; and that when the light extends along a line perpendicular to the surface of the mirror, it will be reflected along this line itself.

[81] As for the cylindrical convex mirror: when it is mounted in the instrument, the centre of the plate will meet the mirror at a point on the straight line imagined to extend in the mirror's surface, i.e., the line extending [lengthwise] in the middle of the face of the post; and the position of this post with respect to the plate will be the same as that of the post that is fitted with the plane mirror. Thus, the straight line drawn from the centre of the plate to bisect the arc that is in it will be perpendicular to the face of this post, for | this can IV 47b be shown as it was shown in the case of the post holding the plane mirror. But the axis of the tube meets the surface of this mirror at the point in the middle of the post, since the line extending on the surface of the mirror is on the surface of the post. Therefore, the straight line issuing from that meeting point to the centre of the middle hole in the thickness of the instrument will be perpendicular to the surface of the post, and it can be shown as was

shown earlier that this line is perpendicular to the face of the post holding the plane mirror. Therefore, the plane through the centres of the holes will be perpendicular to the face of this post. Now, if the light reflected from this mirror has emerged from the hole with the tube, and if the light extending along the axis of the tube is reflected to the centre of the hole corresponding to the one with the tube, then the extending light will be reflected from this mirror along the line that contains, together with the axis of the tube, two equal angles with the perpendicular to the face of the post at the point of reflection; and the
IV 48a three lines, namely the axis of the tube, the line along which the light is reflected, | and the perpendicular to the face of the post at the point of reflection, will be in a single plane, namely the plane through the centres of the holes that is orthogonal to the face of the post. Further, the face of the post is tangent to the surface of this mirror at the reflection point, because it is tangent to the mirror's surface on the straight line that extends in the mirror's surface. Therefore, the light extending on the axis of the tube will be reflected from the surface of the cylindrical convex mirror in the plane orthogonal to the plane that is tangent to the surface of that mirror at the reflection point; and [the reflection will take place] along the line that contains, together with the line along which the light has extended to the reflection point, equal angles with the perpendicular to the tangent surface at that point. And this will also be the position of the axis of the tube and the line of reflection of the light with respect to the face of the post when the post is inclined sideways.

[82] Examination of the middle hole by means of this mirror would [further] show that
IV 48b the light that extends along the axis of this hole | to the mirror would be reflected along the axis itself, which is perpendicular to the plane tangent to the surface of this mirror at the point of reflection.

[83] Again, examination of this mirror by means of the ⟨middle⟩ hole when the post is inclined backwards would show that the light is reflected from the cylindrical mirror only in the plane orthogonal to the plane tangent to the mirror's surface.

[84] The form of reflection from the conical convex mirror can be shown in the same manner. ((For the centre of the plate meets the surface of this mirror)) on the straight line extending in the middle of the face of the post, and thus the face of the post will be tangent to the surface of the mirror at the point of reflection; and all positions of the lines and points in this mirror are the same as those employed in the demonstration for the cylindrical convex mirror. Therefore, the light that extends along the axis of the tube to the conical convex mirror will be reflected in the plane orthogonal to the plane tangent to the surface of this mirror at the point of reflection, and [the reflection will take place] along ((the line that contains, together with the line along which the light has extended to the mirror's surface, two equal angles made with the perpendicular to)) the tangent plane
IV 49a at the point of reflection. |

[85] With regard to the cylindrical concave mirror: it was shown that it meets the centre of the plate at a point on the line that extends along its length, namely the line parallel to the straight line that extended in the middle of the face of the post. Thus, the line drawn in the plate from its centre to the middle of its arc, and [also] the line drawn parallel to it from the reflection point to the centre of the middle hole, will be perpendicular to the plane that is parallel to the face of the post and which passes through the reflection point. ((Further, the plane imagined to be parallel to the face of the post and passing through the straight line that extends along the length of this mirror and through the reflection point)) will touch the mirror's surface at that point, and the light extending along the axis of the tube will be reflected in the plane orthogonal to the plane tangent to the surface of the mirror at the reflection point. [And, finally, the reflection will take place] along the line that contains, together with the line along which the light has extended, two equal angles made with the perpendicular to this [tangent] plane [at the reflection point].

[86] It is in this same way that the form of reflection [takes place] from the conical concave mirror. For the centre of the plate meets the surface of this mirror on | the straight IV 49b
line that extends along its length [i.e., height], parallel to the straight line that extended in the middle of the face of the post. But the positions of all points and lines in this mirror are the same as those corresponding to them in the demonstration for the conical convex mirror. Therefore, the light that extends along the axis of the tube to the conical concave mirror will be reflected in the plane orthogonal to the plane tangent to the surface of the mirror at the point of reflection; and [the reflection will take place] along the line that contains, together with the line along which the light has extended to the mirror, equal angles with the perpendicular to the tangent plane at the reflection point.

[87] As for the spherical convex mirror: it was shown that the point of its convexity on the [plane of] the face of the post is the point at the middle of that face. But the mid-point on [the face of] each one of the posts lies in the plane through the centres of the holes, because it is the endpoint of the tube's axis; and [as we have seen] earlier, the post with the convex mirror is so positioned that, while examining the light reflected from it,
| its face passes through the centre of the plate. Thus, the position of the face of this post, IV 50a
during this examination, is the same as that of the post holding the plane mirror, and the point at the middle of the face of this post is the limiting [point] of the mirror's convexity. Thus, the axis of the tube meets the surface of this mirror at the point in its middle that lies in the plane through the centres of the holes. Therefore, the position of this point on the face of the post with respect to the plate and to the holes is the same as that of the reflection point in the middle of the post in the case of the plane mirror. Therefore, the light extending along the tube's axis to this mirror will be reflected from this point in the middle of the post; and the line going from this point to the centre of the middle hole will be perpendicular to the face of the post; and the line going from this point to the centre of the hole corresponding to the one with the tube, when the latter hole is inclined from [i.e., lies to one side of] the middle hole, will contain with the perpendicular an angle equal to that contained by the axis of the tube | and the same perpendicular; and the three lines will lie in IV 50b
the plane through the centres of the holes, which is orthogonal to the face of the post; and the face of the post is tangent to the surface of the mirror at the endpoint of the tube's axis. Now, it is clear from the experiment with this mirror that the light reaching this mirror along the tube's axis is reflected to the centre of the hole corresponding to the one with the tube. Now, lights only extend along straight lines, and they are reflected only along straight lines; and the point on the mirror where the axis of the tube ends is the point at which the face of the post is tangent to the mirror's surface. Therefore, the light extending along the tube's axis to this mirror will be reflected in the plane orthogonal to the plane tangent to this mirror at the reflection point; [and the reflection will take place] along the line that contains, together with the perpendicular to the tangent plane at the reflection point, an angle equal to that contained by the axis of the tube and this perpendicular.

[88] As for the spherical concave mirror: it was shown that the axis of its concave surface, | when the mirror is set up for experimental examination, lies in the plane passing IV 51a
through the holes; that the extremity of the axis lies at the middle of the plane parallel to

the face of the post; that the axis is perpendicular to this parallel plane; and that the line joining the extremity of the axis to the centre of the plate is perpendicular to the surface of the plate. That being so, the plane parallel to the face of the post, which passes through the extremity of the axis and through the centre of the plate, will have the same position with respect to the plate as that of the face of the post in which the plane mirror is fitted; and the position[s] of all points and lines to be used in the demonstration will be the same with respect to the plane parallel to the face of this post, as was the case with respect to the post fitted with the plane mirror. Now, the axis of the tube is in the plane through the centres of the holes, and ((so also is the axis of the mirror)); and the extremity of this axis is at the middle of the plane parallel to the face of the post. Therefore, the axis of the tube meets this parallel plane at the extremity of the axis of the mirror's concave surface; and the axis
IV 51b of the concave mirror, if extended rectilinearly, | will reach the centre of the middle hole; and the line issuing from the extremity of the axis of the mirror to the centre of the hole corresponding to the one with the tube – if the tube is inclined – will contain, together with the axis of the mirror, an angle equal to that contained by the tube's axis and the axis of the mirror, and the three lines will lie in the plane through the centres of the holes that is orthogonal to the plane parallel to the face of the post.

[89] Since it was shown by experiment that the light extending along the axis of the tube to the surface of this mirror is reflected to the centre of the hole corresponding to the one with the tube, then the light extending on the tube's axis meets the surface of this mirror at the extremity of its axis, and is reflected from this extremity along the line that contains, with the axis of the mirror, an angle equal to that contained by the axis of the tube and the axis of the mirror. Now, the axis of the mirror is perpendicular to the plane parallel to the face of the post and tangent to the mirror's concave surface at the extremity
IV 52a of the axis; and the plane in which the light is reflected, i.e., the plane through | the centres of the holes, is perpendicular to this tangent plane. Therefore, the light that extends along the axis of the tube to the surface of the spherical concave mirror is reflected in the plane orthogonal to the plane tangent to the mirror's concave surface at the reflection point, and [that reflection takes place] along the line that contains, with the perpendicular to the tangent plane at the reflection point, an angle equal to that contained by the tube's axis and this perpendicular.

[90] It is thus clear from all that we have explained regarding the conditions of these mirrors: that the light extending along the axis of the tube to each of the described mirrors will be reflected in the plane orthogonal to the plane tangent to the mirror's surface at the reflection point, [with the reflection taking place] along the line that contains, together with the line along which the light has extended to the reflection point, equal angles with the perpendicular to the tangent plane at the reflection point; that the light is reflected only
IV 52b in the plane orthogonal | to the tangent plane; and that the light extending on the axis of the middle hole, which is perpendicular to each of the planes tangent to these mirrors at the reflection point, will be reflected from each of the mirrors along this perpendicular itself.

[91] Now, the light extending along the axis of the tube and along the axis of the middle hole is not reflected in this manner on account of either axis, because if the tube were to be removed the light would [still] be reflected in the same manner and would only widen. Again, if the holes in the instrument [i.e., ring] were to be changed in width or shape or length, the light would still be reflected in the manner described. Nor does the reflection take place in that manner on account of the line itself along whose extension the light

extended from the luminous body to the mirror's surface at the time of the experimental examination, for the manner of reflection remains the same when the light is examined from many separate positions and from different luminous bodies. Therefore, the reflection of light from the [various] mirrors | in the manner we have determined must be a property IV 53a
that is peculiar to the nature of light, so that every light extending to the specified point on the described mirrors is reflected in the specific manner we have determined.

⟪IV 3.3⟫

[92] Furthermore, the position of every point on the surface of each of the described mirrors, with respect to that surface, is the same as that of any other point on the same surface. As for plane mirrors, their surfaces are the same in regard to all their conditions. This is also the case with the spherical mirror: every point on it has the same position with regard to its sphere as any other point on it. As for the cylindrical mirror, convex or concave, it is the case that every point on the surface of this mirror has the same position with respect to its straight length, ((and to its circular breadth, and to every line across the mirror's surface between the straight and the circular line, and to the plane orthogonal to the plane tangent to the mirror at that point, as the position of every [other] point on the surface of the mirror with respect to the straight line)) drawn from this point along the mirror's length, ⟨and to⟩ the circular line through this point on the mirror's surface, and to the line between the straight and the circular line, which has the same inclination to the | IV 53b
straight line as that of the first line, and ⟨with respect⟩ to the plane orthogonal to the plane tangent to the mirror at this point. Therefore, every light that reaches any point on the surface of the cylindrical mirror will be reflected in the same manner, because the position of every point on the mirror with respect to [the mirror's] length and breadth, and to the plane in which the reflection takes place, is one and the same.

[93] Also, if the two previously described cylindrical mirrors, the convex and the concave, are examined experimentally at a point other than the middle point, the reflection will be found to take place in the same manner. The experimental examination of these [two] mirrors at a point other than the first is made possible by placing in the depression beneath the mirror a thin body with a plane surface. The position of the mirror will then be this [different] position, and the reflection point will be different from the first. Examining the cylindrical convex and [cylindrical] concave mirrors in this manner, it will be found that the light is reflected from them in the first manner already explained. This proves that
| reflection takes place in the same manner from every point on the line extending along IV 54a
the length of the mirror; that all straight lines extending on the cylindrical surface have the same position with respect to that surface; and that light is reflected in the same form from every point on the cylindrical surface.

[94] Furthermore, the form of reflection does not vary with the cylinder being large or small: for if the cylinder described is removed from the post and replaced by a larger or smaller cylinder, the reflection will be found to take place in the same manner. And the same condition will hold for the conical convex and [conical] concave mirrors: for the position of any point on the surface of the conical mirror with respect to the straight line extending along its length, and to the circular line across its width, and to any line cutting the conical surface between the straight and the circular line, and to the plane orthogonal to the plane tangent to the conical surface at that point, is the same as the position[s] of all

IV 54b | [other] points on the mirror's surface. The conditions of the points on the conical surface will vary only in regard to the difference in magnitude of the circles and sections that cut the cone through these points. But whether the circle is large or small does not change the form of reflection.

[95] If the previously described conical mirrors are examined at a point other than the first, previously specified point, by raising the post in the manner indicated earlier, the reflection will be found to take place in the manner shown. If the post fitted with the conical mirror is inclined to one side, as we described with the cylindrical mirror, and the reflected light examined, the reflection will be found to take place in the same manner. Again, the position of the straight line extending along the length of the cone with respect to the surface of the cone is the same as the position[s] of all [other] straight lines along the length of the cone relative to that cone. And if the described conical mirror is removed from the post and replaced by another mirror [cut] from a larger or smaller cone, then
IV 55a examined, the reflection will be found to take place as in that | first manner.

[96] Thus, the reflection of lights from all mirrors, whether plane, cylindrical (convex and concave), conical (convex and concave), or spherical (convex and concave), takes place in the preceding manner; and if that is so, regardless of the different shapes of these mirrors, then the specific manner of reflection cannot be due to the shape of the mirror, but is rather something characteristic to polishedness, and dependent on the perpendicular to the polished surface. If mirrors of different shapes than these are also examined, such as those obtained by conic sections, the light will be found to be reflected from them also in the plane orthogonal to the plane tangent to the mirror's surface at the reflection point; and the line along which the reflection takes place will be found to contain, with the line along which the light has extended to the mirror, equal angles with the perpendicular drawn from that point.

[97] It is also clear from all these experiments, and from all the reasonings we have shown: that the reflection of every light from a polished surface takes place in the plane
IV 55b orthogonal to the plane | tangent to the polished surface at the reflection point; that the line along which the light extends to the reflection point, and the line along which the light is reflected from that point, contain equal angles with the perpendicular to the plane tangent to the polished surface at that point; that if the light extends along the perpendicular to the tangent plane at the point reached on the mirror by the extending light, then that light will be reflected along the perpendicular itself along which it has extended to the polished surface; and that if different lights extend to a single point on the polished surface ((from different points and at different)) inclinations to the surface, then they will be reflected from that single point to different points, and the inclinations of the different, reflected lights will depend on the inclinations of the lights that have extended to the reflection point. That, then, is the manner of reflection of all [kinds of] light from all polished
IV 56a bodies. |

[Reflection from the whole of a polished body]

[98] It was shown previously that, from every point on a luminous body, light radiates along every straight line that may extend from that point through the transparent body adjacent to it. When a luminous body faces a polished body, light will radiate from every point on that luminous body upon every point facing it on the surface of the polished body; and all the lights that radiate upon that surface will be reflected from it in the manner that has been shown. Thus, for any luminous body that faces a polished body, light will radiate from every point on that luminous body upon the whole of the polished surface facing that point; and every point on the polished surface facing the luminous body will be irradiated by the light shining upon it from the whole surface of the luminous body facing that point.

[99] It is clear from this detailed account that when a luminous body faces a polished body, light will shine, in the shape of a cone, from every point on that luminous body upon all of what faces it on that | polished body; and that every point on the polished body will IV 56b receive light from all of what faces it on the luminous body, in the shape of a cone. And from [the fact] that the lines in which light issues are all straight, it follows that light will extend along all parallel lines that can be imagined to extend between the luminous and the polished body. It follows from that also that, for any point imagined [to lie] between the luminous body and the polished surface, if a cone is imagined between that point and the whole of the luminous body, with the point as vertex and the luminous body as base, and if that cone is imagined to extend from the imagined point towards the polished surface, then the light from the luminous body will reach all that falls within this cone on the polished surface through those two opposite cones.[8] And such light will extend to that point,[9] where it will be gathered, then it will extend and expand until it reaches that part of the polished surface that falls within | that cone. Again, if a cone is imagined between that IV 57a point and the whole polished surface, with that point as vertex and the polished surface as base, and if this cone is [also] imagined to extend from that point towards the luminous body, then the light will extend from all [points] that fall within this cone on the luminous body to the polished surface through these two cones. Now, the light that extends through these two cones is the same as that which extends through the first two cones. That being so, and it being the case that every light that extends to a polished surface is reflected along a line whose position relative to the polished surface is like that of the line along which the light has extended to that surface, then [it follows]: that the light shining upon a polished surface will be reflected from it on lines corresponding to all the lines we have described; that the light extending to a point on the polished surface in the shape of a cone will be reflected from that point in the shape of a cone; that the light issuing from a point on | the luminous body to the whole polished surface or to a portion of it will be IV 57b reflected from the whole surface or from that portion; that the light extending from any point between the luminous and the polished surface, through the two opposed cones that have that point as vertex, will be reflected from all [points] on the polished surface that fall within the cone close[r] to that surface; that the light extending along parallel lines will be reflected along lines whose positions with respect to the polished surface are like those of the parallel lines; and that the shapes of the reflected lights will depend on the shapes of the polished surfaces from which they have been reflected, which notion we shall clearly explain afterwards.

8. That is, the original cone and its rectilinear extension.
9. That is, the vertex of the two opposite cones.

⟪IV 3.4⟫

[100] It follows from the extension of light along all straight lines extending from the luminous body, that the lights passing through an opening to a polished surface, provided the opening is fairly large, will extend along every line that can be imagined to stretch inside the opening in such a way that one extremity of it reaches the luminous body while the
IV 58a other extremity reaches the | polished surface. It also follows that, for any point imagined on the circumference at the end of the opening facing the luminous body, if a cone is imagined to be drawn from it to the circumference of the other end of the opening facing the polished surface, and if that cone is imagined to extend until it meets the polished surface, and if that cone is further imagined to extend from the imagined point until it meets the luminous body, then the light will reach the point at the circumference of the hole from all [points] of the luminous body that fall within this cone. Then this light will extend through the other cone and meet all [points] of the polished surface that fall within this [other] cone, provided that the opening is cylindrical and also fairly long. Similarly, for any point imagined on the circumference at the end of the opening facing the polished body, if a cone is imagined to be drawn from it to the circumference of the other end of the opening facing the luminous body, and the cone is imagined to extend from the imagined point to ((the luminous body, and that cone is also imagined to extend from that point until
IV 58b it reaches)) the polished surface, | then the light will reach that point from all [points] on the luminous body that fall within this cone, then it will extend through the other cone until it reaches the polished surface. Likewise, for any point imagined inside the opening, if two [opposed] cones are imagined to be drawn from it to the ends of the opening, so that one of [the cones] will reach the luminous body while the other reaches the polished surface; and two other and similar cones are imagined to extend [from the same point] to both ends of the opening, then light will reach that point from all [points] of the luminous body that fall within the portion common to the two cones next to it, and [this light] will extend to the polished surface through the portion common to the two other cones.

[101] Similarly, for all parallel lines imagined to extend rectilinearly through the opening until they reach the luminous body on one side and [reach] the polished surface on the other, light will extend from the luminous body along these parallel lines to the polished surface. Now, in any cylindrical or non-cylindrical opening, parallel lines of different
IV 59a positions can be imagined, | some of which are parallel to each other and to the length of the opening, while others are parallel to each other but cut the lines that are parallel to each other and to the length of the opening. In regard to all these differently positioned lines, if they end on one side at the luminous body and on the other at the polished surface, light will extend along all of them at one and the same time from the luminous body to the polished surface. That being so, the lights emerging from the openings, if these are fairly large and reach a polished surface, will also be reflected from it on all lines corresponding to all the lines that may extend through the opening so as to reach the luminous body on one side and the polished surface on the other. But if the light is continuous in breadth, then the light reflected from it will also be continuous, since continuous light only extends
IV 59b along compacted lines, and will thus meet the polished surface in a | continuous area; and if the light meets the polished surface in a continuous area, then it will [also] be reflected on compacted lines, and will therefore be continuous after reflection. Therefore, when the light emerging from an opening meets a polished surface, it will be reflected in the

positions of all the lines corresponding to the lines that may have extended through the opening to the luminous body and to the polished surface, ((and the reflected light will still be continuous,)) and the shape of the reflected light as a whole will depend on the shape of the polished surface.

⟨⟨IV 3.5⟩⟩

[102] It has also been shown that light weakens as it recedes from its origin, for the strength of the light that shines from a luminous body depends on its distance from its origin, and the farther it gets from the origin, the weaker it becomes because of its distance. Again, light becomes weaker as it expands and disperses, and it becomes stronger the more it is compacted and gathered. For the strength of the light that issues in the shape of a cone from a point on the luminous body to the whole of the polished surface, or to a portion of the surface, will depend on its distance | from that point. The light at any place in this IV 60a
cone closer to that point will be stronger because of its being gathered and compacted, whereas the light at any place in the cone farther from that point will be weaker because of its dispersion and expansion. Thus, every light that extends from a point in the shape of a cone will become weaker as it gets farther from that point, and this weakening will be due to two causes: one is the distance of the light from its origin, and the other is its expansion and widening. Therefore, the strength of the light that issues from a point on a luminous body to a polished surface will depend on the distance of that surface from that body and on the width of the [illuminated] surface. Similarly, the light reflected from this light, if it widens in the shape of a cone, will become weaker as it recedes from the polished surface, and this weakening will be due to three causes: one is the distance of the light from the point of reflection, another is the light's expansion and widening, and the other is reflection [itself], for it was shown that reflection itself weakens the light.

⟨⟨IV 3.6⟩⟩

[103] If this light gathers after reflection, it | will gain in strength because of its IV 60b
gathering, and it will weaken in accordance with its distance and because of the reflection. If the [resulting] strength and weakness are equivalent, then the strength of the light after reflection will be the same as at the point of reflection. If the strength due to the gathering of the light exceeds the weakness due to distance and reflection, then the light will be stronger after reflection than it was at the reflection point. If, however, the weakness due to distance and reflection exceeds the strength due to gathering, then the light will be weaker after reflection than it was at the reflection point.

⟨⟨IV 3.7⟩⟩

[104] It is similarly the case with the light that extends from the whole of the luminous surface to any point on the polished surface in the shape of a cone that converges at that point. If the strength due to the light's converging exceeds the weakness due to its distance, then the light at that point on the polished surface will be stronger than the light at any point within this cone. I mean that the light that is found at any point on the polished surface | in this manner will be stronger than [the light] at any point in that light that IV 61a

extends along any of the lines extending within this cone, if the light extending along a single line is considered by itself. We single out this light only because light goes to every point on every line inside such a cone from the whole of the luminous surface, i.e., the cone's base. The reason is that, of the light that reaches in this manner a point inside the cone, nothing will extend to the point on the polished surface, i.e., the cone's vertex, except the light extending along the single line that joins that point with the point forming the vertex of the cone on the polished surface. Thus, if the light that extends along every one of the lines imagined to extend within the cone to the point on the polished surface where the light gathers ((is considered singly; and if the strength due to the light's gathering exceeds the weakness due to the distance of that point; then the light)) gathered at the point on the polished surface will be stronger than the light extending along that single line, if
IV 61b the gathered light is compared with [light at] any point on | that line. Therefore, the light gathered at the point on the polished surface will be stronger than any light extending from any point on the luminous surface, as the base of that cone, along a single one of the lines extending inside that cone to reach that point on the polished surface, if the gathered light is compared at any point on that line. If, however, the weakness due to distance exceeds the strength due to gathering, then the light at the gathering point will be weaker than any light extending from any point on the luminous surface as the base of the cone along a single one of the lines inside the cone that reaches that point on the polished surface, if the gathered light is compared with [the light at] any point on that line. And if the strength due to gathering equals the weakness due to distance, then the strength of the light gathered at the point on the polished surface will equal the strength of every light extending from any point on the luminous surface, namely the base of the cone, along a single one of
IV 62a the lines | inside that cone, and reaching that point on the polished surface. If this light gathered at the point on the polished surface is reflected from that point, it will widen in the shape of a cone, and its strength will become weaker as it recedes from the reflection point. This light will therefore be stronger at the reflection point than after reflection, and its weakening after reflection will be due to the three causes previously [noted]: namely, distance, reflection, and expansion.

[105] As for the lights that extend along parallel lines from the luminous to the polished surface: they will be weaker at the polished surface than at their origin or [at a point] close to it; I mean that they will be weaker at every point than the light that has extended from a point on the luminous body from which that light has reached the point on the polished surface, if the light at this point is compared with [the light at] any point on that line along which the light has extended. And the cause of this weakness will only be the
IV 62b distance of the light from its origin. | Again, when this light is reflected from a polished surface, the reflection taking place in parallel lines, it will be weakened, and its weakness will be due to the distance and the reflection only. If the reflected light expands in the shape of a cone, then it will be weakened further, and this will be due to three causes: namely, distance, expansion, and reflection. If the light is gathered after reflection, it will be strengthened by gathering. If the strength due to gathering exceeds the weakness due to distance and reflection, then this reflected light will be stronger at the gathering point than at the reflection point. ((If the weakness due to distance and reflection exceeds the strength due to gathering, then the reflected light will be weaker at the gathering point than at the reflection point.)) And if the strength and the weakness are equivalent, then the strength of this [reflected] light will be the same at the gathering point as at the point of reflection.

[Light shining or reflected along a straight line is not the imagined straight line]

[106] All of this having been made clear, I say that the light extending along a single straight line is not the imagined straight line, I mean that it is not length without breadth. For light can extend only in a body; | and a body, however subtle and thin, must have IV 63a width. And, therefore, the smallest light, than which no subtler light can exist, must have width. But the smallest or most subtle light, than which no subtler light can exist, when it is by itself, will extend along the straight line imagined to go through it along the middle of its length, while it is possible at the same time to imagine throughout its length many straight lines other than that line, some of which are parallel to that line and others [of which] intersect it. But if within the subtlest light, than which no subtler light can exist, two parallel lines are imagined to extend along its length, and the body in which this light exists is divided at a line in the middle between those two lines, then the light will cease and vanish and cease to be light. If, however, the width of the light equals or exceeds double the width of the most subtle light, and that light is divided lengthwise into two halves, then each of the parts will remain to be light. | But if divided lengthwise into IV 63b two unequal parts, so that one part is narrower in width than the subtlest light, then the narrower part will cease, and the wider will remain.

[107] Also, if the light is reflected and a straight line is imagined to cut the parallel straight lines stretching along its length; and, [if] surrounding this line on all sides of it and in the same straight [direction], there exists a light whose entire thickness is not less than the subtlest light; and [further] if the rectilinear extension of this line reaches the luminous body from which that light has radiated; then, if this light is divided, thereby separating the part of it that extends along this line ((in the manner [previously] indicated, then the light that extends along this line will remain)) and continue to be light. That being so, the smallest light that is reflected must have width, and must be reflected from a point on the polished surface that has [a certain] magnitude, even if extremely small. For light must have width, and if so, then it meets the polished surface at a point that has magnitude, and it is reflected through a body that has | width, even if extremely subtle. As a consequence, IV 64a if the light that extends to the polished surface is the subtlest [possible] light, then it will meet the surface in the smallest part in which it can meet that polished surface, and it will be at the moment of meeting it a light that can be reflected. Then the light will be reflected from the polished surface; and the position, relative to the surface, of the line imagined to extend through the middle of this reflected light, will be the same as that of the line imagined to extend through the middle of the first [i.e., incident] light; and these two lines will meet at a point imagined to be at the middle of the small part in which that subtle light has met the polished surface. Then if, within this reflected light, another straight line is imagined to extend from a point other than that imagined in the middle of that [small] part, and if this reflected light is divided by dividing the small part from which it has been reflected, or if the [property of] polishedness were to be abolished from a[ny] portion of that part, then the entirety of this reflected light will cease and vanish.

[108] Further, if the width of the light extending to the polished surface | equals the IV 64b width of two [minimal] parts, [each being] a most subtle light, then [the light] should be imagined to be ⟨reflected⟩ along two lines similar to the two lines imagined along the length of this [incident] light; and the lines along which such light is reflected will either

be parallel to each other, or meet at a point, or diverge in the direction of expansion, as required by the shape of the polished surface. If the place from which such light is reflected is divided into halves, then the reflected light will [also] be divided, but will remain unchanged as reflected light. If, however, the width of the light that extends to the polished surface equals two [minimal] parts, [each being] a most subtle light; and in addition, if it is possible to imagine through this light a line that cuts the two parallel lines imagined along its length and rectilinearly reaches the luminous body; and if this third line is surrounded on all sides ⟨and in the same rectilinear [direction]⟩ by a light that is not less than the subtlest light; and if the part of the polished surface in which this surrounding light meets the surface is not smaller than the smallest surface from which it is possible for
IV 65a light to be reflected; | then the light will be reflected from this part of the polished surface along a line similar to this third line, in addition to being reflected along the two parallel lines extending along the light's length.

⟨⟨IV 3.8⟩⟩

[109] It follows from all this that when light passes through an aperture and meets a polished surface, and it is the subtlest light that can exist, then it will be reflected only along a line similar to that which extends through the middle of that light. But if the aperture is fairly large, so that the light passing through it is wider than the subtlest light, then that light will be reflected along every line similar to one that extends through the aperture and reaches the luminous body. But the reflected light will nevertheless be continuous, and upon meeting an opaque body it will appear in a single place.

[110] It is in this manner, then, that the lights passing through apertures are reflected.

[111] Having shown this, we now say that the light is not [a gathering of] compacted lines, nor is the reflected light [such a gathering of] compacted lines, but that the smallest
IV 65b light must have width. When the light recedes | from the aperture, it is light and it may be reflected if the width of the aperture is not less than the least width through which the light may pass while continuing to be light. But continuous light is a single thing, not lines gathered together, which despite its continuity cannot but extend along straight lines, so that light will extend along every line imagined to be stretching from a point on the luminous body from which that light radiates. Likewise, the reflected light is a single, continuous thing that must extend along straight lines. And to every straight line that can be imagined to extend in the first [i.e., incident] light between the luminous body and the place of reflection, there corresponds, in the reflected light, a similar line along which the light will be reflected from the light that has extended along the first [incident] line. Further, if, through the continuous light that extends to a polished surface, a straight line is imagined to stretch from a point on the luminous body to a point on the polished surface, then the light extending along that line will be reflected along the line corresponding to it, as [was said] previously; and if another straight line parallel to the first line is imagined to
IV 66a extend in | that light, so that the width between them is shorter than the least width through which light can extend, when the light is subtle and separate, then the light extending along this second line will also be reflected, because there will exist on either side of this line too a light that has a fairly large width; and if a third line is imagined between these two lines and parallel to them, then the light extending along this line will also be reflected, because there exists on either side of this line too a light that has a fairly large

width. And continuous light will extend along every straight line that can be imagined in it, and will be reflected along every straight line to which a line may correspond in the first [incident] light. Thus, continuous light is not governed by the rule that governs disjunct light that is most subtle,[10] nor is the character of continuous light the same as that of subtle lights collected together. Rather, it is a property of light that it extends along every straight line that can be imagined to extend through it, and that it is reflected along every line corresponding to any straight line imagined to extend from the luminous body to the polished surface. Thus, the smallest light, than which | no subtler light can exist, will IV 66b
extend along the line imagined in its middle, and upon meeting a polished surface will be reflected from it only along a line corresponding to that extending along its middle. And a light that is fairly large in width will extend from a luminous body on every straight line that can be imagined to stretch from every point on that body through the transparent body contiguous to it. When [that light] meets a polished surface it will be reflected from it on every line corresponding to every line imagined to extend through the first [i.e., incident] light from the luminous body to the polished surface.

[112] We have now explained in detail the manner in which lights are reflected from polished bodies.

[Reason for the reflection of light from polished bodies]

[113] [The reason] why light is reflected from polished bodies but not from rough bodies is that polished bodies resist the lights completely, whereas rough bodies either offer no resistance to the lights, or do not completely resist them. By resistance I mean that [resisting bodies], by virtue of their polish, are not disposed to be affected.

[114] Now, it was shown in Book II that the extension of light along straight lines only takes place | by means of motion, and motion that is extremely fast. When light extends IV 67a
along straight lines by means of the forceful motion that is in it, then encounters a polished body, that body will completely resist it and prevent it from passing through. When the resistance is strong, the light will be reflected from [that body].

[115] [The reason] why polished bodies completely resist the lights, whereas rough bodies do not, is because rough bodies have gaps and separations in them, and thus when the lights encounter these bodies they pass through to the end of these gaps and disperse among the separate parts of the body; and thus, when the lights encounter rough bodies, they are scattered and their parts are dispersed. Polished bodies, on the other hand, have no gaps or separations, and thus when light encounters a polished body it finds no entry into it; and the polished body resists it by all its parts; and the light is thereby reflected. This resistance is due to the polish itself, not to the hardness of the polished bodies, for lights are reflected from polished bodies even when these are not hard, such as water and all moist bodies | whose surfaces are smooth. IV 67b

[116] Light may be reflected from some rough bodies, though this may not be apparent after reflection. That is because some rough bodies have small polished ⟨parts⟩ that, however, are always separated or variously situated. When light encounters such a body, the parts of the light that reach the gaps in it will go through them, whereas those reaching

---

10. Cf. *Optics*, Book II, Chapter 3, paragraph [173] (Sabra, *The Optics of Ibn al-Haytham. Books I–III*, I, p. 191), where the word 'disjunct' (Arabic *munfaṣil*) is used in relation to objects.

the small polished parts will be reflected from them, but will be dispersed and scattered after reflection owing to the separation of the [body's] parts and their differing situations; and when the reflected light is dispersed and scattered, it will not be apparent. Also, if parts of the light that enter the gaps encounter some polished parts there, they will be reflected after entering, but the reflected light will not get out, or, if it does, it will be dispersed and scattered. Thus, the light that may be reflected from such rough bodies will not be
IV 68a apparent. The light reflected from any body will be apparent only if the polished parts | of that body are more than the rough parts in number, and also if the polished parts are similarly situated or nearly so, and the gaps between them are narrow. A body is [counted] among polished bodies if it is of this description.

[117] The reflected light will be strong only if the gaps in the polished body are extremely narrow; the narrower and fewer the gaps, the stronger will be the light reflected from the body. If no gaps exist in the polished body, and if its parts are similarly situated, the strength of the light reflected from it will be at a maximum. For if the polished body has no gaps, the resistance it offers will be complete; and if its parts are similarly situated, the light reflected from it will be continuous; and if the light is reflected with great force and is also continuous, then its strength will be at a maximum.

[The reflection of light has a parallel in natural bodies]

[118] The reflection of light from some bodies but not from others has a parallel in
IV 68b natural bodies. When heavy bodies fall downwards from a high position | and meet at the point of impact a hard body, such as stone or iron or the like, they will immediately turn back with a forceful movement. If they meet at impact a soft body, such as sand or dust or the like, they will adhere to it and will not rebound. If they meet a body with some hardness and some softness, such as gypsum or wood or other, similarly soft, bodies, they will rebound weakly. Again, if someone throws a stone in a direction in which it meets a hard body before the motion in it has expired, it will turn back. If the [initial] movement is forceful, the stone will rebound with great force. If, on the other hand, it meets a soft body, such as wool or cotton or dust, or some other body that is quick to be affected, the stone will adhere to [that body] or [continue to] fall down without returning to the side from which it moved. And if [the stone] meets a body with some hardness and some softness, it will rebound weakly, in accordance with the [degree of] hardness in this body, then descend downwards.

IV 69a **[119]** It is clear from this examination that hard bodies | strongly resist the bodies in motion that encounter them, and when the resistance is strong, these bodies will rebound to the side from which they moved. Similarly, polished bodies strongly resist lights, and when the resistance is strong, the lights will forcefully rebound to the side from which they moved. Rough bodies, on the other hand, do not offer resistance to the lights, or do not resist them strongly. For rough bodies with some polished parts will offer a certain resistance to the lights, and this will result in the reflection of some parts of the light, while ⟨other parts⟩ will adhere to [those bodies]. But [bodies] having no polished parts in them will not resist the lights, so that the lights will not be reflected from them, but rather pass through their gaps. Thus, we have shown the cause on account of which the lights are reflected from polished bodies, but not from rough bodies.

[Why light is reflected in a certain plane and in a certain direction]

[120] [There follows the reason] why the lights are reflected only in a plane orthogonal to the plane tangent to the polished surface, | and along the line that contains, with the IV 69b
line along which the light has extended [in incidence], two equal angles made with the perpendicular [to the polished surface] at the point of reflection, and not along any other line. Light moves with an extremely fast movement, the most forceful movement possible to it. Now, when light shines upon a polished surface, the latter completely resists it. When a thing in motion meets something that resists it, and if the force moving it remains in it at [the moment of] meeting the resisting object, then it will rebound towards the side from which it has moved, and the force of its movement in rebound will depend on the force of its first movement and on the force of the resistance; further, the position of the interval across which [the moving object] moves during the rebound, with respect to the surface of the resisting body, will be according to the position of the line along which it moved with the first [i.e., incident] movement with respect to the same surface; and [further], this second movement is one that the moving object acquires from the resistance itself.

[121] This state of affairs is found also in [the case of] heavy bodies, | both in their IV 70a
natural, downward, motion and in their accidental motions, and can be accurately examined experimentally in the case of heavy bodies [as follows]. [First,] with respect to natural motion, the examination can be conducted by means of a heavy body, made of iron or copper or the like, whose weight does not exceed one *mithqāl*. Let [the body] be spherical and smooth so that it will touch other bodies it may meet at a point. Let someone climb up to a high position carrying this sphere with him. The experimenter should place on the surface of the ground down below an iron mirror, and set the mirror's surface in a horizontal plane, so far as this can be sensibly determined. Let the distance [i.e., height] of the place to which the carrier of the sphere has climbed be not less than twenty cubits; the greater the distance the better and clearer will be the experiment, for as the distance increases the greater will be the force of the [downward] movement through it. Then let [the experimenter] release the small iron sphere, making sure that it falls on the mirror placed below; and let him observe the sphere as it meets the mirror. He will find that it will immediately turn back upwards, then | fall downwards. If this sphere is dropped IV 70b
from a farther distance, its rebound from the mirror will be stronger and through a greater interval, and if dropped from a nearer distance its rebound will be less.

[122] Thus it is shown by this experiment that when a heavy body [naturally] moves downwards, it will rebound in reflection, and its rebound will depend on its force [in descent], because the longer the distance [covered] in descent the more forceful and faster will be its motion. It is also shown by this experiment that the movement in rebound will depend on the motion acquired by the heavy body as it descends from above, and not on the natural motion due to weight.

[123] If the experimenter makes an examination of this matter by means of a sphere weighing more than one *mithqāl*, he should do so from a greater distance [i.e., height] than that from which he examined the smaller sphere, since the acquired motion must depend on the magnitude of the distance and on the magnitude of the weight. This, however, is not the [appropriate] place for a precise account of this matter; what we have determined by examining the movement of the reflection [of light] will be sufficient.

[124] [Now] to examine accidental motion, the experimenter should set up the mirror
IV 71a perpendicularly to the ground surface, | so that the mirror's surface will be parallel to the wall. He should then set the small iron sphere at the tip of an arrow in one of the bows used for shooting pebbles, then shoot the sphere at the mirror that has been set against the wall. He should make sure that the sphere will pass along the perpendicular to the mirror's surface, which he can do by raising the mirror three cubits above the ground; and, standing in front of the mirror, he should horizontally orient the arrow, aiming it towards the mirror in a line sensibly parallel to the ground. The sphere should be projected with great force. Observing it as it reaches the mirror, the experimenter will find that it rebounds along the same perpendicular to the mirror's surface, by perceiving that it returns [along a line] parallel to the horizon; then, soon thereafter, the sphere will descend downwards. The stronger the force with which the sphere has been thrown, the more forceful will be its rebound.

[125] The experimenter should then step aside from facing the mirror, and again project
IV 71b the sphere so that it will pass along a line inclined to | the surface of the mirror and parallel to the horizon. Observing the sphere as it meets the mirror's surface, he will find that it will return from the side facing the thrower, and move at the beginning of its rebound along a line parallel to the horizon and inclined to the mirror's surface as much as the arrow was when it was aimed at the mirror, as far as this can be determined by sense. Then the sphere will soon descend in the downward direction owing to the natural force that moves it downwards. The more forceful the movement [initiated] by the thrower, the more forceful will be the sphere's rebound, as [the experimenter] will find. If, instead of the mirror, the examination is conducted by means of a somewhat softer body, such as wood or the like, the sphere will be found to rebound with less force than at first.

[126] From this experiment, it is clear that when a thing in rectilinear motion meets something that resists its movement, it will turn back, and the force of its rebound will depend on the force of the movement with which it moved at first, and on the force of the resisting thing, or its resistance to being affected. Further, the position of the interval through which [the body] moves in rebound will depend on the position of the interval
IV 72a through which it moved | at first.

[127] Thus, when light meets a polished body, it will be reflected because it is in motion, and because the polished body resists it; and it will rebound with extreme force because its [first] movement was extremely forceful, and because the polished body completely resists it. And a thing in motion will rebound upon being resisted, because it acquires from the resistance a motion in the direction of rebound. That the movement of [the] rebound is produced by the resistance is indicated by the fact that this movement will be in accordance with the [amount of] resistance, and that the stronger the resistance the more forceful will be the rebound. Moreover, the force of the resistance will depend on the force of the first [incident] movement and on [the extent to which] the resisting body is not disposed to being affected.

[128] [There follows the reason] why the position of the [spatial] interval across which the ⟨moving body⟩ rebounds will depend on the position of the interval across which it first moved. When the body moves along the perpendicular to the resisting surface, then the resistance and the motion are mutually opposed, and the resistance is at a maximum; for if the moving body were to pass through by virtue of its first movement, it would do so
IV 72b along the perpendicular extending into the | resisting body itself, and thus the movement produced by this resistance will be produced only along the perpendicular, because the

first movement was along the perpendicular and the resistance is along the perpendicular, and [because] there is no movement [in this case] except along the perpendicular, and no resistance except along the perpendicular. Therefore, such a moving body will rebound along the perpendicular itself. If, however, the line along which the body moves [in incidence] is oblique to the surface of the resisting body, then the resistance and the movement will not be [totally] opposed, and the resistance will not [therefore] be at a maximum; rather, the movement [of reflection] will take place along a line between [on the one hand] the perpendicular to [the surface of] the resisting body, which it goes through at the meeting-point, and [on the other hand] the perpendicular to this perpendicular, which is drawn in the plane containing the first perpendicular and the line along which the first [incident] movement took place. For if this moving body were to pass through [the resisting body], the point on it at which it meets the resisting [body] would extend in the plane containing the perpendicular to the resisting surface and the line of the [first] movement, and the extension of this point would take place in a straight line [continuous] with | the line of the [first] movement. Thus, for every point at which a moving thing meets IV 73a
a resisting body, if [the thing] were to pass through the body with that movement, then the [moving thing's] movement would take place along a line between the perpendicular to the resisting surface at the meeting-point and the perpendicular to this perpendicular, drawn from the meeting-point in the plane of the movement.

[129] Now, if the movement takes place in the direction of the line we have described, then the impact [pressure] of the mover upon the resisting body must be composed of the movement in the direction in which the perpendicular to the surface of the resisting body extends into that same body, and of the movement in the direction in which the perpendicular orthogonal to this perpendicular extends in the plane of the movement. If the impact is composed of these two movements, then the movement produced by the resistance will be composed of the movement along the perpendicular to the surface of the resisting body in the outward direction from that body, and of the very movement that existed in | the direction of the perpendicular orthogonal to this perpendicular and extending IV 73b
in the direction of [that] movement. For if the impact is composed of the two movements mentioned, then the part of the impact due to the movement along the perpendicular that goes into the resisting body will cease, because the resisting body is [facing] in this direction and it resists the mover in this direction. There will thus be engendered, from this part of the impact and from the resistance to this part by the resisting body, a movement along the perpendicular itself along which this part of the impact took place, but in the direction contrary to that of the impact. But the other part of the impact, namely that due to the movement on the perpendicular orthogonal to this perpendicular, will remain as it was; and no opposing movement will be engendered from it, because no resistance exists in the direction of this perpendicular. And if this part remains ⟨as it was⟩, and a movement is engendered from the first part along the perpendicular to the surface of the resisting body in the outward direction | away from the resisting body, then the [actual] movement IV 74a
produced will be composed from the movement along the perpendicular to the surface of the resisting body, and from the movement along the perpendicular orthogonal to this perpendicular and extending in the direction of motion.

[130] That being so, the line along which the reflected movement occurs will be [situated] between the perpendicular to the surface of the resisting body and the perpendicular orthogonal to it, these two [perpendiculars] being the lines of the movements on which

the reflected movement has been engendered. The distance of this oblique line from the second perpendicular will be the same as that of the line along which the movement would take place if the body in motion were to pass through [the resisting body] in a straight line, because the part of the movement along this perpendicular has not been annulled or reduced or augmented. Further, this [oblique] line will be in the plane of the perpendicular ⟨⟨and the line of the [incident] movement⟩⟩, because the two movements from which this [reflected] movement has been engendered are in this plane, and so also is the first [i.e., incident] movement; and this plane is orthogonal to the plane tangent to the polished surface at the
IV 74b meeting-point, the tangent plane being | that in which the second perpendicular extends. And if the distance of this line from the second perpendicular is the same as the distance of the continuation of the line along which the first movement took place from that second perpendicular, then the inclination of this line to the first perpendicular that is orthogonal to the surface of the resisting body will be equal to the inclination of the line of the first movement to this [same] perpendicular.

[The reflection of light versus the reflection of heavy bodies]

[131] However, the case of the reflection of light differs from that of the reflection of heavy bodies. For when a heavy body encounters a resisting [body] and is reflected on account of the resistance, its reflection will take place in a line such as that [previously] described. But [the heavy body] will nonetheless possess the force of weight responsible for the motion downwards, and, therefore, it will not remain after reflection in the [same] condition, but rather incline downwards, owing to the motion due to weight. Its movement after reflection will thus be composed of the movement entailed by reflection and the movement due to weight. If the first [i.e., incident] movement is forceful, and the reflected
IV 75a movement engendered by it is more forceful than the movement due to weight, | then [the heavy body] will move along the interval [i.e., line] entailed by reflection until this force weakens, then it will incline downwards. If, however, the movement due to weight is stronger than the movement required by reflection, then it will, from the beginning of its return, be reflected with a compound movement, and its movement will thus take place along a line other than that entailed by reflection. Light, on the other hand, has no force that makes it move in a special direction; rather, it is a property of light that it should move rectilinearly in all directions open to it, provided that these directions extend through a transparent body. Thus, when the light is reflected by the force it has acquired and is set on the rectilinear direction entailed by reflection, it will [continue to] extend in that direction, and will have no force that makes it move in any other direction, because it is not a property of light to seek a special direction, but rather to move only rectilinearly, whatever the rectilinear direction. Once it has extended in a certain direction, it will not deviate from that rectilinearity, unless it is hindered by resistance. When light moves in reflection, it
IV 75b will have only the force | that moves it along the line entailed by reflection. Its movement after reflection will therefore always be along straight lines, and its rectilinearity will not change after reflection, as happens with the movement of reflection of heavy bodies.

[132] Light is therefore reflected from a polished body, because its passage is prevented by the body; and it extends in the direction of reflection, because the [encountered] resistance engenders in it a motion in the direction of reflection. The reflection of light from every point on the polished surface ⟨takes place⟩ in the plane orthogonal to the

plane tangent to the polished surface at the reflection point, because the movement engendered from this resistance is engendered from two movements, both of which exist in this [same] plane, and because the first movement is in this plane. The line along which the reflection takes place lies between the two perpendiculars, of which one is perpendicular to the plane tangent to the polished surface and the other extends in this tangent plane, because the two movements from which the movement of reflection has been engendered depend on these two perpendiculars: | so the part of this movement that IV 76a
depends on the perpendicular orthogonal to the tangent plane pulls this [reflected] motion in ⟨the direction⟩ of this perpendicular, while the part [of the movement] that depends on the perpendicular extending in the tangent plane pulls it in ⟨the direction⟩ of this other perpendicular. It is for this reason that the line along which this movement [of reflection] occurs lies between these two perpendiculars, and that the inclination of the reflection line to the perpendicular orthogonal to the plane tangent to the polished surface equals the inclination of the line continuous with the first line along which the first [incident] light extended to the reflection point. For the compound movement of reflection is such that the part of it that pulls in ⟨the direction⟩ of the perpendicular extending in the tangent plane will always depend on the inclination of the line rectilinearly continuous with the first line to this perpendicular.

[133] Thus we have shown the cause on account of which reflection takes place in the manner previously determined.

⟪IV 3.9⟫

[Explanation of the manner of reflection of colours from polished bodies]

[134] With the explanation of the manner of reflection of lights from | polished bodies IV 76b
there was also shown the manner of reflection of the forms of colours from these bodies: for the forms of colours always accompany the lights with which they are mixed, so that when a luminous coloured body faces a polished body, the form of the light and the form of the colour in it will shine upon the polished body, and will together be reflected from it, and their reflection will take place in the manner that has been shown. It was shown in the preceding chapter [Chapter 2, [23]] that the forms of colours are reflected from polished bodies together with the lights mixed with them, and how this may be experimentally examined.

[135] The reflection of the forms of colours from polished bodies may be examined by means of the instrument we have described for the examination of the manner of reflection of lights, but examination of the forms of colours in the manner [shown] earlier [i.e., in the preceding chapter] is preferable, because it is clearer. For the forms of the lights and of the colours that emerge from the hole[s] [in the instrument] will be weaker than the forms of the lights and the colours extending in open space, | and the narrower the hole, the weaker IV 77a
still will be the forms of the light and colour emerging from it. For the emerging ⟨forms⟩ of light and colour weaken as they recede from their origin, and the subtler and smaller of these forms will weaken more quickly and after a shorter distance.

[136] This state of affairs becomes clearly manifest upon contemplating the sun's light as it passes through very small apertures: for when sunlight goes through two neighbouring apertures, of which one is so very narrow as to admit only a needle or the like, while the other is fairly wide, and the [passing] ⟨light⟩ reaches the surface of the ground or of a wall, the thin light will be found much weaker than the wider light; often the latter may not show any weakening while the former may become hardly visible, especially when there is a large distance between the aperture and the place where the light shows up.

IV 77b **[137]** The reason for this is that when the light | passing through a wide aperture, or the light extending in open space, shines upon some place, it will reach every point on that place either from a wide portion of the luminous body or from the whole luminous body, and not from a single point on it. But when the light passing through a narrow aperture shines on some place, it will reach every point on that place from a single point or a small part of the luminous body. For this [reason], then, the light passing through narrow apertures will be weaker than the light emerging from a fairly large aperture or extending through open space, even when their distances from their origins are equal.

[138] It is similarly the case with reflected light when the first [i.e., incident] light extends through open space or a wide aperture and, upon reaching a polished surface, it is reflected to some place: for to every point on the [area illuminated by the] reflected light, or to many such points, light will have been reflected from a fairly large portion of the luminous body, and not from a single point on it. If the first light has passed through a
IV 78a narrow aperture, | then, to every point on the reflected light, [some] light will be reflected from a single point or small portion of the luminous body. This, then, is the reason why the light emerging from a narrow aperture will be weaker after reflection than the reflected light that has emerged from a wide aperture or has extended through open space.

[139] The same [conclusion] holds also for the forms of colours: when they are reflected from a form extending through open space or through a wide aperture, the reflected form will be stronger and more manifest for the reason we have mentioned, but when reflected from a form that has extended through a narrow aperture, they will be less apparent and weaker.

[140] From this detailed account, it is clear that to examine the reflection of the forms of colours in the manner we explained in the preceding chapter is preferable and clearer.

[Examination of the forms of colours by means of the described instrument]

[141] And yet, it is possible to examine the forms of colours by means of the instrument described in the present chapter. The experimenter who wishes to examine the reflection of the forms of colours by this instrument should employ a silver mirror, since their examination by means of silver ⟨mirrors⟩ will be clearer, because iron mirrors dim the
IV 78b forms of colours by their [own] colours. The reason why | iron mirrors dim the forms of colours by their own colours is that, as has been shown, when any light is found in some body, there will shine from that body a secondary light in all the straight lines that extend from the body; and the form of the colour of that body will shine, along with that light, along all those straight lines, and will be mixed with it. Thus, when an illuminated coloured body faces a mirror, and the light in that body shines along with the form of the colour upon the mirror, then extends after reflection from [the mirror] along the straight lines required by reflection, the straight lines along which the reflection takes place will be among the straight lines that extend from the place where the light is found on the mirror and along which the secondary light [also] extends. Thus, along the straight lines along which the shining forms of the lights and colours are reflected, there will extend the form

of the mirror's colour, mixed with the secondary light extending from the illuminated place on the mirror; and therefore the form of the colour reflected from the mirror will be mixed with the form of the mirror's colour. For this reason, then, the forms of the colours reflected | from iron mirrors are dimmed. IV 79a

[142] If the experimenter wishes to examine the reflection of the forms of colours by means of the instrument described in the present chapter, he should employ a silver mirror three digits in length and breadth, and avail himself of one of the posts previously used for mounting the mirrors, and mount the silver mirror on it. He should then watch for the sun's light to shine on the wall we specified when we examined accidental light. When that happens, let him open two of the holes in one half of the instrument and close the others with [bits of] white paper, as previously described. He should mount the instrument on the opening in the door, as we explained in connection with examining accidental light. Then, before mounting the upright post in the instrument, let him look through one of the opened holes and mark the place he sees on the white wall facing the door. He should then attach to this place on the wall an object of a bright colour, such as light green, | purple, red IV 79b [*ṣaʿwī*], or the like, and let this object be fairly large. Let him then restore the instrument to its former position, look through the hole he looked through before in order to see the coloured object, then look through the other hole to see the whiteness of the wall. He should note the position of the instrument at this moment and keep it in mind.

[143] Let [the experimenter] then mount the post in which the silver mirror was fitted as previously shown, then mount the instrument in the door's opening, making sure that it has the same position as it did when he looked through the two holes, and contemplate the interior of the instrument. He will find on the interior of the instrument two light [patches] that have been reflected upon the two holes corresponding to the opened holes; one of these lights will be pure white, and the other will have a delicate colour of the same kind [i.e., hue] as the colour of the object; and the coloured light will be located at the hole through which he saw the coloured object. In this experiment we have used | two IV 80a holes, because the forms of colours are more manifest when they are close to a pure light, and this colour will appear in the [patch of] reflected light if the colour of the coloured object is bright and pure. But if the colour of the object is dark, such as dark blue, wine, musk, or the like, [the experimenter] will find the light reflected from it to be dark only in comparison with pure light, but he will not discern the quiddity [i.e., hue] of such a colour, because of its darkness. The appearance of a delicate colour in the light reflected from a body that has a bright colour, and the darkness of the colour reflected from a body that has a dark colour, are manifest proof that the form of the colour is reflected along with the form of the light that is mixed with it, and along the same lines along which the mixed lights are reflected. This experiment thus shows that the forms of colours are reflected in the same manner as in the case of lights.

[144] The cause of the weakness of the forms of colours reflected from narrow holes has been shown. But [these forms] will not be discerned in experimenting with narrow holes for the reason we have explained, namely that they are | in themselves weak, for IV 80b the reason explained previously, and that sight perceives the form of the colour present in the reflected light only from a secondary form extending to it from that form, as was shown in Book I. If the primary form is weak, the secondary form will be extremely weak, and will not therefore be clearly discerned by sight. If, however, the eye is at the place where the light is reflected, then it will perceive the colour of the body whose form has

been reflected from the hole to that place. For the primary form will arrive at the eye in this position, and sight will perceive the colour of that body from a primary form that is stronger than the secondary form that, in another position, would have reached the eye from [the primary form]. ((It is clear from this detailed account that it is possible for sight to perceive colour by reflection from narrow holes if the eye is at the place [upon which the light is] reflected,)) but sight [could] fail to discern the quiddity of the colour at that place if the eye is located elsewhere. This perception is like sight's [direct] perception of
IV 81a all colours of visible objects | in straight lines; but [sight] nevertheless fails to perceive most of the forms of the colours of visible objects on the bodies facing them, because it perceives the colours of visible objects [directly] from the[ir] primary form, whereas it perceives the forms of the colours of visible objects [that shine upon] the bodies facing them from the secondary form that has extended from the primary form to the eye.

[145] We have now completed our explanation of all matters concerning the manner of reflection of the forms [of light and colour] from polished bodies.

CHAPTER 4.
WHAT SIGHT PERCEIVES IN POLISHED BODIES IS PERCEIVED BY REFLECTION
⟨⟨Introduction and One Proposition⟩⟩

⟨⟨Introduction⟩⟩

[1] That sight may perceive visible objects in mirrors is clear. But thinkers have differed as to how this perception happens. Thus, the mathematicians have held that the ray goes out of the eye and is reflected upon meeting the mirror, and that sight will perceive
IV 81b whatever this ray encounters after reflection. | Some of the natural scientists, however, have held that when a polished body faces a visible object, the form of that object will be found in the polished surface, and that sight will perceive the form that is found on the surface of the polished body in the same way as it rectilinearly perceives the visible objects that face it. Each of these two groups has broached the subject, but without either completely investigating it or precisely determining the manner in which this perception really takes place. In this chapter we shall show that this perception happens by reflection, and that sight cannot perceive any visible object on the surfaces of polished bodies except by reflection. We will then show in the next chapter, precisely and in detail, how this reflection happens, and sum up the subject.

⟨⟨IV 4.1⟩⟩

[2] I say that there does not exist in the mirror a definite and fixed form of the visible object that sight perceives [in the same way] as it rectilinearly perceives visible objects. For if a fixed and definite form existed in the mirror, sight would perceive it as it does the
IV 82a visible objects facing it from all | positions. But sight does not perceive the visible object

in the mirror from all positions. Rather, the confusion regarding this matter arises when, upon perceiving his own face in a mirror, the viewer presumes that a form of his face is [actually] there. For when the mirror is placed on the ground, and [when someone] moving around it on all sides perceives from all sides the form of his face in the mirror's surface and in front of his own face, a person of the opinion that this perception is due to a form impressed in the mirror will form the opinion that the form of his face has actually taken shape in the mirror, and that he perceives it from all sides around the mirror, because the form of his face has been impressed in the mirror. But the false apprehension is easily exposed when the person looks at the mirror to perceive an object ((other than himself. For if the mirror is placed on the ground or set against a wall, and the person looks at the mirror to see one of the objects)) facing it other than himself, then, if he moves from his position to one side and looks at the mirror, he will fail to perceive that | object, or fail to IV 82b
perceive the whole of it, or will not perceive it in the place on the mirror in which he [first] perceived it.

[3] This state of affairs becomes clearly manifest if the person places the mirror on the floor, sits beside it, and, looking into it, he sees an area of the ceiling above, or of the wall. If he then moves from his position, and sits on the side closer to that visible [area] and looks at the mirror, he will fail to perceive that [area] and see [instead], in this situation, a different [area] of the ceiling or wall; upon returning to his former position, he will perceive the first area, but not the second. And if he moves a little from his position, he will see the first object at a place in the mirror other than exactly that first place where he saw it. Thus, if the object seen in the mirror were a definite form of the first object, the observer would have seen that form upon changing his position and looking into the mirror; for the occurrence of the object's form in the mirror, in the way | claimed by the contender, IV 83a
would not have taken place on account of the observer, nor would it have had anything to do with the observer. Further, if a definite and fixed form were to be found in the mirror, sight would perceive it from all sides around the mirror, and perceive it in the first place in which it was [formerly] perceived in the mirror, provided that the object and the mirror were fixed in their places with their mutual positions unchanged. But since sight perceives the form of such an object only from a determinate position, and when the eye changes position the object ceases to be perceived, and when the eye moves a little, the object will be seen in a different place in the mirror despite the fixed positions of the object and the mirror, then this is proof that there does not exist in the mirror a fixed and definite form for the object facing it, and that sight's perception of the object in the mirror only takes place in accordance with the position of the eye with respect to the object and the mirror.

[4] This state of affairs may be precisely examined experimentally | by means of IV 83b
the instrument we have described for examining the manner of reflection of forms, as follows. Let the post that is fitted with the plane mirror be mounted in the instrument we have described and in the manner we have determined. Close all holes in one half of the instrument with bits of paper, as already indicated, and open all holes in the other half. Then write on one of the paper pieces on the closed holes a small word like *abjad*, or some other small word, and let it be written at the middle of the hole or around its centre. The experimenter should then place his eye at the hole corresponding to the one with the written word and look at the mirror. He will perceive the word in a reversed position, so that letters on the right will be on the left and vice versa. The experimenter should make sure that the experiment is conducted in a lighted place, and that the light shines on the

interior [concave] surface of the instrument and on the paper pieces that close the holes,
IV 84a and that this shining light | is strong. Having thus perceived the word written on the paper, the experimenter should move his eye to another one of the open holes and look at the mirror. He will neither perceive the word nor any part of it if it is small and positioned on the centre of the hole, or around the centre but not outside the hole's circumference. Moving his eye from one of the open holes to the others one by one, he will not perceive that word. But he will perceive it if his eye returns to the first hole corresponding to the one with the inscribed word.

[5] Now, if a definite, fixed, and distinct form of a visible object were to be found in the mirror when the object faces the mirror, then the form of that written word that faced the mirror would have been found in the mirror, and would have occupied a single, unchanging place in the mirror, because there was no change in position of the word or
IV 84b of the fixed mirror. | If the form of this word had been found in a [definite place] in the mirror's surface, sight would have perceived it from all the open holes through which it looked at the mirror, because the place on the mirror perceived from one of the holes in the instrument is the same as that perceived from all the [other] holes, since the axes of all the holes in the instrument meet in a single point on the mirror's surface. Since, however, sight perceives the inscribed word only by looking through the hole corresponding to the one with the inscribed word; and since sight fails to perceive that word or any part of it when looking through another hole; and, moreover, since upon looking through every one of the holes sight perceives the [same] place on the mirror perceived from the hole corresponding to the one with the inscribed word; then it is clearly manifest that there does not exist at the place on the mirror perceived by sight when perceiving the word in
IV 85a the mirror's surface a definite form that is fixed in | that place. For if a definite and fixed form existed in that place, sight would have perceived it from all the holes, given that it perceives that place from all the holes, and that the position of the eye relative to the mirror and to the inscribed word, when perceived, is the one proper to reflection: I mean the similar positions of the lines drawn from the point on the mirror where the form is seen to the two corresponding holes, on one of which the eye is placed and on the other the word is written.

[6] Similarly, if the experimenter removes that word, closes the hole with a bit of white paper and writes the word on another hole, then looks through the hole from which he perceived the word when it was on the first hole, he will not perceive the word written on the first hole. Then he will perceive the word when he looks through the hole corresponding to the one on which it is written. If he moves his eye to any one of the open holes other
IV 85b than this corresponding hole, | he will neither perceive the word nor any part of it.

[7] Again, if the experimenter replaces the plane mirror with the cylindrical concave mirror and looks through the hole corresponding to the one with the word, he will perceive the word unreversed, that is, letters on the right will appear on the right and left ones on the left. Upon shifting his eye to a hole other than this one he will fail to perceive that word. If the experimenter lifts this mirror from the instrument and looks into it while it stands upright on its base, he will perceive his own form, with his limbs on the right side appearing on the right and his left-hand ones on the left, and mutually exchanged. If he tilts the mirror, letting its straight length [i.e., height] take the transverse position, and looks into it, he will see his form upside down, the upper parts of his face being down and the lower parts up.

[8] Also, if the experimenter looks into the conical concave or spherical concave mirror, he will find his form changed | in shape and size. IV 86a

[9] It is thus clear from these experiments that sight perceives the visible object in the mirror only if [the eye's] position with respect to the mirror and the object is the position proper to reflection, and it fails to perceive the object from another position for the mirror or the object. And from perceiving the form of the word in the mirror when the eye is placed at the hole corresponding to the one with the inscribed word, and from the absence of this perception when the eye is placed at another hole while perceiving that same place on the mirror where the form was [formerly] perceived, it is clear that there does not exist in the mirror a fixed and definite form. It is also clear from the reversal of the form mentioned in the experiment with the plane mirror and its unreversed position in the experiment with the cylindrical concave mirror; and from the previously described manner of reflection of the viewer's form as he looks into this mirror; and from the differing form of the viewer as he looks into a conical concave or spherical concave mirror; that the form[s] | of objects perceived by sight in polished bodies depend on the IV 86b
figure[s] of the surfaces of these bodies.

[10] Furthermore, the objects perceived by sight in plane mirrors are perceived, not on the surface of the mirror, but as if they were behind the mirror. If the object is fairly distant from the mirror in which it is perceived by sight, the distance of the perceived form from the mirror's surface will be fairly large.

[11] It is also clear from this experimental examination that there does not exist in the mirror a definite and fixed form. This being so, and since sight cannot perceive any visible object in mirrors unless [the eye's] position with respect to the mirror and to the object is the position proper to reflection, then sight's perception of visible objects in mirrors can take place only by reflection. | IV 87a

CHAPTER 5.
THE MANNER IN WHICH SIGHT PERCEIVES VISIBLE OBJECTS BY REFLECTION
⟨⟨Introduction and Eleven Propositions⟩⟩

⟨⟨Introduction⟩⟩

[1] It was shown in the preceding chapter that when a coloured body shines with any [kind of] light upon a facing polished body, the form of the light and colour in the shining body will radiate upon the polished body, and will be reflected from the polished surface along the lines proper to reflection; that if some [kind of] light exists at any point facing a polished body, the light will radiate from that point upon the whole polished surface in front of it, and will be reflected from the whole of that surface along the lines proper to reflection; that the form extending from a [given] point to a polished surface will have the shape of a solid cone whose parts are continuous with one another, and will be reflected from the whole of that polished surface as the base of the cone; that the form after reflection will be continuous and coherent; and that when the form that has been reflected in this manner encounters an opaque body, it will continuously and coherently spread over that body.

IV 87b **[2]** Further, if | the form that issues from a single point to a polished surface is reflected as a continuous [form], and it continuously spreads over the opaque body that it has encountered, the lines extending from that point as vertex of the cone to its base will have positions similar to those of the lines along which the form has been reflected from the cone's base to the surface of the facing opaque body upon which the form has spread. And if the positions of these lines are [respectively] similar, then the form extending from the surface of the opaque body, which the reflected form has reached from the base of the cone on the polished surface along the lines of reflection after the extension [of the form] from the point at the cone's vertex to that polished surface, will be reflected along the same lines that extend through the cone along which the form has extended from that point to that place on the polished surface; and the form will [thus] arrive, through that cone, at
IV 88a that | same point that is the cone's vertex. Now, if a cone is imagined between any point opposite a polished surface and any portion of that surface, or between that point and the whole of the polished surface, then if lines are imagined to extend from the base of that cone along lines similar in position to the lines that have [first] extended through the cone, then the forms of bodies that encounter these second lines will extend along them to the base of the cone on the polished surface and will be reflected to that same point, which is the cone's vertex.

[3] And it has been shown that the form of the light and the colour radiates from every point on the coloured and shining body upon every polished surface facing the shining body, and that the form extends along every line imagined to extend from every point on the surface of the coloured and shining body to every point opposite that point on the polished surface.

[4] All that being so, when a shining coloured body faces a polished surface so that
IV 88b the form of that body extends to | that surface on every straight line that may extend from every point on the body to every opposite point on the surface, then the lines that extend from the body to a portion of the polished surface, whose positions relative to the surface are similar to those of the lines that extend from that portion to a single point outside the shining coloured body, will be the same lines along which ((the form of the light and colour in that body)) will extend to that portion and be reflected to that point where they all come together and where the form will [thus] be collected. But if the form that extends in this manner is reflected to a single point, then, if the eye is placed at that point, and the form of the shining coloured body is reflected to it, sight will perceive that shining coloured body.

⟪IV 5.1⟫

[5] Now, since it was shown that sight perceives visible objects in polished surfaces only by reflection, and in accordance with the position proper to reflection, and since sight
IV 89a perceives its objects only through the straight lines | that issue from the centre of the eye, then the position proper to reflection, according to which sight perceives the object by reflection, is that in which the lines extending from the visible object to the place on the polished surface where sight perceives the object have positions similar to those of the

lines that extend from that place on the surface to the centre of the eye. Further, if the form of a visible object facing a polished surface radiates upon that surface and is reflected from it to a single point, then, when the eye faces a polished surface, and the lines through the cone that extends from the centre of the eye to that surface or to a portion of it are reflected from the cone's base to a visible object, then the form of that object will be reflected from that surface and will reach the eye along the straight lines that extend through the cone and meet at the centre of the eye.

[6] And it was shown in Book I that sight can perceive the light and the colour only through this form; and it | perceives the forms only through perception of the light and IV 89b
colour that exist in the object. But if the form of the colour and light in the object are reflected to the eye, and it has been shown that sight can perceive the colour and light in the object only from this form, and that [sight] perceives the object only through perception of the light and colour in the object; then sight will perceive the visible object by reflection only from the form of the light and colour that extends from the object to the surface of the polished body and is reflected to the eye on the straight lines that meet at the centre of the eye.

[7] It was also shown in Book I that sight perceives the visible objects only through something that reaches the eye from the object, whether or not a ray goes out from the eye. And it was similarly shown that sight does not perceive a visible object by reflection except through something that reaches the eye from the object, whether or not a ray goes out from the eye. Therefore, if the form of the object reaches the eye by reflection, and sight always perceives its objects from this form, then the issuing of a ray from the eye is useless and redundant.

[8] Sight, then, | perceives an object by reflection only from the form that extends IV 90a
from the object to the polished surface and is reflected on the straight lines that meet at the centre of the eye. The position in which sight perceives its object by reflection is that according to which the cone imagined to be drawn from the centre of the eye to the polished surface, or to a portion of it, is such that lines drawn from its base will reach that object if their positions are similar to those of the lines that have extended through that cone. Thus, when the form of the object radiates upon the polished surface and is reflected from the surface to a single point, then the positions of the lines along which the form is reflected to that point will be similar to the positions of the lines along which that form has extended to the reflection point on the polished surface. And, therefore, if the eye is located at that point, sight will perceive that object from that form, and will perceive it by reflection. That is the manner in which sight perceives | its objects by reflection. IV 90b

[9] This notion was not disclosed to any one of the ancient mathematicians, nor do I know anyone who has mentioned it. Yet it does not contradict what mathematicians have held with regard to reflection. For they believe that the lines of the [visual] rays are reflected from polished bodies at equal angles, and that an object is perceived by reflection along these lines. But they [also] believe that a visual ray goes out along these lines to reach the object, and that perception of the object takes place by means of this ray. We have shown, however, that this [latter] notion is useless and redundant, and that nothing goes out from the eye other than the imagined lines. We have used the opinion of the mathematicians in a number of our writings that we composed before the present Book. However, nothing of what we previously used is in contradiction with what we have now shown; we merely followed then the way of the mathematicians. But we only used there

the positions of the lines of the [visual] rays, so that if that is put together with what we have now shown, no contradiction will be found between them; rather, there is, in what
IV 91a we have | just explained, added confirmation of those notions without contradicting any of them.

⟪IV 5.2⟫

[10] Now that it has been shown how sight perceives visible objects by reflection from polished bodies [I say]: when a single object faces a polished surface in a position such that straight lines can be imagined to extend from it to different locations on the polished body; and there extend from each of these locations straight lines that meet in a single point; and these lines have similar positions [with respect to the polished surface] as the lines that have extended from the object to those locations; and, further, if at every one of the [meeting-]points there is an eye looking at that polished surface; then all these eyes will simultaneously perceive that object at that location on the polished surface. That is why several viewers may perceive one and the same object in the same mirror at the same time. Similarly, if the eye [of one viewer] moves to different locations and looks at the
IV 91b mirror from each of them while the mirror remains fixed, it | may perceive from these different locations one and the same object from among the objects facing the mirror. But each of those viewers will perceive that object from a location on the mirror's surface other than that from which the object is perceived by another viewer. Similarly, if a single eye perceives a single object in a single mirror from different locations, it will perceive it from different locations on the surface of that mirror.

⟪IV 5.3⟫

[11] Again, it was made clear that when a shining coloured body faces a polished surface, light will extend from every point on that body to every point facing it on the polished surface. That being so, when a visible object faces a polished surface, the form of every point on that object will go forth in the shape of a cone to the whole polished surface facing that point; and to every point on the polished surface there will proceed the form of the whole surface of the object facing that point, in the shape of a cone. From this it follows that the form of every point on the object's surface will be found in the whole
IV 92a polished surface, provided that | the whole polished surface faces the whole ⟨surface⟩ of the object. And from this it follows that there will be found at every point on the polished surface the form of every point on the object's surface, and that the form will be reflected from every point on the polished surface. But the reflection of the forms of different points, at any single point on the polished surface, will take place along different lines; and the form of the whole surface of the object that has extended to a single point on the polished surface in the shape of a cone will be reflected in the shape of a cone.

⟪IV 5.4⟫

[12] It also follows from this situation that, if the polished surface is flat, every portion of it whose periphery has the shape of the periphery of the facing object will contain the form of that object, whether that portion is small or large, provided [only] that the shapes

of the peripheries are similar. For if the form of a point on the object is found in the whole of the polished surface, then it will be found within the portion that has the shape of the object at the point | corresponding to that point on the object; and the form of a point on IV 92b
the object next to the first point, if it is also found upon the whole of the polished surface, will be found at the point corresponding to it within the similarly shaped portion, i.e., next to the first point. Similarly, the form of every point on the object will be found at the point corresponding to it in the portion that is similar to the object. For the form of every point on the object will be found in the whole of the polished surface and in the whole portion of this surface that is similar to the object. Thus, every portion of the polished surface will contain the form of the facing object if this portion is similar to the object. But this form that is found in the portion of the polished surface will not be separate from the other forms, but will be continuous with them; and every point of this form will be continuous with the form that has spread over the whole of the polished surface, issuing from the point on the object that corresponds to that point in the form, | and will be neither distinct nor IV 93a
discrete. Thus, there will be found in the polished surface an infinity of potential forms of the facing object, but none of these forms will be distinct or discrete; and thus, when the form is reflected from a portion of the polished surface and gathers at the centre of the eye, sight will perceive the form of that object from that portion of the polished surface from which the form has been reflected.

[13] Therefore the form perceived by sight is in the polished surface, and it is one of the forms previously described, and every point of this form will be continuous with the remainder of the same form that has spread upon the polished surface; but ((sight will perceive of the form of [any] one point on the object)) only that single point in the particular area from which the form of that point [on the object] has been reflected to the eye. And, therefore, the form that sight perceives of the object from the polished surface facing it will [appear] to sight to be separate and distinct from the forms continuous with it, as a result of being reflected along the lines that come together at the centre of the eye. For no [part of] the form of [any] one point | on the object that has spread upon the polished surface IV 93b
will be reflected to the centre of the eye, except from a single point on the surface, if the surface is plane, so sight will perceive the point from which the form has been reflected, but the remainder of the form that has spread over the polished surface will be reflected to other locations and will not be perceived by sight. And the same holds for all points on the object. The form that is found in that portion of the polished surface from which the form has been reflected to the eye will thus be separate from all forms continuous with it as a result of being reflected and gathered at the centre of the eye and perceived by sight. ((And sight's perception of [this form])) will thus be due to its particular position with respect to the eye. The occurrence of the form in a [certain] shape on the polished surface, and sight's perception of the form from that surface, will therefore be due only to the position proper to reflection, on account of which sight perceives that form.

[14] It is thus clear from this situation that no definite and discrete form exists in the polished surface, and that of the form of every point | that has spread over the polished IV 94a
surface sight will perceive only a single point, with the result that it will perceive from the whole continuous form a discrete form with a certain shape because it occupies a position at which a discrete form of that point is gathered, not because a discrete point exists in the polished surface. And that is the case with all differently shaped mirrors, but the example of the plane mirror is clearer, because, in the case of the other mirrors, accidents of visual

error occur that bring about changes in the forms perceived from them. It is in the manner we have [just] explained that the forms perceived by sight take shape in mirrors.

[15] Furthermore, when the eye faces a mirror, there is formed between the centre of the eye and what faces it of the mirror's surface an imagined cone, and there will exist between every point on the facing surface of the mirror and the centre of the eye an imagined straight line. If this line is perpendicular to the plane tangent to the mirror's
IV 94b surface, there will be reflected from that point on the mirror | the form of the point on the eye's surface through which this line passes. If the line is not perpendicular to the tangent plane, then it will contain an acute angle with the perpendicular to the tangent plane at that point. But there issue from that point an infinite [number] of imagined straight lines. Thus, if another one of the straight lines issuing from that point contains with the perpendicular to the tangent plane at the [same] point an angle equal to the angle contained by this perpendicular and the line extending between the point and the centre of the eye, and if both lines lie with the perpendicular in the same plane, then the two such lines will be among the similarly situated lines along which the forms may be reflected. If the second line that issues from that point reaches a visible object, then the form of the point reached on the object will extend along that line and arrive at the point on the mirror's surface, and
IV 95a will be reflected along the other line that extends between this | point and the centre of the eye; and having been reflected along this line, the form will reach the eye, and sight will perceive it. The second line along which the form has extended to the mirror's surface will belong to the cone through which the form of that point on the object has extended to that surface; and the line between that point [on the mirror] and the centre of the eye will be among the lines along which the form of that point [on the object] will be reflected from the mirror's surface. That is the manner in which sight perceives, by reflection, all that it perceives in mirrors.

[16] That being so, when the eye faces one of the mirrors, an imagined straight line will exist between every point on the facing surface of the mirror and the centre of the eye. Then, if it is possible for another line to extend from that point on the mirror so as to make, with the perpendicular to the plane tangent to the mirror's surface at that point, an angle equal to the angle contained by that perpendicular and the line extending between
IV 95b that point and the centre of the eye, | and if the three lines are in a single plane, then it is possible for the form of some point on one of the visible objects to be reflected from that point [on the mirror] to that eye.

[17] This having been clarified we [now] say: when the eye faces one of the mirrors previously described in the preceding chapter, then it is possible that a straight line should extend from every point on the facing surface of the mirror so as to contain, with the line joining that point and the centre of the eye, two equal acute angles made with the perpendicular to the plane tangent to the mirror's surface at that point, and both lines will be with the perpendicular in the same plane.

⟨⟨IV 5.5⟩⟩

[18] That should be clear with regard to the plane mirror. For it is possible to draw a line perpendicular to the surface of such a mirror at any point on it, and the plane tangent to the mirror's surface coincides with the surface; and every perpendicular to the surface of the plane mirror ((will be parallel to the perpendicular drawn from the centre of the eye to that

surface or to the plane continuous with it; and the perpendicular to the mirror's surface)) at any point on it and the perpendicular drawn from the centre of the eye will both lie | in [one IV 96a
and] the same plane, because they are parallel to each other. Further, the line that joins the centre of the eye and that point is in the same plane with the two perpendiculars; and the line that joins that point to the foot of the perpendicular drawn from the centre of the eye to the mirror's surface contains a right angle with the perpendicular, and it lies in the plane of the two perpendiculars. Therefore, the line joining the centre of the eye ⟨to that point⟩ on the mirror's surface will contain with ⟨the perpendicular at that point an acute angle⟩, since the eye is located above the mirror's surface. That being so, it is possible to draw from that point on the mirror's surface another line that extends in the plane of the two parallel perpendiculars, and which will contain with the perpendicular at that point an angle equal to the acute angle contained by this perpendicular and the line joining that point with the centre of the eye, the two lines being on either side of the perpendicular. Now, the plane of the two perpendiculars is orthogonal to the surface | of the mirror, IV 96b
because it is drawn through the perpendicular to that surface; and the line joining the feet of the two perpendiculars lies in the mirror's surface, ((because it is the intersection between this orthogonal plane and the mirror's surface. Therefore, if the two lines drawn from that point extend in this orthogonal plane,)) and they contain two equal angles made with the perpendicular to the mirror's surface at that point, then these two lines are among the lines along which the forms may be reflected. Thus, if the second of these two lines ends at a visible object, the form of the point that it has reached on the object will extend along that second line to the point on the mirror's surface, and [then] it will be reflected along the line joining this point to the centre of the eye, and sight will perceive that point on the object from this point on the mirror. If the point at the foot of the perpendicular drawn from the centre of the eye lies on the [actual] surface of the mirror, then there will be reflected from that point the form of the point on the surface of the eye through which this perpendicular passes.

[19] It is clear from this demonstration | that, when the eye faces a plane mirror, it is IV 97a
possible for any point on the facing mirror to reflect the form of some point on the visible objects. Thus, when the eye faces a plane mirror, what it may perceive of the visible objects in the mirror in this situation will be all [the forms] that exist at the extremities of the lines that have been reflected from all [points] that face the eye in this situation on the mirror's surface. Furthermore, sight's perception of all objects at the extremities of the reflected lines will take place at one and the same time. Thus, when the eye faces a plane mirror it is possible for it to perceive visible objects by reflection from the whole surface of the mirror, and its perception of all that it perceives from the whole surface of the mirror will take place at one and the same time. And all the planes orthogonal to the mirror's surface, in which the forms are reflected from the surface of this mirror to the eye, will intersect at the perpendicular drawn from the eye to the mirror's surface or to the plane continuous with it. For every perpendicular | to the surface of this mirror lies IV 97b
in a single plane with the perpendicular from the centre of the eye. Now, every plane in which the form of one point or other on the visible objects is reflected from the surface of this mirror is a plane in which there lie at once the centre of the eye, the visible point, and the foot of the perpendicular drawn from the visible point to the mirror's surface. And all the intersections between [on the one hand] the planes in which the forms are reflected from this mirror and [on the other] the mirror's surface, will be straight lines, all of which

intersect at that point on the mirror's [actual] surface, or on the plane continuous with it, namely the foot of the perpendicular drawn from the centre of the eye.

⟪IV 5.6⟫

[20] With regard to the spherical convex mirror: the [portion] of its surface that may face the eye at one and the same time is the segment cut off by the circle drawn by the [straight] line extending from the centre of the eye and tangent to the mirror's surface. For the eye to be facing this mirror is to be outside the [actual] surface of this mirror and [outside] the spherical surface [imagined to be] continuous with it. Now, if the eye lies
IV 98a outside | the surface of this mirror and outside the surface [imagined to be] continuous with it, then, if we imagine a straight line drawn from the centre of the eye, tangent to the mirror's surface, and imagine this line to move in a circle on the mirror's surface until it returns to its [initial] position while being tangent to the surface throughout this motion, this line will draw by this motion a circle on the surface of the spherical mirror; and that [circular portion] cut off from the mirror's surface is what the eye will perceive. For what lies beyond this circle on the surface of the mirror will not be seen, because it lies beyond the tangent line, and what lies beyond the tangent line cannot be seen. Therefore, that [portion] of the surface of the spherical convex mirror that may face the eye at one and the same time is the segment cut off by this circle from the mirror's surface. Now, the diameter extended from the centre of the mirror to the centre of the eye will pass through the middle
IV 98b of this segment, and the point at which | this diameter cuts the mirror's surface is the pole of the circle that bounds this segment facing the eye. And, for every plane drawn through this diameter and cutting the mirror, the arc of this circle produced by this plane in the mirror's surface, namely [the arc] between the diameter passing through the eye's centre and the circumference of the circle that bounds the segment facing the eye, will be less than a quarter of a circle. For the line drawn from the centre of the mirror to the extremity of the tangent line, which [extremity] is the point of tangency that has produced the circle containing the segment facing the eye, will make an acute angle with the diameter through the centre of the eye, because it will contain a right angle with the tangent line.

[21] This having been made clear, I say: it is possible for every point on this segment, except [the points on the circumference of] the surrounding circle, that the form of one visible point or other should be reflected from that point [on the segment] to the eye, and this reflection will take place from the whole of this surface at one and the same time. For it is possible to draw a diameter from the centre of the mirror to any point on this segment, and this diameter will be perpendicular to the plane tangent to the mirror's surface at that
IV 99a point, it being possible to draw through this diameter | many planes that cut the mirror. If two such planes are drawn that intersect at this diameter, then there will be produced on the surface of the mirror two circles that intersect at this diameter and have the centre of the sphere as their centre. If from the extremity of this diameter two lines are drawn tangentially to these two circles, then the diameter will be perpendicular to the two tangent lines and to the plane in which they lie; and this plane will be tangent to the sphere, because every line drawn in this plane will contain a right angle with the perpendicular that has been drawn to the point. For the plane in which this line lies [together] with the diameter through the point of tangency produces a circle in the mirror's surface, and so this line will be tangent to the circle, meeting the circle only at the point of tangency. Thus,

this plane will meet the surface of the mirror only at the point of tangency, and therefore it is tangent to the mirror's surface.

[22] If we imagine a line from the centre of the eye to the centre of the mirror it will be a diameter of the mirror, and it will lie in the same plane as the first diameter; | the IV 99b intersection between this plane and the mirror's surface will be a circle whose centre is the centre of the mirror. If we imagine a line from the centre of the eye to the point of tangency,[11] it will lie in the plane of this circle. If we imagine the diameter through the point of tangency to extend rectilinearly outside the circle, it will contain, with the line joining its extremity and the centre of the eye, an acute angle that lies outside the circle. For the plane tangent to the mirror's surface at that point does not pass through the centre of the eye, but rather cuts the line from the centre of the eye to the mirror's centre [somewhere] between the centre of the eye and the mirror's surface. This is so because the line from the centre of the eye to that point [of tangency] cuts the surface of the mirror, since it is [situated] before the tangent line that has cut off the segment [of the mirror] facing the eye. If this line cuts the surface of the mirror, then it will cut the [circumference of the] circle that passes through that point and in the plane of which this line lies; and if it cuts this circle, then the line tangent to the circle at the intersection point, | which lies in the tangent IV 100a plane, will cut this line that cuts the circle, and it will be closer to mirror's surface than this [latter] line. Thus, the tangent line at the intersection point will cut the line [segment] from the centre of the eye to the centre of the circle [somewhere] between the centre of the eye and the mirror's surface, and therefore the centre of the eye will be higher than the tangent plane. That being so, the line joining the centre of the eye and the point of tangency will contain, with the diameter through this point, an acute angle outside the mirror. And if it does, then it is possible that there may be drawn from the point of tangency, in the plane passing through this diameter and the centre of the eye, another line that contains, with the diameter, an angle equal to the angle contained by the diameter and the line joining that point and the centre of the eye; the angle [so produced] will be outside the tangent plane, and the line [so produced] will be higher than this plane. But the plane passing through the centre of the eye and the point of tangency and the centre of the circle is orthogonal to the plane tangent to the mirror's surface, because it is drawn | through the perpendicular to the IV 100b tangent plane. Thus, if the second line, as it rectilinearly extends, reaches a visible object, then the form of the point reached on the object will extend along this line and arrive at that [intersection] point on the mirror's surface, and will be reflected along the line joining this point and the centre of the eye, and sight will perceive [this form].

[23] The same is clear with respect to every point on the segment cut off from the mirror's surface by the circle drawn by the tangent line, other than the [points on the circumference of the] circle itself. As for the point at which the mirror's surface is cut by the diameter drawn from the centre of the eye to the centre of the mirror, given that this point is located on the [actual] surface of the mirror, there will be reflected from it the form of the point on the surface of the eye through which this diameter passes, since this diameter will [in this case] be perpendicular to the mirror's surface and to the plane tangent to the mirror's surface at the intersection point; and the segment determined on the surface of the mirror is that which may face the eye.

---

11. That is, the given point on the mirror, which is the endpoint of the first diameter.

[24] Therefore, when the eye faces a spherical convex mirror, it is possible, for every
IV 101a point on the facing surface | of the mirror, that the form of one point or other on visible objects should be reflected from it ⟨to the eye⟩; and sight's perception of all that is reflected from all of what faces it on the surface of this mirror will take place at one and the same time; and all planes orthogonal to the planes tangent to the surface of this mirror, [namely those] in which the forms are reflected to the eye, will intersect at the diameter drawn from the centre of the eye ((to the centre of the mirror; and every plane, in which the form of one point or other on the visible objects is reflected from this mirror, will be a plane in which are collected the centre of the eye,)) that visible point, the centre of the mirror and the reflection point; and all intersections between, on the one hand, the planes in which the forms are reflected, and, on the other, the surface of this mirror, will be circles that pass through the centre of the mirror, and all will intersect at the point at which the line drawn from the centre of the eye to the mirror's centre cuts the mirror's surface.

⟪IV 5.7⟫

[25] With regard to the right-cylindrical convex mirror: the [portion] of its surface that may face the eye is the segment cut off between the two planes drawn from the centre of the
IV 101b eye tangent to the cylinder's surface. For the eye, to be facing this mirror is | to be located outside the [actual] surface of this mirror and outside the cylindrical surface [imagined to be] continuous with the mirror's [actual] surface. Let the eye be located outside the surface of a right-cylindrical convex mirror and outside the cylindrical surface continuous with that surface. If we imagine a plane through the centre of the eye and parallel to the base of the cylindrical mirror, it will produce a circle in the mirror's surface. If, from the centre of the eye, two lines are drawn tangent to the circle on either side of the diameter joining the centre of the eye and the circle's centre, and two lines are drawn through the points of tangency along the length of the cylinder, then the two planes through a line tangent to the circle and a line extending tangentially along the length of the cylinder's surface ((are the planes that mark off, on the cylinder's surface, the segment facing the eye)). For if either of these two planes is not tangent to the cylinder, then it cuts the cylinder. But the straight line that extends along the length of the cylinder through the point of tangency lies in this plane, and any plane that is drawn through a straight line extending along the length of
IV 102a the cylinder, | and which cuts the cylinder will [also] cut the cylindrical surface at another straight line parallel to the first line. Thus, if either of these two planes cuts the cylinder, then it will produce in it a straight line parallel to the line extending along the length of the cylinder through the point of tangency. Now, the line drawn from the centre of the eye tangentially to the circle parallel to the cylinder's base lies in this plane, and it meets the other line that extends along the length of the cylinder through the point of tangency. If, therefore, this line [from the eye] extends rectilinearly, it will meet the other straight line that the plane has produced in the cylindrical surface, since the first line is parallel to the produced line. But the tangent line lies in the same plane as these two parallel lines, and the point on the other, produced, line, which is reached by the tangent line drawn from the centre of the eye, lies in the cylinder's surface, since the entire produced line lies in that surface. Therefore, the line drawn from the centre of the eye tangentially to the circle
IV 102b parallel | to the cylinder's base will meet the cylindrical surface at two points; and this line lies in the plane parallel to the cylinder's base. Therefore, the two points at which this line

meets the cylindrical surface will lie on the circumference of the circle that is parallel to the cylinder's base. Therefore, this straight line will meet the circumference at two points and will thus cut the circle. But this line is tangent to this circle at the first of these two points. And this is impossible. Therefore, the plane defined by the tangent line from the centre of the eye and the straight line extending along the length of the cylinder through the point of tangency will not cut the cylindrical surface; this plane will therefore be tangent to the cylinder's surface at the line drawn from the point of tangency along the length of the cylinder. And so the two planes previously defined will be tangent to the cylinder's surface. Likewise, for every two lines drawn from a point on a cylindrical surface, such that one of them extends along the length of the cylinder | and the other is tangent to the IV 103a
circle passing through that point, the plane in which these two lines lie will be tangent to the cylinder's surface. And, therefore, the segment cut off between the two tangent planes is the [portion] of the surface of the right-cylindrical convex mirror that may face the eye at one and the same time.

[26] Now, these two tangent planes intersect, because the centre of the eye is on both of them and therefore it lies on the intersection between them, and this intersection is parallel to the axis of the cylinder, since the axis is perpendicular to the plane passing through the centre of the eye parallel to the cylinder's base. And the two lines that extend along the length of the cylinder are parallel to the axis and are therefore perpendicular to the plane parallel to the base. Therefore, the two tangent planes are perpendicular to the plane parallel to the base, and so is their intersection. But the axis is perpendicular to this plane, and therefore the intersection is parallel to the axis, and so it is parallel to every line that extends along the length of | the cylinder; and the centre of the eye lies on this plane. IV 103b
But every line drawn from the centre of the eye to ⟨a point on⟩ the segment of the mirror's surface that lies between the two tangent planes will cut that surface. [This is] because, if from that point a straight line is drawn along the length of the cylinder, it will end at the circumference of the circle in the plane of which the centre of the eye lies, and this line will be parallel to the intersection between the two tangent planes; and, therefore, this line will be with the intersection in one and the same plane, and the line drawn from the centre of the eye to that point will [also] lie in this plane.

[27] If a line goes out from the centre of the eye to the point on the circumference of the circle that is reached by the line extending along the length of the cylinder, then that line will cut the circumference, because it will lie between the two lines that are tangent to the circle; and if it cuts the circumference of the circle, then it will go inside the cylinder. But this line is in the plane through the intersection and the line extending along the length | of IV 104a
the cylinder, which are parallel. Therefore, the plane through the intersection and the line along the length of the cylinder will cut the cylinder's surface. But the line from the centre of the eye to the point on the cylinder's surface lies in this plane, and it cuts the line that extends along the length of the cylinder. Therefore, this line, if extended rectilinearly, will go inside the cylinder. Therefore, any line that goes from the centre of the eye to a point on the segment of the mirror that is contained between the two tangent planes ((will cut the mirror's surface. Likewise, for any point on the intersection between the two tangent planes, any line drawn from it)) to a point on this segment will cut the segment.

[28] That being so, then any plane tangent to the cylinder's surface at a straight line in this segment will cut the two tangent planes, and will not meet the intersection between these two planes, because, if it did, then the line from the meeting point to a point on the

IV 104b straight line that extends along the length of the cylinder, where | the tangency occurs, will cut the mirror's surface. But this line will lie in the third plane that is tangent to the cylinder at the line extending along the length of the segment facing the eye, and thus the tangent plane will cut the cylinder. And this is impossible. Therefore, any plane tangent to the cylinder's surface at a line extending along the length of the segment between the two tangent planes will cut these two tangent planes and will not meet the intersection between them. But the centre of the eye lies on the intersection between these two planes; and no plane tangent to the mirror's surface at a line extending along the length of the segment facing the eye will pass through the centre of the eye, and [any such plane] will be located between the mirror and the centre of the eye. Therefore, any line drawn from the centre of the eye to a point on the segment of the mirror's surface that faces the eye will lie above [i.e., on the near side of] the plane tangent to the mirror, at the line passing through this point.

[29] Having made that clear, I say that when the eye faces a right-cylindrical convex
IV 105a mirror, then for every point on the [segment] of the mirror's surface that is marked off | between the tangent planes that pass through the centre of the eye, other than [the points on] the two lines that bound this segment, ((it is possible that the form of one point or other)) on the visible objects should be reflected from that point [on the surface] to the eye. For let a straight line be drawn along the length of the mirror through any point on this [marked-off] surface; through that point draw [i.e., imagine] a plane that cuts the mirror parallel to its base, thus producing in it a circle; draw from the point a straight line tangent to the circle produced, and a diameter to the centre of the circle; the diameter thus drawn will then be perpendicular to the tangent line. But this diameter will contain, with the axis of the cylinder, a right angle, since the axis of a right cylinder is perpendicular to any circle in it. And the line through the length of the cylinder is parallel to the axis, and it is also perpendicular to the plane of the circle. The diameter will then be perpendicular to this line and to the tangent line, and thus perpendicular to its surface [i.e., the tangent plane]; and this plane, [in which the longitudinal line and the tangent line lie,] is tangent to the cylinder's surface, as was shown earlier. Therefore, it is possible for the point at which
IV 105b | the circumference of the circle drawn in the plane through the centre of the eye is cut by the line joining that centre and the centre of the circle, if this point lies on the [actual] surface of the mirror, that there will be reflected from it ⟨to the eye⟩ the form of the point on the surface of the eye through which this diameter passes. And the line drawn from any other point on the segment facing the eye perpendicularly to the tangent plane, will lie with the centre of the eye in a single plane; and the tangent plane through the extremity of the perpendicular will not pass through the centre of the eye, but will rather lie between the cylindrical surface and the centre of the eye. Therefore, the line joining the centre of the eye and the extremity of this perpendicular will be located above [i.e., on the near side of] the tangent plane, and in the plane through the perpendicular and the centre of the eye. Thus this [latter] plane, in which the perpendicular and the centre of the eye lie, will cut the tangent plane. But the perpendicular lies in that plane; that plane is therefore orthogonal to the tangent plane.

[30] If the perpendicular is rectilinearly produced outside the mirror, it will contain,
IV 106a with the line joining the perpendicular's extremity and the centre | of the eye, an acute angle, since the perpendicular is orthogonal to the tangent plane, and the joining line is located [on the near side of the eye] above this plane. That being so, then it is possible to

draw from the perpendicular's extremity another line behind the perpendicular, and in this plane orthogonal to the tangent plane, that will contain, with the perpendicular outside the mirror, an acute angle equal to that between the perpendicular and the line from its extremity to the centre of the eye. And, if that [other] line extends rectilinearly to reach a visible object, then the form of the point reached will extend to the point on the mirror's surface and be reflected from this point along the line drawn to the centre of the eye; and when the form reaches the eye, sight will perceive it from this line.

[31] The same can be shown with regard to every point on the bounded segment on the surface of the cylindrical mirror, excepting the [points on the] two lines that bound the segment. Therefore, when the eye faces a right-cylindrical convex mirror, it is possible, for every point on the mirror's segment | facing the eye, that the form of one point or another IV 106b on the visible objects should be reflected from it to the eye. And sight will simultaneously perceive all the forms that are reflected [to it] from the whole of this segment.

[Concerning the intersections between the surface of the cylindrical mirror and the reflection planes]

[32] However, the intersections between the surface of the [right-]cylindrical [convex] mirror and the planes [in] which the forms are reflected will differ. Thus, for the plane through the centre of the eye and the whole axis of the cylinder, the intersection between it and the mirror's surface will be a straight line, since any plane drawn through the cylinder's axis produces a straight line on [the cylinder's] surface. That is because the right cylinder is produced by a rectangular parallelogram when it is turned about a fixed side until it returns to its [initial] position: the fixed side will be the axis of the cylinder, and the moving side will produce the cylinder's surface. But any plane drawn from the cylinder's axis and cutting its surface will coincide with the plane of the imagined parallelogram that produced the cylinder by its rotation; | thus the side of the parallelogrammic plane IV 107a will coincide with the intersection between this plane and the cylinder's surface. Thus, the intersection between the cylindrical surface and the plane through the centre of the eye and the whole axis of the cylinder will be a straight line; and the plane through the centre of the eye parallel to the cylinder's base will produce a circular intersection [with the cylinder's surface]. And all other planes in which the forms are reflected will meet the axis and cut the cylindrical surface: for the perpendiculars drawn in these planes from the reflection points will meet the axis, since each of these perpendiculars is a diameter of the circle through the perpendicular's extremity parallel to the cylinder's base. But only one of these planes passes through the whole axis, because all of them pass through the centre of the eye, and only one plane can be drawn from a single line through a single point outside that line, and therefore only one plane will pass through the whole axis of the cylinder and the centre of the eye. | Therefore, all the planes in which the forms are reflected, with the IV 107b exception of the plane through the whole axis, will cut the axis; and only one plane will cut the axis and produce a circle on the surface of the cylinder, namely the single plane parallel to the cylinder's base.

[33] Therefore, all the planes in which forms are reflected from all points on the mirror's surface outside the line along the length of the cylinder that lies together with the eye's centre in the plane through the whole axis, ⟨and⟩ outside the circumference of the [parallel] circle, will cut the axis of the cylinder and the planes parallel to the base of

the cylinder. The intersections between those planes and the cylinder's surface will be sections of the type that occurs in cones. For it has been made clear that, for any point on the surface of the mirror, a plane will pass through that point and the centre of the eye which is orthogonal to the plane tangent to the surface at that point and in which forms will
IV 108a be reflected from that point. Now, all [these] planes | orthogonal to the tangent planes will cut the cylinder, because the perpendiculars to the tangent planes, which [perpendiculars] are in these [orthogonal] planes, will cut the cylinder, since they meet the cylinder's axis. Therefore, all planes in which the forms are reflected will cut the cylinder and meet the cylinder's axis; and they pass through the centre of the eye. And, of all the planes that pass through the centre of the eye and cut the cylinder and meet the axis, only one will produce a straight line in the cylinder, namely the one that passes through the whole axis. [This is] because only planes passing through the whole axis will cut the cylinder, meet the axis, and produce a straight line in the cylinder's surface. For all planes that cut the axis will cut all the straight lines along the length of the cylinder, since these lines are parallel to the axis; and only one of these planes will pass through the centre of the eye and cut the
IV 108b cylinder's surface in a circle, namely the one parallel to | the base of the cylinder.

[34] Thus, all the planes in which forms are reflected from all points on the surface of this mirror that are outside the straight line forming the intersection between the surface and the plane through the whole axis, and outside the circumference of the circle produced by the plane parallel to the cylinder's base and passing through the centre of the eye, will cut [the cylinder] in [conic] sections. For only one of the intersections on the surface of this mirror is a straight line, and only one is a circle. All other intersections are ellipses of the type of sections that occur in cones, for mathematicians have shown that the sections of a cylinder are the same as elliptic conic sections.

[35] All the forms reflected from points on the straight line [extending along the length of the cylinder] will be in the plane passing through the axis in which this line lies. For the plane tangent to the mirror's surface at this line is unique; and any perpendicular to
IV 109a the tangent plane [at this line] will meet | the axis; and any perpendicular drawn from this line will lie in the plane passing through the centre of the eye and the axis, which [plane] is orthogonal to the tangent plane; and, therefore, all forms reflected from [points on] this line will be in this [orthogonal] plane.

[36] Now, with regard to the circle: let any plane touch the surface of the cylinder, passing through a point on the [circle's] circumference; then the perpendicular at this point to the tangent plane will be a diameter of the circle and will therefore lie in the circle's plane. But the plane of this circle is orthogonal to every plane tangent to the cylinder at a point on the circle's circumference; therefore, the forms reflected from the circumference of this circle will all be in the plane of the circle.

[37] With regard to [conic] sections: no form will be reflected from the periphery of any such section in the convex [surface] of this cylinder to a given eye, except from a single point on the periphery; for these planes [producing the sections] cut the cylinder's axis and the planes parallel to the cylinder's base, and the axis will thus not be perpendicular to any
IV 109b of those planes, | and the plane of each one of these sections will be inclined to the axis, and each will cut the axis only in a single point. Now, the line drawn from the periphery of one of these sections, perpendicular to the plane tangent to the cylinder's surface, will contain a right angle with the line extending along the length of the cylinder, parallel to the axis, and will therefore meet the axis, containing with it two right angles. Thus, when the

forms are reflected in one of these [orthogonal] planes, the perpendicular drawn from the point of reflection will lie in it. But this perpendicular meets the axis at right angles; and if the perpendicular lies in that plane and it meets the axis, and the plane meets the axis in a ((single)) point, ((then this perpendicular will meet the axis)) at the point in which the axis is cut by that plane. This perpendicular is orthogonal to the axis that is inclined to the plane of the section; and only one line can be drawn from the extremity of a line inclined to a plane so as to lie in the plane and contain two right angles with the inclined line; therefore,
| only one line can be drawn in the plane of the section perpendicular to the tangent plane. IV 110a

[38] Again, this perpendicular is a diameter of the circle that cuts the cylinder through the reflection point, and thus this perpendicular is an intersection of the surface of the [cylinder's section] and the plane of the circle. But the intersection of two planes can only be a single line; and, therefore, only one diameter can be drawn from the centre of this circle in the plane of that section. If, from the periphery of that section, a perpendicular other than the first perpendicular, namely the diameter of the circle, is drawn to a tangent plane, then it will be a diameter to a circle other than this circle: for any perpendicular to a plane tangent to the cylinder will be a diameter to one of the circles in the cylinder, but only one diameter can be drawn in the first circle so as to cut the periphery of the section; and thus the other perpendicular will be a diameter of another circle. But the circles in the cylinder are parallel to one another, and, therefore, the point on the axis that is reached by
that other perpendicular will be other than the centre of | the first circle. And only one point IV 110b
of the axis lies in the plane of the section, namely the centre of the first circle; therefore, the other point on the axis, that reached by the other perpendicular, will be other than the point that lies in the plane of the section; and, therefore, that ((other point on the axis)) will lie outside the plane of the section; and the other perpendicular will therefore cut the plane of the section and will not lie in it. But if that [other] perpendicular cuts the plane of the section, then the form reflected from the extremity of that perpendicular will not be in the plane of the section, but will rather lie in another plane; and, therefore, the first perpendicular, i.e., the intersection of the plane of the [elliptical] section and the plane of the circle, will have only one of its two extremities in the segment of the cylindrical surface that faces the eye, since this segment is less than half the cylinder, and the perpendicular will be the diameter of the circle that cuts the cylinder. That being so, no form of visible objects will be reflected to a given eye from the periphery of a single [elliptical] section of
a cylindrical convex | mirror, except from a single point on that periphery. IV 111a

[39] Now, for any object seen by reflection in this mirror, the form of any point on it will be reflected to the eye in a plane other than that in which the reflection of another point on that object will take place, excepting two lines. One of these is the intersection between the surface of the object and the plane through the whole axis of the mirror and the centre of the eye, if this plane cuts the object's surface. For the form of this whole line will be reflected to the eye in a single plane, i.e., the plane passing through the whole axis and the centre of the eye. And the other [of the two lines] is the intersection between the surface of the object and the plane parallel to the mirror's base and passing through the centre of the eye, if this plane cuts the object's surface. For the form of this whole line on the object will be reflected in a single plane, namely the plane parallel to the mirror's base passing through the centre of the eye. Thus, only the form of the whole of each of these two lines
on the object may be reflected | in a single plane. But, for any point on the surface of the IV 111b
object perceived in this mirror, with the exception of [the points on] these two lines, the

form of that point will be reflected to the eye in a plane in which no form will be reflected to the same eye from any other point on the object.

[40] It is clear from what we have shown that, for any perpendicular to a plane tangent to the surface of a right cylinder, the rectilinear extension of this perpendicular will meet the axis of the cylinder at a point forming the centre of the circle that cuts the cylinder through the point from which the perpendicular is drawn. For the diameter drawn from the centre of the circle passing through that point, to that same point, will be perpendicular to the tangent plane, since it will contain a right angle made with the straight line in which this plane touches the mirror's surface, and will also contain a right angle with the line tangent to the circle, which [line] is in this plane. Thus, this diameter will be continuous
IV 112a with the perpendicular to the tangent plane. For if it were not, it would cut it, and | two perpendiculars would have been drawn from the same point in a single plane, which is impossible. Therefore, if any perpendicular to a plane tangent to the surface of a right cylinder is produced rectilinearly, it will meet the axis of the cylinder at the centre of the circle that cuts the cylinder passing through the point from which the perpendicular has been drawn.

[41] And that being the case, all planes in which the forms of visible objects are reflected from this mirror will meet the axis; and any plane in which the form of one point or other on the visible objects is reflected will be a plane that contains the centre of the eye, the visible point, the point of reflection, and the point on the axis at which the perpendicular drawn from the reflection point meets this axis.

⟨⟨IV 5.8⟩⟩

[42] With regard to the right-conical convex mirror: when the eye faces such a mirror, the [segment] facing it on the mirror's surface will depend on the position of the eye with respect to the mirror. And, for the eye, to face such a mirror is to lie outside the mirror's surface and outside the surface of the cone continuous with the surface of the mirror. If the
IV 112b eye is located outside | both [this surface and its continuation], the line joining the centre of the eye and the vertex of the cone will lie outside the surface of the mirror and the conical surface continuous with it. For that line will be located outside the cone's surface; and, therefore, if it contains with the cone's axis an acute angle next to the cone, then, if we imagine a plane through the centre of the eye, parallel to the cone's base, it will cut the surface of the cone, producing a circle in it. Now, if two lines are drawn from the centre of the eye tangent to this circle on either side of the diameter joining the centre of the eye and the centre of the circle, and two lines are drawn from the points of tangency to the vertex of the cone, then the two planes, in each of which there are a line tangent to the circle and a line extending along the length of the cone, will be tangent to the cone's surface. For if one of these planes were to cut the surface of the cone, then, if any point of intersection
IV 113a outside the line between the vertex of the cone | and the point of tangency is joined to the vertex by a straight line, and if this line is rectilinearly produced, it will meet the circle's circumference in the plane through the centre of the eye at a point other than the point of tangency. That point will then lie in the plane that cuts the cone, because this line is in the plane, since the cone's vertex and the intersection point are both in the plane. But that point is in the plane of the circle in which the centre of the eye is located. Therefore, that point is on the intersection between the plane of the circle and the plane cutting the cone.

But the intersection between these two planes is the line drawn from the centre of the eye tangent to the circle, because this line lies in the plane and in the plane of the circle. Therefore, the point on the circle's circumference that is reached by the line drawn from the vertex of the cone to the point of intersection would be in a straight line with the line drawn from the centre of the eye tangent to the circle. Therefore, the tangent line would meet the circle | at two points, which is impossible. This plane will not, therefore, cut the IV 113b surface of the cone, so it will be tangent to it. And the same holds for the other plane. Likewise, for any two lines drawn from a point on a conical surface, such that one of them extends along the length of the cone and the other is a tangent to the circle passing through that point, the plane, in which these two lines lie, will be tangent to the cone; and therefore, when the eye is in this [specified] position, the segment of the mirror's surface that faces the eye, namely the [segment] cut off between the two planes tangent to the cone's surface, will fall short of half the cone, since the two lines drawn from the centre of the eye, tangent to the circle, will cut off less than half the circle.

[43] If the line issuing from the centre of the eye to the vertex of the cone is perpendicular to the axis, [and] we then draw a plane parallel to the cone's base and cutting its surface somewhere at random, then the intersection between this plane and the plane passing through the [whole] axis and the centre of the eye will be a straight line; and this intersection | will contain a right angle with the axis. But the line drawn from the centre of IV 114a the eye to the vertex of the cone lies in the plane passing through the whole axis ((and the centre of the eye, and it contains a right angle with the axis, therefore this line is parallel to the intersection between the plane passing through the whole axis)) and the circle. [If] we draw from the centre of the circle a diameter at right angles with the intersection, then produce it in both directions, and draw from both of its extremities two lines tangent to the circle; and [if we] also draw from the extremities two lines to the vertex of the cone; then the two planes, in each of which there lie a tangent line and a line extending along the length of the cone, will be tangent to the cone's surface, as was shown earlier. These two planes will pass through the centre of the eye: for the two lines tangent to the circle at the diameter's extremities are parallel to the intersection drawn from the circle's centre; and this intersection is parallel to the line joining the centre of the eye and the vertex of the cone; therefore, the two tangent lines will be parallel to the line joining the centre of the eye and the cone's vertex. But each of the tangent lines will lie, with the line joining the centre of the eye and the vertex, in a | single plane; and the lines extending from the two IV 114b points of tangency to the vertex are in these two planes, and they lie [respectively] with the tangent lines in the tangent planes; therefore, the line joining the centre of the eye and the vertex will lie with each of the tangent lines in the plane containing the tangent line and the line extending along the length of the cone. But this [latter] plane is the plane tangent to the cone; therefore, the line joining the centre of the eye and the vertex of the cone is the intersection between the two planes that are tangent to the cone. Now, the two planes tangent to the cone pass through the centre of the eye; therefore, the segment cut off from the mirror's surface between the tangent planes is what faces the eye if the eye's position is the one [specified above]; and this segment of the cone's surface will be half the cone.

[44] If the line joining the centre of the eye and the vertex of the cone contains, with the axis of the cone, an obtuse angle next to the cone, and if | produced rectilinearly IV 115a this line does not cut the cone and [remains] outside the cone, then let us also imagine a plane parallel to the cone's base and cutting the cone somewhere at random. This plane

will produce a circle in the cone's surface, and produce in the plane drawn from the axis through the centre of the eye an intersection ((that will be orthogonal to the axis)) and which will meet the line drawn from the centre of the eye to the vertex behind the surface of the mirror that faces the eye. If from the meeting-point two lines are drawn tangent to the circle, and two straight lines are drawn to join the tangent points to the vertex, then the two planes, in each of which there lie a tangent line and a line extending along the length of the cone, will be tangent to the cone's surface. Now, these two planes will pass through the centre of the eye, because each of the lines tangent to the circle, and the line from the
IV 115b centre of the eye to the cone's vertex, will intersect at | the point from which the tangent lines are drawn, and thus the tangent planes will pass through the eye's centre. Therefore, the segment cut off between the tangent planes towards the eye will be [the portion] of the mirror's surface that faces the eye; and this segment will be larger than half the cone, because these planes tangent to the cone meet behind this facing segment of the cone's surface.

[45] If the line joining the centre of the eye and the cone's vertex extends along the cone's surface, then the whole of this surface will face the eye, excepting only the line extending along the length of the cone. For if from this extending line a plane is imagined to be drawn tangent to the cone's surface, then only this plane will not cut the surface; all other planes drawn from that line ((will cut the cone)). That being so, if any point on the
IV 116a cone's surface is joined to the centre of the eye with a straight line, and it is [also] joined | to the cone's vertex with a straight line, then the plane containing these two lines and the line joining the eye and the vertex will be one of the planes that cut the cone. But nothing of this plane will meet the cone, except the line that extends along the middle of the cone's surface and joins the cone's vertex and the point to which the line was drawn from the centre of the eye; and the line drawn from the eye's centre to that point lies in this plane, and it cuts the line extending from that point to the vertex; therefore, if the line drawn from the eye's centre to that point is rectilinearly produced, it will cut the cone and go through it. But this line will not meet the line intersecting it at another point, nor will it meet the line extending from the eye's centre to the vertex, namely the [line] along the surface of the cone, at a point other than the centre of the eye. Now, this line lies in the cutting plane
IV 116b in which these two lines lie; | and nothing of the cutting plane lies on the surface of the cone, except these two lines only, I mean the lines extending along the length of the cone; therefore, the line drawn from the centre of the eye to that point will not meet the cone's surface in any other point. Likewise, the line drawn to any point on the surface of the cone from the centre of the eye will not meet the surface at any other point. And, that being so, no part of the cone's surface will be hidden from the eye by any part of the cone's surface. The whole surface of the cone will therefore face the eye, except the single line that extends along the surface of the cone.

[46] If the line joining the centre of the eye and the vertex of the cone enters the cone upon being rectilinearly produced, then the whole surface of the cone will face the eye. For every plane drawn from [i.e., through] this line will cut the cone, and, therefore, the
IV 117a line drawn from the centre of the eye to any point on the cone's surface | will cut the cone, because it lies in a plane cutting the cone; and [such a line] will cut ((the intersection between)) this plane and the cone's surface, which [intersection] joins that point and the vertex of the cone. Now, this plane that cuts the cone does so in two lines drawn from the cone's vertex, one of which is the line cut by the line drawn from the centre of the

eye to that point on the cone's surface, and the other [of which] cuts this line that goes from the centre of the eye to that point ⟨somewhere⟩ between that point and the centre of the eye; for the line from the centre of the eye to the cone's vertex will lie between the two intersections in the cone's surface; and, therefore, the second of the two lines will cut the angle contained by this line that is drawn from the centre of the eye to the cone's vertex and the line from that point to the vertex, [namely the angle] subtended by the line IV 117b
drawn from the centre | of the eye to that point; and if it cuts the angle, then it cuts the [subtending] chord. Therefore, if the line drawn from the centre of the eye to the point on the cone's surface crosses the surface and enters the cone, then it will not meet any [point] on these two lines that lie in the cone's surface. But only these two lines on the cutting plane will lie in the cone's surface; therefore, the line from the centre of the eye to the point on the cone's surface will not meet ((the cone's surface)) beyond that point. That being the case, no part of the cone's surface will be hidden from the eye by any other part of the surface. Thus, if the line from the centre of the eye to the cone's vertex cuts the cone, then all planes tangent to the cone will cut this line at the vertex, since the straight line at which the plane touches the cone's surface cuts this line. Therefore, no plane tangent to the cone's [surface] will pass through the centre of the eye; | and any plane touching the cone IV 118a
will be situated between the surface of the mirror and the centre of the eye.

[47] If the line from the centre of the eye to the vertex of the cone extends along the cone's surface, then only one of the planes tangent to the cone will pass through the eye, namely the plane tangent to the cone at that line. Not one of the other planes tangent to the cone will pass through the eye's centre, because all cut the line extending along the length of the cone at the cone's vertex; and, therefore, all planes tangent to the cone's surface will be situated between the mirror's surface and the centre of the eye.

[48] If the line from the centre of the eye to the vertex of the cone lies outside the cone and in two planes tangent to the cone, | then any line drawn from the centre of the eye to IV 118b
⟨a point on⟩ the segment of the cone's surface, between the two tangent planes, will cut the segment. For the line from the vertex to that point will lie in the cone's surface, and it will be in one plane with the line joining the vertex and the centre of the eye, and this plane will be situated between the two tangent planes; this plane will then cut the cone, and the line drawn from the vertex to that point will be an intersection between this plane and the cone's surface. But the line from the centre of the eye to that point cuts this intersection, and it lies in the plane that cuts the cone; therefore, this line, if rectilinearly produced, will enter the cone. And, therefore, any line drawn from the centre of the eye to a point on the segment between the two tangent planes will cut the surface of the segment. That being so, no plane tangent to | this segment will then meet the centre of the eye. For if it did, and IV 119a
a straight line were drawn from the centre of the eye to a point where the tangency occurs, then this line would lie in the plane passing through the centre of the eye and would be tangent to the mirror; but this line cuts the surface of the mirror, and therefore that plane would cut the surface as well as being tangent to it, which is impossible. Therefore, no plane tangent to the segment that is cut off from the mirror's surface between the two tangent planes will meet the centre of the eye. Now, any plane tangent to a cone will pass through the cone's vertex; and the vertex lies on the intersection ((between the two tangent planes that cut off the segment; and the centre of the eye lies on this line and outside)) the cone's surface; therefore, any plane tangent to this segment will be situated between this segment and the centre of the eye; and, therefore, any line drawn from the centre of the eye

to a point on this segment will be situated above the plane tangent to the segment, passing through that point.

[49] Having made that clear, I say that, for any point on the surface of a right-conical
IV 119b convex mirror that faces | the eye, it is possible for the form of one point or another on the visible objects to be reflected from that point [on the mirror] ⟨to the eye⟩. For if, from any point on the mirror's surface, a line is drawn along the length of the mirror, and a plane is drawn parallel to the mirror's base so as to produce a circle in the cone; and [if] a line is drawn tangent to the circle at that point; then the plane in which the tangent line and the line along the length of the cone lie will be tangent to the cone; and the centre of the circle will be on the axis of the cone, so that, if drawn, the axis will pass through the circle's centre. If, from that point [on the mirror], a diameter is drawn to the centre of the circle, it will contain a right angle with the axis, since the axis is perpendicular to the plane of the circle, and the cone is right; and so the angle contained ⟨by⟩ this diameter and the longitudinal line on the cone will be acute. Then, if, from the extremity of this diameter, a line is drawn in the plane that passes through the axis so as to contain a right angle with the longitudinal line, then it will meet the axis of the cone; and this line will be
IV 120a perpendicular to | the tangent plane. For if a line is drawn from the centre of the circle and in the plane of the circle so as to contain a right angle with the diameter, then this line will be parallel to the line tangent to the circle; and this line, being in the plane of the circle, contains a right angle with the axis; this line then contains a right angle with [both] the axis and the diameter, and, therefore, it is perpendicular to the plane in which the diameter and the axis lie. Now, the line tangent to the circle, which is parallel to this line, is also perpendicular to this plane; and the line that contains a right angle with the line along the length of the cone lies in this plane; so the tangent line contains a right angle with this line. But this line contains a right angle with the line along the length of the mirror; so this line is perpendicular to the tangent line and to the line along the length of the mirror, and is, therefore, perpendicular to the mirror's surface. But the plane in which the tangent line and the line extended along the length of the cone lie is tangent to the cone's surface; and this
IV 120b perpendicular, | if produced rectilinearly, will either end at the centre of the eye or will not pass through it; if it does, then there will be reflected, from the point of this perpendicular that lies on the mirror's surface, the form of the point on the eye's surface through which the perpendicular passes. Now, any line issuing from a point on the mirror's surface without passing through the centre of the eye will lie in a single plane with this centre; and the plane tangent to the cone and passing through that point will not pass through the centre of the eye, but will be situated between the eye's centre and the surface of the cone, as was shown earlier; so the line joining the centre of the eye and the diameter's extremity will be situated above the tangent plane, and above the plane containing the perpendicular and the eye's centre; therefore, this plane containing the perpendicular and the centre of the eye will cut the tangent plane. But the perpendicular lies in this plane and is orthogonal to the tangent plane, and this plane is orthogonal to the tangent plane.

IV 121a **[50]** Therefore, the perpendicular, | if produced rectilinearly outside the mirror, will contain an acute angle with the line joining the perpendicular's extremity and the centre of the eye, the latter line being higher [i.e., nearer the eye] than the tangent plane, and the perpendicular being orthogonal to the tangent plane. That being the case, it is possible to draw from the extremity of this perpendicular another line in the plane orthogonal to the tangent plane and behind the perpendicular, so as to contain with the perpendicular

an acute angle equal to that contained by the perpendicular and the line drawn from the perpendicular's extremity to the centre of the eye. If that line rectilinearly extends to reach one of the visible objects, the form of the point reached on the object will extend along that line and reach that point on the mirror's surface, then will be reflected along the line drawn from that point, namely the perpendicular's extremity, to the centre of the eye, whereupon the eye will perceive the form. And the same holds for every | point on the surface of the IV 121b
mirror that faces the eye.

[51] Therefore, when the eye faces a right-conical convex mirror, then it is possible, for any point on the segment of the mirror's surface facing the eye, that the form of one point or another on the visible objects should be reflected from that point [on the mirror] ⟨to the eye⟩; and the eye will simultaneously perceive all the forms reflected from all that faces it of the mirror's surface. Likewise, it can be shown by this argument that, if the whole surface of the cone faces the eye, it will be possible, for any point on the mirror's surface, that the form of some point or another on the visible objects should be reflected from that point [on the mirror] ⟨to the eye⟩; and that the eye will ⟨simultaneously⟩ perceive all that is reflected from the mirror's surface.

[52] It is [also] clear from this argument that, if any perpendicular to a plane tangent to the surface of a right cone is produced rectilinearly, it will meet the axis of the cone. For, let a line be drawn from the point [on the cone's surface] through which the perpendicular passes, and in the plane containing the axis and that point, so as to make a right angle with the straight line drawn from | the cone's vertex to that point; then [that line] will meet the IV 122a
axis, and will be orthogonal to the tangent plane, and continuous with the perpendicular to the tangent plane. For if it were not continuous with it, it would cut it; and, then, two perpendiculars to a plane would have been drawn from a single point, which is impossible. And, therefore, any perpendicular to a plane tangent to the surface of a right cone will, if produced rectilinearly, meet the cone's axis.

[53] That being the case, the planes in which the forms of visible objects are reflected from this mirror will meet the mirror's axis; and any plane in which the form of one point or another on the visible objects is reflected will be a plane in which there lie the centre of the eye, the visible point, the reflection point, and the point on the mirror's axis where the perpendicular from the reflection point meets the axis.

[Intersections of the cone's axis with the reflection planes]

[54] With regard to the intersections between the surface of the conical mirror and the planes in which the forms are reflected: if the line from the centre | of the eye to the cone's IV 122b
vertex is continuous with the cone's axis, then all the intersections will be straight lines, because any perpendicular drawn from a point on the mirror's surface will meet the axis; and the reflection plane passing through the extremity of that perpendicular is one that contains that perpendicular and the centre of the eye; and if the centre of the eye lies on the axis, and the perpendicular meets the axis, then the reflection plane is that which contains the perpendicular and the whole of the axis. But any plane drawn from the axis so as to cut the cone will produce a straight line as an intersection; and all planes drawn from the centre of the eye without passing through the whole of the axis will not meet the axis inside the cone. Therefore, if the centre of the eye lies on the axis of the cone, then reflection will take place from the whole of the cone's surface, and all the intersections will be straight lines.

[55] If, however, the centre of the eye lies outside the axis, then all the intersections will
IV 123a be [conic] sections,[12] | except one intersection, which will be a straight line, namely the intersection between the plane through the whole axis and [through] the eye's centre and the surface of the cone, there being only one plane [through the axis and the eye]. But all other intersections will be [conic] sections. For all planes in which the forms are reflected meet the axis of the cone, since the perpendiculars drawn in them from the reflection points meet the axis; and the whole axis can only be in one of these planes, since only a single plane can be drawn from a [whole] straight line through a point outside the line; ((and, therefore, if all these planes meet the axis, and only one of them will contain the whole axis,)) then all other planes will cut the axis. Now, no forms will be reflected in the plane drawn through the centre of the eye, parallel to the base of the cone, so as to produce a circle in the cone, because all the lines extending along the length of the cone will be
IV 123b inclined to this plane, the axis being perpendicular to it; | and, hence, the lines extending along the length of the cone will contain acute angles with the lines drawn in this plane; and, therefore, no line can be drawn in this plane perpendicularly to a plane tangent to the cone's surface. And, therefore, the intersections between [on the one hand] all planes in which the forms are reflected from the surface of the right-conical mirror, other than the plane through the whole axis, and [on the other hand] the surface of the cone, will be inclined to the cone's base and will not pass through the cone's vertex. Such planes will produce [conic] sections on the surface of a cone; and the intersections produced by the reflection planes that cut this mirror, other than the plane through the axis, will therefore be [conic] sections.

[56] Now, the reflection of forms from any point on the straight line forming the intersection between the mirror's surface and the plane through the whole axis and the centre of the eye will take place in this plane, because any perpendicular drawn from a
IV 124a point on this line will lie in | this plane, since it meets the axis, and thus the reflection point, the point in which the perpendicular meets the axis, the centre of the eye, and the visible point will [all] be in the plane passing through the axis; and [because] the plane containing [all] these points is the reflection plane; therefore, for any point whose form is reflected from a point on the straight line that lies in the plane through the whole axis, the plane in which that form is reflected is the plane passing through the whole axis. And, for any point on that line, it is possible that the form of one point or another on the visible objects should be reflected from it ⟨to the eye⟩, because from any point on this line a perpendicular can be drawn to the plane tangent to the cone at this line.

[Conic sections as intersections]

[57] With regard to [conic] sections [as intersections]: some of them are such that forms can be reflected from their periphery at only one point, while for others the forms can be reflected from the periphery at only two points; and, for any [conic] section produced
IV 124b in the | right-conical convex mirror, it is not possible for the forms to be reflected from its periphery at more than two points. For the perpendicular drawn from the reflection point will meet the axis of the cone at an acute angle towards the vertex of the cone. If

12. That is, ellipses, parabolas, or hyperbolas.

from the perpendicular's extremity a plane is drawn parallel to the cone's base, it will produce in the cone's surface a circle whose centre will lie on the axis; the axis will be orthogonal to the plane [of the circle]; the centre of the circle will lie between the vertex and the point on the perpendicular where it meets the axis; and any line drawn from this point on the axis to the circumference of the circle will be equal to the perpendicular. Thus, if this point on the perpendicular is imagined to be fixed while the perpendicular moves round the circumference of the circle, a right cone will be produced with that point as vertex and that circle as base. Then, any plane in which some form or another is reflected from the extremity of that perpendicular will either be tangent to this [produced] cone ((at that perpendicular, or it will cut the surface of the cone at the perpendicular. Let it be tangent to the cone,)) then any | point on the periphery of the [conic] section, IV 125a
namely the intersection between that plane and the mirror's surface, other than the extremity of the perpendicular, will be outside the plane of the circle and placed behind the circle, next to the [original] cone's base. Now, every line drawn from the vertex of the cone to a point on the periphery of this [conic] section, other than the extremity of the perpendicular, will cut the circumference of the circle before it reaches that point; so that, if a straight line joins the point of intersection and the point on the axis, namely the vertex of the produced cone, it will be equal to the line joining the cone's vertex and the perpendicular's extremity; and thus, if a straight line joins the vertex of the cone, i.e., the mirror, and the point of intersection, it will be equal to the line joining the vertex of the cone and the extremity of the perpendicular; and, therefore, the angle contained by these two lines at the point of intersection on the circle's circumference will be right, because it will be equal to the angle at the | perpendicular's extremity. If, then, the point IV 125b
on the axis, i.e., the vertex of the produced cone, and the point on the periphery of the conic [section], which [point] is reached by the line cutting the circle, and which [point] lies outside the circle, are joined by a straight line, the angle contained by that line and the line extending to it from the vertex of the conical mirror, i.e., the line that has cut the circle's circumference, will be acute. Thus, if to any point on such a [conic] section other than that on the circle's circumference a line is drawn from the point on the axis, then it will contain, with the line drawn to that point from the vertex of the conical mirror, an acute angle. And, therefore, none of these lines will be perpendicular to a plane tangent to the cone. But these lines are those drawn in the plane of the [conic] section from the point on the axis to the periphery of the [conic] section; and only one point of the axis lies in the plane of the [conic] section; and, therefore, every perpendicular drawn from a point on the periphery of the [conic] section, other than the point on | the circumference IV 126a
of the circle, will meet the axis at a point other than the endpoint of the first perpendicular, because [that perpendicular] will contain a right angle with the line drawn from the vertex of the cone of the mirror to that point. But if the perpendicular meets the axis at a point other than that point, then it will lie outside the plane of that [conic] section; and, thus, if some form is reflected from the extremity of that perpendicular, the reflection will take place in a plane other than ⟨that⟩ of the first [conic] section. Therefore, if the reflection plane is tangent to the surface of the cone produced by the movement of the perpendicular on the circle's circumference, then there does not exist on the periphery of the [conic] section, namely the intersection between that plane and the mirror's surface, a point from which some form will be reflected in that plane other than the single point on the circle's circumference.

[58] If, however, the reflection plane cuts the surface of the cone produced by the perpendicular's movement, it will cut it in two lines, one of which is that perpendicular;
IV 126b and one part of the | intersection, or [conic] section, will be next to the cone's vertex, and the other part next to [i.e., behind] the cone's base. The part next to [i.e., behind] the cone's base will be such that the line drawn to any point on it from the point on the axis, i.e., the vertex of the produced cone, will contain, with the line from the vertex of the conical mirror to that point, an acute angle, as was shown for the case of the tangent plane. The part next to the cone's vertex will be such that the line drawn to any point on it from the point on the axis will contain an obtuse angle with the line from the vertex of the conical mirror to that point. The two lines in both of which this plane cuts the surface of the produced cone will [respectively] contain two right angles with the two lines drawn from the cone's vertex to their extremities. Thus, if the reflection plane cuts the surface of the cone produced by the movement of the perpendicular drawn from the reflection point on the circle's circumference, then there will be drawn in that plane two
IV 127a lines [respectively] perpendicular to two planes tangent to the surface of | the cone. And, therefore, if the extremities of these two perpendiculars lie in the segment of the mirror's surface that faces the eye, the forms of two points will be reflected from the extremities to the eye. But only two lines can be drawn in that plane that are perpendicular to a plane tangent to the cone, namely the two intersections between this plane and the surface of the cone produced by the perpendicular's movement, i.e., [the intersections] of which one is the perpendicular producing the cone. That being so, no forms will be reflected from the periphery of this [conic] section, except from two points only, provided that both points lie in the segment facing the eye. Therefore, of the planes in which forms are reflected from the right-conical convex mirror whose sections are [conic] sections, some are such that the forms are reflected from one point only on the periphery of the [conic] section produced by that plane, and others are such that ⟨the forms⟩ will be reflected from only two points on the periphery. There is no [conic] section from the periphery of which reflection can take
IV 127b place | from more than two points. And the angles contained by the perpendiculars and the axis will all be equal, because the angles contained by the axis and the lines extending from the vertex of the cone along its length are equal, and the angles contained by the perpendiculars and the lines along the length of the cone are right.

⟪IV 5.9⟫

[59] As for the spherical concave mirror: it is possible for its whole surface to face the eye. If the eye lies inside the spherical surface [i.e., inside the sphere], then [this will hold] for all cases without qualification. If the eye lies outside the spherical surface, it will be possible for the whole mirror to face the eye, provided that it is not larger than the larger [i.e., concave] segment cut off by the circle produced by the line drawn from the eye tangent to the sphere's surface.

[60] Let the eye face such a mirror. If the eye lies at the sphere's centre, then it will not perceive any visible object other than the part of the eye's surface that is cut off by
IV 128a the cone | having the centre of the eye as vertex and ⟪what faces the eye on⟫ the mirror as base. For any diameter drawn from the mirror's centre to one point or other on its surface will be perpendicular to the plane tangent to the mirror's surface at that point; thus, if the eye's centre coincides with the centre of the sphere, the eye will not perceive any of the

visible objects outside the eye, because no form will be reflected in this case from any point on the mirror's surface other than the form of the point on the eye's surface through which the line from the eye's centre passes to that point.

[61] If, however, the centre of the eye lies outside the sphere's centre, then it is possible for any point on the mirror's surface that the form of some point or other on the visible objects outside the eye should be reflected from it [to the eye]. For the diameter through the centre of the mirror and the centre of the eye will reach a point on the mirror's surface from which there will be reflected the form of the point on the eye's surface through which this diameter passes; and any | other point [on the mirror's surface] is such that IV 128b
the diameter drawn from it to the mirror's centre will lie with the centre of the eye in one and the same plane. And this diameter will be perpendicular to the plane tangent to the mirror's surface at that point, because every line drawn from that point in the plane tangent to the mirror's surface will be tangent to the circle in whose plane this line lies, and this diameter will be a diameter to this circle, and thus any line drawn in the tangent plane [at that point] will contain two right angles with this diameter. This diameter will therefore be perpendicular to the tangent plane, and the diameter drawn from the mirror's centre to the centre of the eye will lie, together with this diameter, in one and the same plane. Moreover, the intersection between this plane and the mirror's surface will be the circumference of a circle, so the line drawn from the centre of the eye to that point [of reflection] will contain, with this diameter that passes through that point, an acute angle inside the circle in which this diameter lies. For, with regard to any line | drawn from the centre of the eye to a point IV 129a
on the spherical surface, this is manifest if the eye lies within the spherical surface or on the surface itself; but if the eye lies outside the spherical surface, then [the reason is] that all these lines will be within the circumference of the circle produced by the line drawn from the centre of the eye tangent [to the sphere], and therefore the tangent plane will lie behind the mirror's surface.[13] It will, then, be possible to draw from the extremity of the diameter, namely the perpendicular, another line in the plane containing the diameter and the centre of the eye, so as to include with the diameter an acute angle equal to that included between the diameter and the line drawn from the diameter's extremity to the centre of the eye. This line will lie behind the diameter, and inside the mirror, because the angle contained by the diameter and the circumference of the circle on the mirror's surface is greater than any acute angle [between] two straight lines. Then, if that [other] line is produced rectilinearly to reach a visible object, the form of the point reached on that object will extend along that line and reach the point on the mirror's surface, and be reflected from this point to the centre of the eye; and upon reaching the eye, it will be perceived. | IV 131b
And the same can be shown for every point on the mirror's surface. Thus, for any point on the surface of ⟨this⟩ mirror, it is possible for the form of one point or other on the visible objects to be reflected from that point; and sight's perception of all that is reflected from the surface of this mirror will take place simultaneously.

[62] Moreover, all planes in which the forms are reflected to the eye will intersect at the diameter passing through the centre of the eye and the centre of the mirror, for this

13. In MS F, the continuation of the text at IV 129a, line 6 is at IV 131a, line 3; this and the subsequent breaks must have been caused by misplacement of four pages in an archetype, which the scribe copied in the wrong order without noticing. See further nn. 15–18.

diameter lies together with any line drawn ⟨from the centre of the eye⟩ to a point on the mirror's surface in the plane orthogonal to the plane tangent to the mirror's surface at that point. And every plane in which the form of one point or other on the visible objects is reflected from the surface of this mirror is a plane that contains the centre of the eye, the visible point, the reflection point, and the centre of the mirror. Further, all intersections between these planes and the mirror's surface will be circles that pass through the centre of the sphere, and all will intersect at the point on the mirror's surface where the diameter
IV 132a through the centre of the eye and the centre of | the mirror will meet the mirror's surface.

⟨⟨IV 5.10⟩⟩

[63] With regard to the right-cylindrical concave mirror: it is possible for its whole surface to face the centre of the eye. This is so without qualification if the eye lies inside the cylindrical surface [i.e., inside the cylinder] or on the surface itself. If the eye lies outside the cylindrical surface, it will be possible for the whole surface of the mirror to face the eye, if the mirror is not larger than the larger [i.e., concave] segment cut off between the two planes tangent [to the mirror's surface] that pass through the centre of the eye.

[64] When the eye faces such a mirror, it will be possible, for any point on the mirror's surface, that the form of one point or other on the visible objects should be reflected from it [to the eye]. For, let the plane drawn from the centre of the eye parallel to the mirror's base cut the mirror's surface; the point on the circumference of the circle produced by this plane, in which there intersect the mirror's surface and the diameter drawn from the centre of the circle to the centre of the eye, will then be such that the form of the point on the
IV 132b eye's surface through which this diameter passes will be reflected [to the eye]. | Any other point on the mirror's surface will be such that the perpendicular drawn from it to the plane tangent to the mirror's surface at that point will lie with the centre of the eye in one and the same plane; and this plane will be orthogonal to the tangent plane, and this perpendicular will be a diameter of the circle passing through this point, and the perpendicular will meet the axis, as was all shown to be the case for the cylindrical convex mirror. But the line drawn from the centre of the eye to the extremity[14] of the perpendicular will contain, with the perpendicular, an acute angle, the tangent plane being situated behind the mirror's surface; and the line issuing from the centre of the eye will be inside the cylindrical concave surface; and, therefore, it is possible to draw from that point a line that contains, with ⟨the perpendicular⟩, an angle equal to that contained by the perpendicular and the line drawn from the centre of the eye to that point. And, therefore, ⟨it will be possible⟩ for the form of one or another of the points on the visible objects ⟨to be reflected from that point⟩. And the same holds for any point on the surface of this mirror.

IV 133a **[65]** The situation of the planes | in which the forms are reflected from this ⟨mirror⟩[15] will be the same as the situation of the planes in which the forms are reflected from the cylindrical convex mirror; all of those planes will meet the axis; each of them will contain the centre of the eye, the visible point, the reflection point, and the point in which

14. The text of Book IV ends here in MS K.

15. In MS F, the continuation of the text at IV 133a, line 1 is at IV 129a, line 6.

the perpendicular drawn from the reflection point meets the axis; the plane through the whole axis and [through] the centre of the eye will cut the mirror's surface in a straight line; and the plane parallel to the mirror's base will cut the mirror in the circumference of a circle. The intersections produced by all the other planes will be | [conic] sections, in IV 129b
each of which it will be possible to draw only one line perpendicular to a plane tangent to the cylinder's surface. But some of the [conic] sections produced in this mirror will be such that it will be possible for the forms to be reflected from the periphery at both extremities of that line. For that line, which we determined in the case of the [cylindrical] convex mirror, is a diameter of one of the circles produced in the cylinder. And only one extremity of that diameter can lie in the segment of the convex mirror that faces the eye, since [the segment] facing the eye on the convex mirror is less than half the cylinder. But it is possible for both extremities of the diameter to lie in the segment facing the eye on the concave mirror, because what faces the eye in the case of this mirror is larger than half the cylinder. And, therefore, it is possible for the forms to be reflected from the periphery of these [conic] sections at two points. But no forms can be reflected from the periphery of the [conic] sections produced in | this [cylindrical concave] mirror at more IV 130a
than two points.

⟨⟨IV 5.11⟩⟩

[66] With regard to the conical concave ⟨mirror⟩: it is possible for the whole of its surface also to face the eye. This is so without qualification when the eye lies inside the conical surface, whether the mirror is a complete cone or the segment of a cone, and when the eye lies outside the conical surface, if the mirror is not greater than the segment [of the cone] cut off between the two planes tangent to the cone's surface that pass through the centre of the eye. The eye will perceive nothing through this mirror if it is situated next to the cone's vertex, and if the line joining the centre of the eye and the vertex cuts the cone ⟨or⟩ extends in its surface, because every line from the centre of the eye to a point on this mirror will only meet the external surface of the cone, ⟨⟨either cutting it or being tangent to it⟩⟩, and thus sight will perceive nothing of the surface of this mirror unless | the line IV 130b
from the centre of the eye to the vertex of the cone, if produced rectilinearly, lies outside the cone.

[67] But it will be possible, for any point on the [segment of the] surface of this mirror that faces the eye, that the form of one point or other on the visible objects should be reflected to the eye from that point. For, [consider first] the point at which the perpendicular to the plane tangent to the mirror's surface will reach the centre of the eye: there will be reflected from that point the form of the point on the surface of the eye through which this perpendicular passes. As for the other points on the mirror's surface: the perpendicular to the plane tangent to the surface at any one of these points will lie with the centre of the eye in one and the same plane, as was shown in the demonstration concerned with the conical convex mirror. But every plane in which the form of one visible point or other is reflected will be a plane that contains the centre of the eye, the visible point, the reflection point, and the point on the mirror's axis in which the perpendicular at the reflection point meets the axis; and the sections | produced in this mirror will be like those produced in the conical IV 131a
convex mirror.

[68] If, however, the centre of the eye lies on the mirror's axis and inside the conical surface of the mirror,[16] then all the sections will be straight lines; and if the centre of the eye lies outside the axis, then only one of the sections will be a straight line, and all others will be [conic] sections, whether the centre of the eye is situated inside, outside, or on the
IV 135a surface of the cone. And for each one | of these [conic] sections, it will not be possible for the forms to be reflected from the periphery at more than two points.

[69] If, however, this mirror is a complete cone, then the eye, when facing it [from a point outside the cone], will in most cases perceive only the mirror's surface. The reason is that the perpendiculars at whose extremities the forms are reflected will be inclined to the axis, and each of them will contain an acute angle with the axis, and thus the extremity of each perpendicular, where the reflection takes place, will be closer to the vertex of the cone, while the centre of the eye, when the eye faces the vertex, will be closer to the cone's base; and thus, if a line is drawn from the centre of the eye to the perpendicular's extremity, the second line, that reflected from this line, will incline from the perpendicular towards the vertex; and this second line, if extended rectilinearly, will in most cases meet the mirror's surface; and, therefore, sight will in most cases fail to perceive in this mirror
IV 135b any of the visible objects situated outside | the mirror, unless some object is placed close to the cone's vertex inside the mirror.

[70] If the conical concave mirror is a segment [cut] along the length of a cone, then it is possible for sight to perceive in it many visible objects, because the second, reflected lines, if extended rectilinearly, will get out of the cone.

[71] Similarly, if the mirror is a circular segment of a cone, in the shape of a ring that has been cut off close to the cone's vertex, sight will perceive in it many visible objects. For many of the second lines will in this case exit the cone on the side opposite the location of the eye. The eye will perceive more [objects] when it is close to the narrower end [of the conically shaped ring], because ⟪more⟫ of the second lines will then exit from the wider end.

[72] That, then, is the manner in which visual perception takes place in this mirror. The situation [just described] may obtain in the cylindrical concave mirror as well as in this conical concave mirror, ⟪that is to say, if the cut-off ⟨segment of the cone⟩ has the shape of a ring⟫.

IV 136a **[73]** Thus it is the case, for | any point on the surface of any mirror, that only the form of a single point on the visible objects will be reflected to a single eye at any one time. For the plane that contains the eye and the perpendicular to the plane tangent to the mirror's surface at that point must be a single plane, and the line from the centre of the eye to that point must be a single line, and the angle contained by this line and the perpendicular must be a single angle. And, therefore, only a single line can be drawn from that point in the plane passing through the centre of the eye, and orthogonal to the tangent plane, so as to contain with the perpendicular an angle equal to that angle, while being situated behind the perpendicular outside the tangent plane; and it is the form extending only along that line that can be reflected to that eye at that time. For it is not
IV 136b possible for any form to be reflected from that point to that eye | at that time, except along the line that extends between the centre of the eye and that point, which is a single line.

---

16. The text at IV 131a, line 3 continues at IV 134b, line 9.

And, therefore, ⟨only the form of a single point on the visible objects⟩ will be reflected from a single point on the surface of one of the mirrors to a single eye at one and the same time.

⟨⟨Recapitulation⟩⟩

[74] And[17] only the form of a single point will extend along a single line to a single point on a single mirror at [any] one time. For if the form of a point on the single line extends along the line to reach a point ⟨on⟩ the mirror's surface, then no form of any other point on this line will arrive at that point on the mirror's surface, if there exist on this line ⟨the forms⟩ of other points on the visible objects. The reason is that the first point will hide the other points that lie behind it, so that the form of the other points that extend on this line at the same time will not reach the point on the mirror's surface.

[75] Therefore, only the form of a single point on the visible objects will be reflected from a single point on a given mirror to a single eye at one and the same time. It is, IV 133b
however, possible for the form[s] of a number of | points on visible objects to be reflected at one and the same time from a single point on the mirror's surface to a number of eyes, because it is possible for an infinite number of planes to intersect at a single perpendicular, each of which will be orthogonal to the plane tangent [to the mirror at that point on its surface], and it will be possible to draw from the extremity of that perpendicular in each of these planes two lines that contain, with the perpendicular, two equal acute angles.

[76] What we have now explained is the detailed account of what needs to be detailed concerning the properties of mirrors and what pertains to the reflection of forms from them.

[77] Now, all that sight perceives by reflection from mirrors is perceived [to be] facing it, like what it perceives in visible objects rectilinearly, without [involving] reflection; [sight] will not sense that its perception of the object perceived in the mirror is perception by reflection. The reason is that sight does not perceive any visible object, except along the straight lines that meet at the centre of | the eye. When the form of the visible object IV 134a
radiates upon the mirror and is then reflected along the lines that meet at the centre of the eye, sight will perceive each point of that form only along the line that ends at the centre of the eye.

[78] And when sight perceives the form along the lines ⟨that⟩ meet at the centre of the eye, it will perceive that form [to be] facing it and at the extremities of those lines, just as it perceives the forms of the objects perceived rectilinearly, namely [the objects actually] facing it, at the extremities of those lines, because it perceives the reflected and the rectilinear forms along the same identical lines. It is for this [reason], then, that sight perceives the object visible in the mirror [to be] facing it, [thus] taking the object to be at the extremities of the lines along which the [object's] form has come to it. Sight does not sense the reflection of the reflected form, because the lines along which the object's form has extended to the mirror's surface do not reach the eye, and [hence] the eye [*or* sight] does not sense them, nor is it concerned with them. Nor does the extension of forms along these | lines take place on account of the eye or on account of the reflection, but rather IV 134b
forms always [tend to] extend along every straight line along which they are [already]

---

17. The text at IV 136b, line 4 continues at IV 133a, line 1.

extending. Sight, then, perceives visible objects in mirrors by reflection and, moreover, it perceives them [to be] facing it, because it perceives the reflected forms along the lines through which it perceives the forms of facing objects perceived without reflection, and because sight perceives no visible object, except along these lines. ⟨Sight⟩ does not sense the reflection of the forms it perceives by reflection, because the lines along which the form extends to the mirror do not reach the eye, nor is the eye concerned with them,[18] and because the extension of the form along these lines does not take place on account of the eye or of the reflection. ⟪Rather, that [perception by reflection] is due to the nature of the lights that encounter polished surfaces.⟫

[79] Moreover, the visible objects that sight perceives in mirrors are perceived [as being] behind the mirror, but it often perceives them [to be] in front of the mirror and often on the surface of the mirror. The location of the form will vary with the shape of the mirror and with the position of the eye with respect to it; but sight will always perceive the form in a particular location that does not change, as long as the position of the eye with respect to the mirror remains unchanged. Thus, the distance between the mirror's surface and the location at which sight perceives the form will vary with the shape of the mirror. And this location, at which sight perceives the form of the object perceived by reflection, is called
IV 137a 'the location of the image', and the form | that sight perceives in the mirror is called [the] 'image'. And, thus, sight will only perceive a visible object by reflection at the location of the image.

[80] It now remains for us to determine this location for each one of the mirrors that have already been enumerated, and state the reason why sight perceives the object in that location. We should therefore thoroughly explain this matter in the next Book.

[Colophon]

[81] Book IV of the work on optics has been completed. Praise and all due thanks to God. From Him alone I seek help and in Him alone I trust.

18. The text at IV 134b, line 9 continues at IV 136b, line 4. This is the last break in MS F.

ADDENDUM[19]

Commentary by Kamāl al-Dīn al-Fārisī Concerning the Construction of the Experimental Instrument, as also Explained by Ibn al-Haytham in Book IV

I [Kamāl al-Dīn] say: He [Ibn al-Haytham] described an instrument for examining experimentally what he set out, but without a diagram. Seeing that a diagram would help to understand the instrument, I have added a diagram to the description. Let us describe it in order, because it consists of arranged parts. [Diagrams IV.1, IV.2]

He said: The Plate. As to how to proceed to achieve certainty, let the experimenter obtain a copper plate not less than twelve digits in length, and half that in breadth, and let it be thick enough to avoid being bent or corrugated. He should make both surfaces as even as possible. Let the plate be ABGD [Diagram IV.1], then draw a straight line lengthwise close to one edge and parallel to it, let it be EZ. Mark the middle of this line at H, and, with H as centre and half the line as radius, draw the semicircle ETZ. Then draw, from the centre, the perpendicular HT at right angles to the diameter; it then bisects arc ETZ at T.

I say: This is the line that [Ibn al-Haytham] often calls 'the line at the middle of the plate'.

He said that he then divides one of the two arcs as many times as he wishes, then divides the other into parts equal to the first in number, proportion, and order, and let the divisions begin at the middle. Let one set of the divisions be TY, YK, KE, and the other TL, LM, MZ. Then he joins the centre and the points of division by means of HY, HK, HL, HM. And let all radial marks be made with an iron implement so that their impression in the wood may last.

Then draw, with H as centre, a smaller semicircle than the other, let it be the one through N, S, O, and let the distance between the two circles be one digit.

Then mark off from perpendicular HT, and next to the centre, the length of one digit, let it be HF. Then draw, through point F, chord CFQ, parallel to diameter EZ.

Then cut off from the plate what remains between the chord and the end of the plate's length on the side of the diameter, ending the cutting at lines HY, HL; then cut off what remains of the plate next to the circumference of the curve, leaving the figure HRCTQJ. There will remain a small triangle in the middle of the plate next to the centre. He should be careful to save the point at the centre as the tip of the triangle.

He should then cut off slantwise from the thickness of the plate whatever remains directly below the centre, thus sharpening the tip of the triangle, namely the centre.

I say: The honing should be towards the back of the plate, not the face of it.

He said: The Ring. Having finished with this plate, [the experimenter] should obtain a piece of soft wood, such as juniper, pine, or jujube; it should be quadrilateral, its length being more than that of the plate by two digits, and six digits in height. He should make

19. The text of Kamāl al-Dīn's Commentary may be found in Sabra, *Al-Ḥasan ibn al-Haytham, Kitāb al-Manāẓir. Books I–III*, II, pp. 225–30. As noted in the Introduction, there are no diagrams in MS F, which is copied from the fourth volume of the 'Askarī Set. Indeed, neither Ibn al-Haytham, nor Kamāl al-Dīn in his *Tanqīḥ*, found it necessary to include such diagrams, contrary to what would be the case later with the Latin translation of the *Optics*. Kamāl al-Dīn, however, found it useful to draw two diagrams associated with the construction instrument, which we have reproduced here from MSS TA and Leiden, Or. 201.

its top as even as possible. He should then mark point A [Diagram IV.2] at the middle and draw around it a circle with a diameter exceeding the radius of circle ETZ [in Diagram IV.1] in the plate by one digit, let it be circle BGD. He should then draw about A another circle equal to circle NSO in the plate, let it be circle EZH; then the distance between circles BGD and EZH will be two digits.

He should then mark off from circumference EZH [Diagram IV.2] arc TZK equal to NSO [in Diagram IV.1], and divide TZK at [points] L, M, N, S as the division of arc NSO [Diagram IV.1] by the mentioned radii, with regard to number, proportion, and order.

He then should draw from centre A the radii to the points of division and extend them to circumference BGD, and let the drawing be made with an iron implement so that the lines remain in the body of the wood.

Then he should set up this wooden [block] on the lathe at the centre of the circle, and work it to remove the redundant parts around circle BGD all the way to the circumference. The lathe should then be applied again to the middle of the ring to remove its whole interior and the working be stopped at the circumference of circle EZH; this can be achieved by attaching the wooden piece to another wooden piece, and determining the point [on the latter] opposite the centre of the first piece. The lathe should then be applied to this point until it reaches the first piece, and the wood worked until it reaches the circumference of the smaller circle, thereby cutting from the [whole] wooden piece a ring measuring two digits in width. There will remain on the surface of the ring the extremities of the radii that divide both arcs into equal divisions like those of both arcs of the plate.

Having finished the wooden ring, he should draw on the outer [i.e., convex] surface of its height straight lines stretching from the extremities of the lines across its width, meaning its face, which are orthogonal to the surface of its width. He can do this by placing upon [the ring's] height a straight sharp-edged ruler. He should place the ruler's edge at the extremity of the straight line across its width, and adjust the ruler upon the surface of the height until the edge coincides with the circumference of the other width, that is, the base of the ring. He should then draw the line on the height at the [edge of] the ruler's line; [the line] will then be straight and perpendicular to the surfaces of both widths of the ring.

If he gently makes the extremity of the straight line on the [top] surface of the ring a centre; and [if], while pressing the leg of a light, sharp compass at the extremity of the line on the top of the thickness of the ring, he draws a semicircle about it on the surface of the cylindrical semicircle; and [if he] divides the arc in two halves and, applying the ruler's edge upon the points of the centre and the division, draws the line, it would be better, for the edge of the ruler would then be certainly applied upon the cylindrical surface.

He should then also draw on the concave surface the lines corresponding to those drawn on the convex surface from the ends of the lines perpendicular to the two surfaces of the width [i.e., the two bases] of the ring. He then should remove from the ring the piece defined by the two [small] arcs and the lines through points T and K [Diagram IV.2].

Then he should draw on the inner [i.e., concave] surface of the ring an arc parallel to the other arcs, so that the distance between [the drawn arc] and the end of it [i.e., the concave surface] on the side of the [top] face of the ring is four digits.

Diagrams to Book IV, Addendum

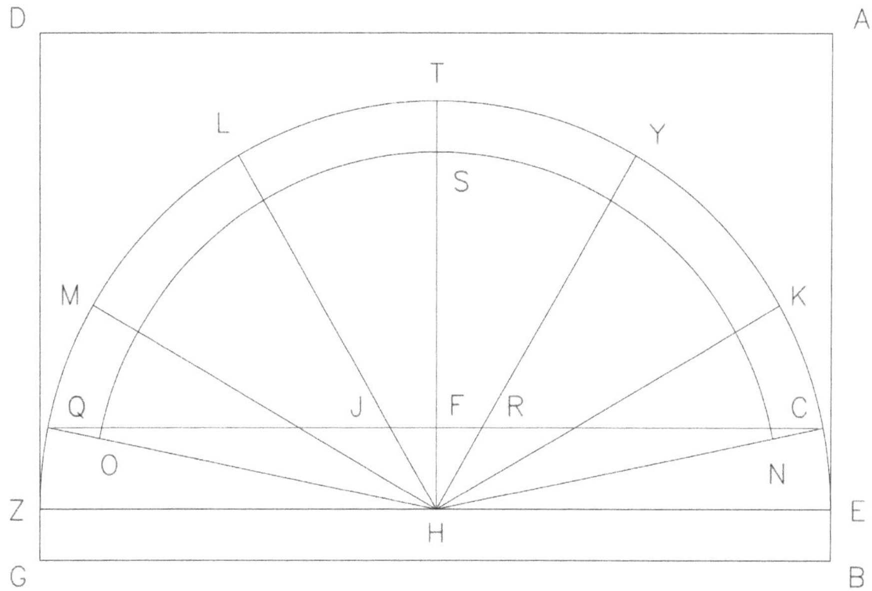

Diagram IV.1

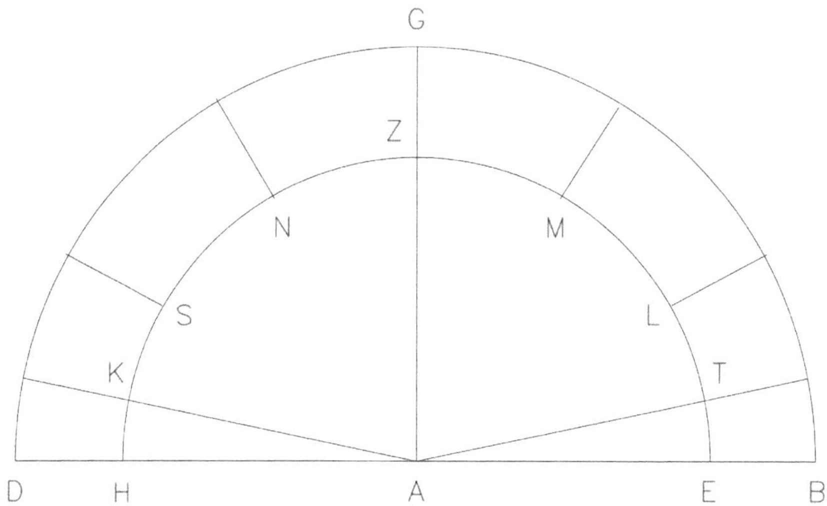

Diagram IV.2

| *In the Name of God the Compassionate and Merciful* V 138b

THE FIFTH BOOK OF THE *OPTICS*

OF AL-ḤASAN IBN AL-ḤASAN IBN AL-HAYTHAM

CHAPTERS OF THIS BOOK: THEY ARE TWO

CHAPTER 1. PREFACE

[1] In Book ⟨IV⟩ we showed that the forms of visible objects are reflected from polished bodies, and that sight perceives visible objects in polished bodies | [only] from V 139a
the forms reflected from the polished bodies; and we have shown how sight perceives visible objects in the polished bodies from the[ir] reflected forms. Now, sight perceives the visible object in the polished body at a certain location with respect to the polished body, the position of the form with respect to the polished body being invariable. And the form that is perceived in the polished body is called the 'image' [*khayāl*]. We shall now show in this Book the locations of images in polished bodies, and show how these locations may be examined experimentally, and how they can be determined and located by reasoning and demonstration.

CHAPTER 2.
ON IMAGES [PERCEIVED BY REFLECTION]
⟪Consisting of [an Introduction and] Nine Inquiries⟫

[Introduction]

[1] An image is the form of the visible object that sight perceives by reflection from the surface of a polished body. The location of an image is the place where sight perceives
V 139b this form, and every point of the form of an object | perceived by reflection is the image of the corresponding point of that object that is perceived by reflection. And the image of any point of an object perceived by reflection is at the point where the line along which the form of that point is reflected from the object to the eye, that is [the line] on whose rectilinear extension sight perceives the point that belongs to the object, meets the ⟨line⟩ drawn from that point on the object perpendicularly to the line tangent to the intersection between the surface of the polished body, i.e., mirror, and the plane in which the reflection takes place, or to the continuation of this intersection. This state of affairs can be perceived by experiment, and we shall [now] show how it can be examined in such a way as to achieve certainty.

⟪V 2: First Inquiry.
Experimental Examination of the Location of Images,
Consisting of Seven Experiments⟫

⟪V 2.I.1: Experimental examination of the location of the image in the plane mirror⟫

[2] To examine experimentally the location of images in plane mirrors, the experimenter should take a thin and perfectly straight rod, and let it be pure white. Let him take a plane
V 140a mirror of even surface; using a large mirror will make the [results of the] experiment | more manifest. He should place the mirror on the surface of the ground, ensuring that the mirror's surface is parallel to the horizon. Let him mark a black dot somewhere on the rod; the dot should be conspicuous and close to one end of the rod. He should then place the rod upright on the mirror's surface, so that the end near the dot is on the mirror, and let him look into the mirror. He will see the form of the rod behind the mirror, while also observing the rod itself outside the mirror. Let him ensure that he sees the whole rod in the mirror, including the upper end of the rod, for if he sees the upper end, no error will occur regarding the form, but an error regarding it may occur if he does not see that end.

[3] Upon perceiving the form of such a rod in the plane mirror, the experimenter will [also] see the form of the rod as rectilinearly continuous with the rod, and the dot marked on the rod will be visible on the form seen in the mirror. From this experiment, it is clear
V 140b | that the form of the point that sight has seen behind the mirror lies on the rectilinear extension of the perpendicular drawn from the assumed dot to the mirror's surface.

[4] While perceiving the form of the dot on the rod, sight will see it to be as sensibly distant from the mirror as the dot itself from the mirror's surface. That is to say, it will see the form of the segment of the rod between the form of the dot and the mirror's surface

⟨to be equal to the segment of the rod that lies between the assumed dot and the mirror's surface⟩. Then if [the experimenter] inclines the rod to the mirror's surface, he will see its form inclined in the same direction, and will see the segment of the form of the rod between the assumed dot and the mirror's surface to be equal to the segment itself of the rod that lies between the assumed dot and the mirror's surface.

[5] Then let the experimenter set up another thin rod on the surface of the mirror while the first rod ⟨⟨remains inclined⟩⟩, ⟨so that the marked dot on the first rod⟩ coincides with the tip of this second rod, ensuring that the second rod is perfectly perpendicular to the mirror's surface. He will see the form of this rod also to be continuous | with the rod itself; V 141a
the form of the dot on the first rod will be found at the tip of the form of the second rod; the form of the second rod will be sensibly equal in length to the second rod itself; and the second rod and its form, which is continuous ⟨with it⟩, will be found to be at right angles with the mirror's surface.

[6] It is clear from these experiments that the image of every point on the first rod, whatever the positions of the points, is [located] behind the mirror, on the rectilinear extension of the perpendicular drawn from that point to the mirror's surface; that the perpendicular to the surface of the plane mirror is orthogonal to the intersection between the mirror's surface and the plane orthogonal to the mirror's surface, [the latter plane being that] in which the reflection takes place; and that this point, I mean the image, lies on the straight line along which the form is reflected to the eye, for sight perceives the point, namely the image, on the rectilinear extension of this line.

[7] Similarly, if the experimenter fastens the rod beside the mirror, while ensuring that it makes right angles with the mirror's surface, and looks into the mirror so as to see the whole | rod, including its tip, then examines the form, [and] if [further] he inclines the rod V 141b
as we described earlier and examines its form, he will find the form in both cases to be as [indicated] earlier.

[8] Again, if the experimenter lifts the mirror and, holding it [vertically] in his hand, sets the rod upon it and examines its form [i.e., image], he will find the image of the rod and the image of the dot assumed on it to be as [described] earlier. And, again, if he assumes several dots on the rod and sets up the rod upon the mirror's surface, he will find all the dots to be on a single straight line.

[9] To make this experimental examination precise, let the experimenter take a circular cone made of wax or the like, which should be as correctly [shaped] as possible, and let it be a right cone. Having made the surface of the cone's base as even as possible, he should place it on the mirror's surface, fasten the base to the mirror's surface, and look into the mirror. He will see the form of this cone to be that of a right cone opposed to that cone, and will see the base of the [mirrored] cone to be coincident with the base of | that cone. V 142a
The proof that the imaged cone is a right cone is that the distances between its vertex and the periphery of its base are equal.

[10] When the experimenter moves his eye nearer to the cone, so that the reflection of the cone's vertex takes place from a point on the mirror's surface closer to the cone's base than the first point, he will find the form of the cone to be as he found it before, i.e., he will find it to be the form of a right cone whose base coincides with the base of the cone placed upon the mirror's surface.

[11] Now, for every two opposed right cones with a single circle as their common base, the line joining their vertices will pass through the centre of the base at right angles. Thus,

the line joining the vertex of the cone that is set up on the mirror's surface and the vertex of the cone that is perceived by reflection, i.e., the form of the cone placed upon the mirror, will pass through the centre of the cone's base and will be perpendicular to the surface of
V 142b the cone's base, which coincides with the mirror's surface. | Therefore the line joining the point at the vertex of the cone that is mounted on the mirror's surface with the image of that point, i.e., the vertex of the form [of the cone], will be perpendicular to the mirror's surface, and as such, this line will be perpendicular to the intersection between the surface of the mirror and the plane orthogonal to it, ⟨which is⟩ the reflection plane.

[12] And similarly, for every point on the surface of the cone mounted on the mirror, the line joining it to its form, i.e., the point corresponding to it on the form of the cone, will be perpendicular to the mirror's surface. But any point on any visible object perceived by reflection in a plane mirror is the vertex of an imagined right cone whose base coincides with the surface of the mirror or with the surface continuous with the mirror's surface. Therefore, the point from which the form of the cone's vertex is reflected will lie outside the cone's base, because every point perceived by reflection from any [type of] mirror can be the vertex of infinitely many cones, for all of which that point is a common vertex.

V 143a **[13]** Thus, let that cone, | outside whose base the reflection point is located, be a certain body, and let the surface of the mirror so extend that the base of that conical body may lie upon the mirror's surface, then the form of that cone will be a right cone opposed to it, ⟨and⟩ their base will be a single, common circle, and the form of the visible point will be a vertex of the imaged cone.

[14] Now, for any point on any visible object that sight perceives by reflection in a plane mirror, the image of that point will be located behind the mirror on the line drawn from that visible point to the surface of the mirror perpendicularly to the intersection between the mirror's surface and the reflection plane. And, for any point on any object that is perceived by reflection, sight will perceive that point on the straight line along which the form of that point is reflected to the eye. For sight perceives visible objects only from the forms that reach it from them; it perceives the form only along radial lines; it perceives
V 143b visible objects in mirrors only by reflection; | forms are reflected only along straight lines; and reflected forms reach the eye only along reflection lines.

[15] Now, sight perceives reflected forms along radial lines if reflection lines are among the radial lines. Thus, if sight perceives the form of a point perceived in the mirror along a line other than the reflection line, then the form perceived in this case will have arrived at the eye along a line other than the reflection line. But this is impossible, for it has been shown that sight perceives forms reflected to the eye only along the determinate lines of reflection. Therefore, sight can perceive the forms of the visible objects it perceives in mirrors only along the reflection lines along which the reflected forms reach the eye.

[16] Therefore, for any point on any visible object perceived by sight in a plane mirror, the image of that point will be [located] on the rectilinear extension of the line along
V 144a which the form of that | point has been reflected to the eye. And for any point on any visible object that sight perceives by reflection in a plane mirror, the image of that point will coincide with the point ⟨in which⟩ there meet the line along which the form of that visible point has been reflected from the object to the eye, and the line drawn from that visible point perpendicularly to the intersection between the plane of the mirror's surface and the plane orthogonal to it in which the reflection takes place. That is the position we

determined [earlier], for the imagined straight line, tangent to the intersection in the case of these [plane] mirrors, will coincide with the intersection itself, since the intersection in the case of these mirrors is a straight line.

⟪V 2.I.2: Experimental examination of the location of the image in the spherical convex mirror⟫

[17] To examine experimentally the location of the image in the spherical convex mirror, let the experimenter take a spherical mirror that is part of a fairly large sphere; using a fairly large spherical mirror will make the [results of the] experiment more manifest. Let him set up the previously described rod on the mirror's surface, ensuring that the rod stands perpendicularly [on the surface]; and let him look into the mirror so as to see the form of the rod | and the dot marked on it. He will see [that] the form of the V 144b rod is rectilinearly continuous with the rod itself, if the rod is extremely thin, and the dot marked on the rod will be visible on the form of the rod. From this experiment, it is clear that the image of the dot on the rod lies on the rectilinear extension of the perpendicular from that dot to the mirror's surface. ⟪But the form of the rod⟫ will be found shorter than the rod itself. Similarly, every form perceived by sight in a spherical convex mirror will be smaller than the visible object itself; and if the rod is in the least thick, its form will be conically shaped, that being one of the errors of sight, which we will explain when we discuss errors of sight.

[18] Let us [now] precisely examine experimentally the location of the image in this mirror. Let the experimenter take the previously described cone of wax, and let him hollow out the base of it so that the circumference of the resulting concavity, i.e., the circumference of the cone's base, will be perfectly circular; let him then mount this cone upon the mirror's surface, fasten it to the mirror, and look into the mirror. He will | see the V 145a form of this cone as one of a right cone, and will find the form smaller than the cone itself; but he will [also] find the distances between the vertex [of the mirrored cone] and [points on] its base to be sensibly equal, and that the base of the form coincides with the base of the cone itself that has been mounted on the mirror's surface.

[19] If he tilts the mirror so that the cone's base partly occupies the portion of the mirror's surface facing the eye while partly lying outside [that portion], he will also see the form of the cone as that of a right cone opposite the cone itself, with its base being coincident with the cone's base. If the experimenter brings his eye nearer to the cone so that the reflection of the form of the cone's vertex takes take place at a point on the mirror's surface closer to the cone's base than the first point, he will find the form of the cone to be the same as he found it before, as long as the reflection point lies outside the cone's base; that is to say, he will find the form to be that of a right cone whose base coincides with the base of the mounted cone itself.

[20] If, keeping the mirror fixed, the experimenter | moves his eye away from the V 145b mirror, while ensuring that this motion takes place along the line along which the form of the cone's vertex is reflected to the eye, he will find that nothing in the form has changed, and [also] that nothing has changed regarding the position and distance of the cone's vertex with respect to the mirror. The motion of the eye in this manner can be controlled [as follows]: the experimenter should mark the point on the mirror's surface from which he perceives the form of the cone's vertex, then move his eye away while looking at the cone's form, ensuring that in [this] second situation he perceives the form of the cone's vertex and the point marked on the mirror's surface together in the same position in which he perceived them earlier. For if he perceives the cone's vertex – that is, its form – in the second situation, together with the [marked] point as he perceived them [together] at first, then the centre of sight in the second situation must be on the same line on which it was located in the first situation.

[21] Now, for any two opposed right cones whose base is a single, common circle, the line joining their vertices will pass through the centre of the common base and will
V 146a be | perpendicular to the base. If the cone is mounted on the surface of a spherical convex mirror, then the circular base of the cone will coincide with one of the circles on the sphere, and thus the line joining the vertices of the two cones will pass through the centre of the circle on the sphere and be perpendicular to the circle's surface. And if the line is perpendicular to the surface of the circle on the sphere and passes through the circle's centre, then it will pass through the centre of the sphere, and as such this line will be perpendicular to the surface of the sphere and to the plane tangent to the sphere's surface at the point through which the line passes, and [perpendicular] to the intersection between this tangential plane and the reflection plane.

[22] For the diameter through the vertices of the two cones lies in the reflection plane, since the visible point and its image both lie in the reflection plane. But the intersection between these two planes is a straight line, and a perpendicular is orthogonal to every line that extends from its extremity in the plane to which it is perpendicular. Therefore, the line
V 146b that joins the vertices of the two cones will be perpendicular | to the intersection between the reflection plane and the plane tangent to the mirror's surface at the point of intersection between the mirror's surface and the line joining the two vertices. Now, this intersection is a straight line tangent to the circumference of the circle, which is the intersection between the mirror's surface and the reflection plane. Therefore, the image of the vertex of the mounted cone will be on the perpendicular drawn from the cone's vertex to the line tangent to the intersection between the mirror's surface and the reflection plane; and the point at the vertex of the imaged cone lies on the straight line along which the form of the cone's vertex has been reflected to the eye, as was shown earlier. Therefore, the point at the vertex of the imaged cone, namely the image of the vertex of the mounted cone, lies at the point of intersection between the line along which the form has been reflected to the eye and the perpendicular drawn from the point at the cone's vertex to the line that is tangent
V 147a to the intersection between the mirror's surface and the reflection plane. |

[23] Now, any point on any visible object seen by reflection in this mirror is a vertex of an infinity of imagined right cones, with the vertex of each cone being that point, and with the base being the surface of the mirror facing the eye, or a portion of this surface, or with part of the base being a portion of the facing surface, while the other part lies outside the surface. Each point on any visible object seen by reflection in this mirror is, then, the vertex of an imagined right cone whose base, or part of whose base, ⟨is a portion of the surface⟩ facing the eye, in which the form of that point is seen; and the reflection point from which the form of the vertex of the cone has been reflected to the eye will lie outside that portion ⟨of the surface⟩ that faces the eye as ⟪the centre⟫ of the eye recedes from the mirror along the line of reflection; or, rather, the visible point is the vertex of many such cones.

[24] That being the case, ⟨and⟩ assuming the imagined cone, whose vertex is the visible
V 147b point, to be some visible body, then | the form of it seen by reflection in this mirror will be

the form of a right cone opposite that cone, and the form of the visible point on the visible object seen by reflection will be the vertex of the imaged cone. Therefore, the image of any point on any object seen by reflection in a spherical convex mirror will be at the point of intersection between the line along which the form of that point is reflected to the eye and the perpendicular drawn from that point to the line tangent to the intersection between the mirror's surface and the reflection plane.

⟪V 2.I.3–4: Experimental examination of the location of the image in the cylindrical convex and conical convex mirrors⟫

[25] With regard to the cylindrical convex mirror: if the object perceived through it is orthogonal to the mirror's surface, then its form will be inclined, and also, when the object is straight, not straight but curved. This state of affairs in the case of this mirror is one of the errors of sight. That being the case with the shape of the form perceived in this mirror, the experimental examination of a single point on the object must be conducted in isolation [from other points]. | To conduct a precise examination of the images of objects V 148a
perceived in this mirror, the experimenter must employ the instrument we described for the examination of the reflection of lights. To do so, let the experimenter take a piece of board with parallel surfaces, and place it under the copper plate in the previously described instrument. The height of the board should equal the elevation of the plate from the base of the instrument, so that the board will sit neatly and securely beneath the plate.

[26] The experimenter should ensure that part of the board should extend beyond the plate, so that the small triangle at the middle of the plate's edge rests on this board. He should then surround the board with some wax, which should be pressed in at the level of the plate's surface. Then let him set up the post with the cylindrical convex mirror on this board, so that the side of the post's base that runs the width of [the base] at the edge of the board on which the mirror is mounted coincides with the edge of the plate whose middle is occupied by the small triangle's base. He should also ensure that the extremity of | what V 148b
remains of the line that extends along the length of the post lies at the point of intersection between the line extending along the length of the plate and the line across the breadth of the plate on the plate that passes through the middle of the small triangle. This position being precisely secured, the face of the post will be orthogonal to the surface of the plate, and the line in the middle of the plate will be perpendicular to the face of this post. Having precisely secured this position, the experimenter should surround the base of the post with some wax and securely fasten the post to the board and to the plate, so that they remain fixed in position.

[27] Having set up this post, the experimenter should take a thin ruler and sharpen its straight edge. Let him cut away one end of it slantwise, so that the angle of the ruler at the extremity of the sharp edge becomes like a point. He should then apply this ruler to the top surface of the ring in the instrument, ensuring that the sharp edge of the ruler coincides with the straight line drawn in the middle of the ring's top surface, then gently move the ruler forwards while in | this position, until its sharp tip meets the mirror's surface, V 149a
whereupon he should mark the meeting-point on the mirror's surface. This point will lie on the straight line imagined to stretch along the middle of the mirror and the middle of the ruler's face that is tangent to the mirror's surface ⟨at that point⟩.

[28] That point being marked on the mirror's surface, [the experimenter] should remove the sharp-edged ruler, and place a thin needle on the top surface of the ring, along the straight line drawn in the middle of it, then firmly fasten the needle in this position with wax in such a way that if this needle is imagined to extend rectilinearly towards the mirror it will reach the point marked on the mirror's surface, and the straight line imagined to extend through the middle of the needle will rectilinearly reach the middle of that point, and will be perpendicular to the plane tangent to the mirror's surface at that point.

V 149b **[29]** Then let the experimenter take a very small body; | let it be white or of a pale colour, like a sesame seed, or some white body smaller than a sesame seed, and stick the tip of the needle in it. Let him position one eye behind the ring, at its middle and in the plane that cuts the mirror at the middle of its length. He should raise his eye above the top surface of the ring, cover the other eye, and look into the mirror, so as to see the form of the small body that is attached to the needle's tip. Once the form of this body is visible, he should contemplate it, together with the body itself and the point marked on the mirror's surface. He will see the three of them on one straight line, as far as appears to sense. Now, this line is perpendicular to the surface of the post on which the mirror is mounted, since the line passing through that small body and the point on the mirror's surface is perpendicular to the face of the post, and as such it is perpendicular to the straight line imagined to stretch along the length of the mirror, namely the intersection between the mirror's surface and the reflection plane. And this reflection will take place from a point
V 150a on the imagined straight line | along the length of the mirror higher than the point marked on the mirror's surface.

[30] Having realised this, the experimenter should move his eye and position it in the plane of the ring's upper surface [at a point] close to the edge of the ring, then look into the mirror so as to see in it the form of the small body. Upon seeing it, he should contemplate the form, together with the body itself and the point marked on the mirror's surface. He will see all three on a single straight line, as far as appears to sense. But this line is perpendicular to the face of the post, and thus perpendicular to the line tangent to the circle forming the intersection of the mirror's surface with the [plane] of the upper surface of the ring, if this plane is imagined to extend until it cuts the mirror's surface. Now, this [latter] plane is orthogonal to every plane tangent to the mirror's surface at a point on the circumference of this circle, and therefore it is one of the reflection planes; and in this situation, the centre of the eye and the small body are [located] in this plane; and in this situation, therefore, the reflection point is a point on the circumference of this
V 150b circle that is parallel to the mirror's base | and passes through the point marked on the mirror's surface. We shall later on show that when the eye and the visible object lie in the plane parallel to the base of this mirror, the form of that object will be reflected to that eye only in that same plane; and we shall show this by a demonstration that does not depend on anything we have stated or will state in this [present] chapter. Thus, in this situation, the image of the small body that is perceived lies on the perpendicular drawn from the visible body to the line tangent to the intersection between the mirror's surface and the reflection plane.

[31] Having realised this, the experimenter should remove the needle from its position, place the small, sharp-edged ruler on the top of the ring with the sharp edge coinciding with the straight line drawn in the middle of the ring's upper surface, and slide the ruler gently along this line until its sharp angle reaches the point on the mirror's surface. This done,

let him securely fasten this sharp-edged ruler to the ring's upper surface by | surrounding V 151a
it with wax. Having fixed the ruler, the experimenter should remove the post holding the
mirror from its position, and place on the face of this post holding the mirror a sharp-edged
ruler with its sharp edge coinciding with the two remaining segments of the line drawn
in the middle of the face of the post; in this situation, the ruler's sharp edge will coincide
with the mirror's surface.

[32] ⟨The experimenter⟩ should then draw on the surface of the mirror a straight line along the ruler's edge in black. He should then take a small triangle of wax, of which one side equals the width of the post holding the mirror; let the triangle be fairly thick and let its surfaces be made as even as possible. He should then stick the triangle under the base of the post, so that the side of it that equals the width of the post coincides with the side of the post's base, and the surface of this triangle coincides with the base of the post. He should securely attach this triangle to the post's base.

[33] Once this triangle is attached to the base of the post, [the experimenter] should
make the surface of the triangle, next to the face of the post, as level as possible | with the V 151b
post's surface, which can be done precisely by applying to [the triangle's surface] and the
face of the post a light board with an even surface. The surface of the triangle being made level with the surface of the post in which the mirror has been fitted, [the experimenter] should apply a sharp-edged ruler to the end of the triangle, and cut the triangle's end at the edge of the sharp ruler, so that this end now forms a straight line, which in this case will be the base[line] of the [face of the] post that holds the mirror.

[34] Having finished with the triangle in this manner, [the experimenter] should set up the post on the surface of the board in the instrument, allowing the edge of the post, namely the side of the wax triangle, to coincide with the line extending along the length of the plate, as he did before. In this situation, as before, the surface of the post holding the mirror will be orthogonal to the surface of the plate. But [the plane of] this surface [of the post] will cut the plate at the line extending through the length of the plate, and it is tangent to the surface of the mirror at the line drawn in the mirror's surface, since this plane is
[the plane of] the surface of the post on which the mirror is attached, and, therefore, in this
situation, the sharp angle of the ruler | that is attached to the upper surface of the ring will V 152a
lie in this plane. Further, the mirror will in this situation be inclined to the side on which the vertex of the triangle lies, since the post has been raised on one side by the body of the triangle, but the other side that is next to the triangle's vertex lies on the surface of the plate, and therefore, in this situation, the line drawn in the middle of the mirror will be inclined.

[35] When the side of the wax triangle falls on the line extending along the length of the plate, [the experimenter] should slide the post holding the mirror, while the side of the triangle remains in touch with the line stretching along the length of the plate, i.e., the edge of the plate, moving the post forwards and backwards until the sharp angle in the ruler that is in contact with the ring's upper surface meets a point on the line drawn on the mirror's surface. He should then fix the post that holds the mirror in this position by pressing some wax around its sides to keep it firmly in place.

[36] Once the ruler is fixed in this position, he should wipe off the line drawn on the
mirror's surface and mark on the mirror a small dot at | the extremity of the sharp-edged V 152b
ruler that is attached to the ring's upper surface. He should then lift the sharp-edged ruler
from this surface, and return the needle to its former position, applying it lengthwise to

the line drawn in the middle of the ring's upper surface and firmly fastening it with wax. Thus, if the line imagined to extend along the middle of the needle is produced, it will meet the point marked on the mirror's surface. And this line will be perpendicular to the surface of the post, which is tangent to the mirror's surface at the point marked on the mirror, and perpendicular to every straight line drawn from that point in the plane tangent to that surface, and, therefore, perpendicular to the straight line tangent to the intersection between the mirror's surface and [the plane of] the ring's upper surface.

[37] The experimenter should position his eye at the upper surface of the ring, close
to the ring's end, and look into the mirror, so as to see the form of the small body at the
tip of the needle. Upon seeing it, let him in this situation observe the body itself, the point
V 153a marked | on the mirror's surface, and the form that he perceived in the mirror. He will
find all three on a single straight line, as far as appears to sense. Now, the line imagined
to pass through the small body and the marked point lies in the [plane of] the surface of
the ring's top, and it is perpendicular to the plane tangent to the mirror's surface at the
point marked on the mirror. Therefore, this plane, i.e., the upper surface of the ring, is
one of the reflection planes. And the small visible object is in this plane in this situation,
as is the centre of the eye, and the point of reflection therefore lies in this plane, as this
will be shown later. Therefore, the image of the small body will in this situation also
lie on the straight line drawn from the small body perpendicularly to the plane tangent
to the mirror's surface. But this line is perpendicular to the straight line tangent to the
intersection between the mirror's surface and the reflection plane, namely [the plane of]
the ring's upper surface, and the reflection plane in this situation is one of the inclined
V 153b planes that cut the cylinder between the straight lines that stretch through | the length of
the cylinder and the circle parallel to the cylinder's base, since the post and the mirror it
holds in this position are inclined; therefore, the intersection between this plane and the
mirror's surface will be a [conic] section.[1] In a similar manner, the location of the image
can be shown for the case in which the position of the post holding the mirror is changed
by changing its inclination to the plate's surface, as described, by increasing or decreasing
the inclination.

[38] It is thus shown by this experimental examination that the images of objects seen
in cylindrical convex mirrors will always lie on the perpendiculars drawn from these
objects to the straight lines tangent to the intersections between the mirror's surface and
the planes of reflection. Every point seen by reflection will lie on the rectilinear extension
of the line along which the form is reflected to the eye, as was shown earlier; and the image
of every point seen in a cylindrical convex mirror will be [located] at the point where the
V 154a reflection line meets the perpendicular drawn from the visible point | to the straight line
tangent to the intersection between the mirror's surface and the reflection plane, that being
the position we have determined. Moreover, the perpendicular drawn from the visible
point to the line tangent to this intersection is the shortest line that can be drawn from that
visible point to the cylinder's surface in the reflection plane. For any line drawn from the
visible point in the reflection plane to reach the cylinder's surface will cut the line tangent
to the intersection before it arrives at the cylinder's surface, and thus what is cut off from
this line by the tangent line will be larger than the perpendicular, and, therefore, the whole
of this line will be larger than the perpendicular.

1. Ellipse, parabola, or hyperbola.

[39] [In addition,] similarly to what we have shown in this chapter, we can examine the images belonging to the conical convex mirror that occur in the planes of [conic] sections, and which [also] occur in the plane passing through the axis of the cone and cutting the cone's surface in a straight line. When the images of this mirror are examined experimentally, they will also be found | at the intersection of the perpendicular with the line of reflection. V 154b

⟨⟨V 2.I.5: Experimental examination of the location of the image in the spherical concave mirror⟩⟩

[40] With regard to the spherical concave mirror: the images [produced by it] differ – some of them are [located] behind the mirror, some in front of the mirror, and some on the surface of the mirror; and of all these, some are perceived distinctly, and some are not. Moreover, in the case [of this mirror], the image of a distinctly perceived point will be located at the place where the line along which the form is reflected to the eye meets the line from the visible point to the mirror's centre, [this latter line] being perpendicular to the line tangent to the intersection between the mirror's surface and the reflection plane.

[41] The experimenter wishing to examine the images belonging to this mirror should use a cone of white wax: let it be a right cone; let its conical surface be perfectly shaped; let the diameter of its base be shorter than the radius of the mirror's sphere; and let [the cone] be longer than the radius of the mirror's sphere. He should mark off a part of the cone [near its vertex], so that the straight | line extending along its length equals the radius V 155a
of the mirror's sphere, and, at this mark, draw round the surface of the cone a circle parallel to the cone's base; [he should then] cut the cone at this circle and draw a circle of equal size at the middle of the mirror's surface. Drop some wax in the middle of the concavity of this circle and let it fill the whole area [marked by the drawn circle]; the surface of the wax should be made even, with a circumference equal to that of the circle drawn on the mirror's surface.

[42] Having precisely [spread] the wax, the experimenter should expose the base of the cone to a gentle flame in order to warm it up a little. He should then place the base upon the circle in the mirror, thus binding the waxes together, and ensure that the circumference of the cone's base coincides with that of the circle on the mirror. With the cone thus attached and fixed, its vertex will be at the centre of the mirror. Having precisely secured this position for the cone, a straight line should be drawn on its surface from the point at its vertex to its base; the line should be black and conspicuous. The experimenter should then set the mirror below | his eyes, with the eyes so raised above it that the light can reach the V 155b
cone inside the mirror; the experiment should be performed in a lit area. He should ensure that the distance of one of his eyes from the straight line imagined to continue the cone's axis, if the latter is imagined to be produced rectilinearly, is sensibly longer than the radius of the spherical mirror.

[43] With one eye at this precise position, [the experimenter] should cover the other eye, and look into the mirror, moving it backwards and forwards until he sees the form of the cone. He will see it behind the mirror, and the form will be that of the segment of a cone whose smaller circle coincides with the base of the cone placed inside the mirror; and he will see the surface of that [mirrored] segment as rectilinearly continuous with the cone's [actual] surface, so that the cone and its form behind the mirror will look as if they

were a single, continuous cone. He will [also] find the straight line drawn along the length of the cone to be rectilinearly continuous with its form, as if they were a single line. But if
V 156a the surface of the form is continuous with the cone's surface, then any straight line | drawn from a point on the cone's surface to the centre of the mirror will extend along [both] the surface of the cone and the surface of its form that is continuous with it, and the form of the straight line drawn from any point on the cone's surface along its length will be found to be rectilinearly continuous with that line.

[44] From this experiment [the following] is clear, given the position we have specified for the eye:

- that the form of any point on the cone's surface will be located on the straight line continuous with the line on which the point lies;
- that the form of any point on the cone will be located on the line drawn from the point through the centre of the mirror, since the vertex of the cone is identical with this centre;
- that this line lies in the reflection plane, since the visible point and its image are both in this plane;
- that the line through the centre of the sphere is perpendicular to any line tangent to any circle through the line's extremity on the mirror's surface, with the centre of the mirror as its own centre;

and, therefore,

- that any point on the cone's surface will be located on the perpendicular drawn from it to the line tangent to the intersection between the mirror's surface and the reflection
V 156b plane; |
- that every point seen by reflection is located on the straight line along which the form is reflected to the eye;
- that the form of this cone will be located behind the mirror;
- that any point on any object seen by reflection in this mirror, if it is located between the mirror's centre and its concave surface, will lie on an imagined cone whose vertex is the mirror's centre and whose base is a circle on the mirror's surface;

and, therefore,

- for any point on any object seen by reflection in a spherical concave mirror, if it lies between the centre of the mirror and its concave surface, and its image lies behind the mirror, then the image [of that point] will be located at the intersection between the reflection line and the perpendicular drawn from that point to the line tangent to the intersection between the mirror's surface and the reflection plane.

As for the points whose images lie in front of the mirror, between the eye and the mirror, they comprise all points from which the line drawn to the mirror's centre will run transversally between the eye and the mirror.

[45] To examine the images of such visible objects, the experimenter should lift the
V 157a cone | from the middle of the mirror and set it up on one side of the mirror in the manner

we determined, [namely] so that the vertex of the cone coincides with the centre of the sphere. Then he should look into the mirror, ensuring that the distance between his eye and the mirror's surface exceeds the mirror's radius. Let him take a thin, white rod and bring it close to the mirror, so that the mirror's centre lies between the top of the rod and the centre of the eye; then look into the mirror with his gaze directed to the point on its surface opposite the cone's vertex, and such that the line imagined to extend between this point and the vertex is situated between the eye and the top of the rod. He should then search the mirror until he sees the rod and its top. He will find the form of the rod's top to be located in front of the mirror, and will find it closer to the eye than the vertex of the cone. Having perceived this form, he should examine it and the vertex of the cone, i.e., the mirror's centre, and compare it with the top of the rod itself. He will find that the top of the rod, the vertex of the cone, and the form of that top are always located on a single straight line, as far as appears to sense. Now, this line | is perpendicular to the line tangent to the V 157b
intersection between the mirror's surface and the reflection plane, because the reflection plane passes through [both] the [mirror's] centre and the visible point, and the line passing through those two points will thus lie in the reflection plane. But the intersection between the reflection plane and the mirror's surface is the circumference of a circle. Therefore, that line will meet the circumference of this circle and be a diameter to it, the centre of this circle being the centre of the sphere; and, being a diameter to the circle, that line will be perpendicular to the line tangent to the circle at its extremity and will pass through the visible point and its image.

[46] Examining any visible object perceived in a spherical concave mirror, when the image of the object lies in front of the mirror, the experimenter will find that any point on the object will lie sensibly on a straight line [that also] includes the mirror's centre and the image of that point. Every point perceived ((by reflection [in this mirror] will be located on the straight line along which the form is reflected to the eye, and therefore every point perceived in this mirror)) whose image is located between the eye and the mirror will have its image at the point of intersection between the reflection line and the perpendicular drawn from | that [perceived] point to the line tangent to the intersection between the V 158a
mirror's surface and the reflection plane. Objects distinctly perceived in this mirror will be those whose images lie behind the mirror ⟨or⟩ between the eye and the mirror. All other objects will not be perceived distinctly by sight, the reason being that these [objects] and their images follow [i.e., behave like] the distinct images; and of this category [also] are the images located on the surface of the mirror, as we shall make clear later, [in Book VI,] when we discuss the errors [of vision by reflection]. Therefore, for every point perceived distinctly by sight in a spherical concave mirror, the image of that point will be at the point of intersection between the reflection line and the perpendicular drawn from that [perceived] point to the line tangent to the intersection between the surface of the spherical mirror and the reflection plane.

⟪V 2.I.6–7: Experimental examination of the location of the image in the cylindrical concave and conical concave mirrors⟫

[47] With regard to the cylindrical concave mirror: the images [produced by means of it] also differ – some will be located behind the mirror, some in front of the mirror, and some | on the surface of the mirror; and of all these, some will be distinctly perceived, and V 158b
others not.

[48] To examine the images of the cylindrical concave mirror, the experimenter should proceed as previously described for the examination of the images of the cylindrical convex mirror. Set up the post that holds the cylindrical concave mirror in the instrument as explained earlier; attach the needle to the upper surface of the ring, having inserted the tip of the needle into a very small body; position the eye opposite the middle of the mirror at the middle of the ring; raise the eye above the ring's upper surface and look into the mirror to see the small body. [The experimenter] will see the form of the body, together with the body itself and the point marked on the mirror's surface, on a single straight line, as far as appears to sense. The form will be located behind the mirror, and the reflection will take place from a point on the straight line extending lengthwise [i.e., parallel to the cylinder's axis] at the middle of the mirror's surface.

[49] The experimenter should then position his eye [just] above the upper surface of the
V 159a ring, | next to [the ring's] edge or between the edge and the middle, and search the mirror until he sees the form of the small body. He will find the form of the body located in front of the mirror, and find the form, along with the small body itself and the point marked on the mirror's surface, to be located on a single straight line. This line will be perpendicular to the line tangent to the circle parallel to the mirror's base at the point marked on the mirror's surface. The plane of this circle in this position will be the reflection plane and the plane of the upper [surface] of the ring, and the reflection point will be a point on the circumference of this circle.

[50] The experimenter should then take another fine needle and insert its tip into another small body; [then] advance this body towards the mirror's surface until it occupies a position between the mirror's surface and its axis, ensuring that this body together with the first small body and the point marked on the mirror are on a single straight line, as far as appears to sense. Let him position his eye [just] above the upper surface of the ring [at a point] between the ring's edge and its middle, and search the mirror until he sees the form
V 159b of this body. He will find the form located behind the mirror, and find | the form, together with the body itself and the point marked on the mirror, to be on a single straight line, as far as appears to sense.

[51] The experimenter should then remove the post holding the mirror, place beneath it the small wax triangle, and mount it again in the instrument as previously explained. In this situation, the mirror will be inclined. Let him place his eye at the middle of the upper surface of the ring and search the mirror in order to see the small body at the tip of the needle. He will find the form of the body, together with the body itself and the point marked on the mirror, to be on a single straight line; and he will find the form located in front of the mirror. Reflection in this case will take place from the periphery of a [conic] section, because in this case the mirror is inclined.

[52] Employing the other needle and the other small body, let the experimenter advance this [second] body towards the mirror's surface, ensuring that it is located on the straight line extending from the first body to the point marked on the mirror and occupying a point between the mirror's surface and its axis; then let him place his eye [just] above the upper
V 160a surface of the ring, at its edge or between the edge | and the middle, and search the mirror until he sees this [second] body. He will see the form of this body behind the mirror, and find the form, together with the body itself and the point on the mirror's surface, to be on a single straight line, as far as appears to sense. Further, the form of any point perceived by reflection will be located on the line of reflection.

[53] It is clear from this experiment that the image of any point that is distinctly perceived by sight in a cylindrical concave mirror is located at the point where the line of reflection meets the perpendicular drawn from that point to the line tangent to the intersection between the mirror's surface and the reflection plane. The images of point[s] not perceived distinctly will follow [i.e., behave like] the distinct images.

[54] [As for] the images of the conical concave mirror, they may be examined in the same manner. When the images of this mirror are examined experimentally, they will be found to be located at the point where the lines of reflection meet the perpendiculars.

[55] Therefore, the image of every point distinctly perceived by reflection in any one of | the [types of] mirror specified in this Book will be located at the point where the V 160b
reflection line meets the perpendicular drawn from that point at right angles to the line tangent to the intersection between the mirror's surface and the reflection plane. That point, namely the location of the image, is also called the 'image'.

⟪V 2: Second Inquiry. Explanation of Vision by Reflection, in Two Propositions⟫

⟪V 2.II.1: The reason why the object seen by reflection is perceived to be at the location of the image⟫

[56] We shall [now] state the reason why sight perceives objects visible by reflection at the locations of the image, and why the image is located on the perpendicular drawn from the object to the mirror's surface. That sight perceives the form of an object by reflection as being opposite to it [the object] has already been made clear. As for sight's perception of this form at the location of the image, which [location] is on the perpendicular, the reason is [as follows]. When sight perceives an object, it perceives it immediately, at a glance, as being at some distance [from the eye], and, at the moment of perception, it perceives the object's distance by conjecture; sight may then contemplate the magnitude of the object's distance and ascertain it, or it may be satisfied with its glance-perception. This notion we have thoroughly made clear in Book II [of the *Optics*]. | It was also made clear in Book V 161a
II that sight may perceive the magnitude of the object's distance through inference by comparing the object's [estimated] size with the angle subtended by this size at the centre of the eye; and that in most cases sight perceives the distances of familiar objects in this manner. Sight may also perceive the distances of unfamiliar objects in this manner, if it assimilates the size of the unfamiliar object to the size of a similar but familiar object. Thus, when sight perceives an object by reflection in a mirror, it will at the same moment perceive the distance of the object's image only by conjecture; then, if it contemplates [the image], it will perceive its distance by comparing its size with the magnitude of the angle contained by the cone through which the form of the object has been reflected to the eye. If the object is a familiar one, sight will perceive it by reflection to be at the distance from which the same object, when perceived directly and without reflection, would subtend at the centre of the eye an angle equal to that contained by the cone through which the form has been reflected to the eye; | and sight will perceive [moreover] the distance of the form, V 161b
i.e., image, as it would rectilinearly perceive the distances of [real] objects in the manner described previously.

[57] Similarly, when sight also perceives an unfamiliar object by reflection, it assimilates the size of its image to the size of a similar, familiar object, and perceives the distance of this image by comparing the [image's] magnitude with the magnitude of the angle subtended at the centre of the eye by this image at the moment of reflection; and [it perceives] the location [of the image] to be that from which the form of an object perceived by reflection would rectilinearly extend to the eye through the same cone through which the reflected form has reached the eye, i.e., the cone containing the form [assumed to be] at that location of the image. Thus, sight perceives the object by reflection to be at the location of the image [only] because the form perceived by reflection will be similar to the
V 162a form that sight would perceive | rectilinearly if it were located at the image and contained by the cone through which the reflected form has reached the eye. That is the reason why sight perceives a visible object by reflection to be at the location of the image.

⟨⟨V 2.II.2: The reason why the object seen by reflection is perceived to be on the perpendicular drawn from the object to the mirror's surface⟩⟩

[Case of the plane mirror]

[58] There [now] follows the reason why the image is located on the perpendicular drawn from the object to the mirror's surface, and why sight does not perceive any point on the object, except at the point on the perpendicular drawn from that object-point to the line tangent to the intersection between the reflection plane and the mirror's surface, provided that no perceptual error occurs. The object-point perceived by sight must have magnitude and is not an imagined point, and it can only be perceived as facing the eye on the line along which the reflected form reaches the eye. But the point seen by reflection will not produce in the part of the eye where the point is seen ⟨an impression identical⟩ to that produced when
V 162b the point is perceived rectilinearly, unless that point is located where the image is, namely | on the specified perpendicular. For, in the case of the plane mirror: let the form, namely the image, of the point visible by reflection, and located on the perpendicular, be not so located; and let it be on the line along which the reflected form has extended to the eye; and [now] assume that it occupies [a location] farther than the point on the perpendicular. Its form at the reflection point would then be smaller than the form [actually] produced at that place, and it would [consequently] occupy a portion of the eye smaller than that that it occupies when located on the perpendicular. For the form of the visible point [on the object] has [a certain] magnitude, and the form extending from it to the eye does so only through [the shape of] a cone whose vertex is the centre of the eye; and the farther the visible form is from the eye, the smaller will be the portion of the eye where this visible form occurs.
V 163a Thus, if the form of the point perceived by reflection, which | is a [point] that has [a certain] magnitude, were farther from the eye than the location of the image on the perpendicular, then the form of it that occurs at the reflection point would be smaller, and the form of it that occurs on the surface of the eye would be smaller, and also weaker.

[59] Now, assume the form, namely the image, to be closer to the eye than the [actual] location of the image. The form of it that then occurs at the reflection point would be larger than the form that occurs at that point when perceived by reflection, and it would occur in a portion of the eye larger than that in which it occurs when perceived by reflection, and also [be] clearer.

[60] Therefore, the object-point perceived by reflection from a plane mirror, if perceived rectilinearly while occupying the [assumed, farther or closer] location of the image on the perpendicular, would not impress the same area of the eye in which that point occurs through reflection, nor would the impression [produced rectilinearly] be equal in clarity to that produced by reflection, unless that point were at the location of the image on the perpendicular. The reason is | that the point on the perpendicular, namely the image, is V 163b
as distant from the reflection point as the visible point, of which this image is an image, is from the reflection point. But there is no reason in the plane mirror's surface that would cause a change in the form that occurs in it, and, therefore, the visible [object-]point, as it reaches the reflection point, is at this [latter] point identical in every respect to the form that would rectilinearly reach this same visible [reflection] point, if the latter were at the location of the image – I mean identical to it in shape, position, size, clarity, and effect [made] in the eye when it reaches the eye.

[61] Similarly, let the visible object be the [surface of the] eye itself, and let the line along which the form [of a point on the eye's surface] is reflected be the perpendicular itself: then the image-point can only be located on that perpendicular, behind the mirror, and it will be as distant from the foot of the perpendicular as the [point on] the eye is, since the distance of this image from the reflection point, i.e., the foot of the perpendicular, equals that of the visible point from | the reflection point. For when the viewer perceives V 164a
his eye in the mirror, the image of the point at the middle of the eye['s surface] will be at the middle of the image of the whole eye.

[62] This then is the reason why the image of a point seen by reflection from plane mirrors is located on the perpendicular drawn from the visible point to the mirror's surface.

[Case of the spherical convex mirror]

[63] With regard to the image of a point seen in a spherical convex mirror: the image is located on the perpendicular, i.e., the line drawn from that visible point to the centre of the mirror. This is because the form extending from the visible point to the reflection point on the surface of the spherical convex mirror will, upon reaching the surface, suffer an effect due to the centre. For, let the visible point be the point on the surface of the eye at the intersection of this surface with the perpendicular drawn from the mirror's centre to the centre of the eye; the reflection point ⟨will be⟩ then the intersection of this perpendicular with the mirror's surface. But the image-point is not a determinate point, because all | points on the diameter joining the mirror's centre and the centre of the eye are V 164b
of identical condition. Further, the image of [the portion of] the eye's surface surrounding that point will be located behind the mirror's surface, and thus the image of that point will be continuous with the image of what surrounds [the point] on the surface of the eye, because the parts of this surface are continuous with one another; and the image of that point on the surface of the eye, which is located on the diameter joining the mirror's centre and the centre of the eye, will therefore be surrounded by the image [of that portion] of the eye's surface that encircles that point on all sides. Therefore, the line from the mirror's centre to that point will not be a straight line, but will have the shape of a cone, the part of which forming the intersection between it and the mirror's surface will have [a certain] magnitude; and the form that occurs at this place [i.e., the intersection] will have [a certain] magnitude, because the visible point itself has magnitude while the centre of the

mirror is an imagined point; and the form that occurs at the reflection point, which point
V 165a is on the perpendicular, | will be as a form that has extended from the mirror's centre to this point. This is because the surface of the spherical convex mirror curves inwards on all sides around the central point; and, similarly, all parts and all extremities of parts of this surface curve inwards about the centre. Thus, the extremities and all parts of the form that occurs in a part of this surface and that extends along the lines reaching the mirror's centre will curve about the centre.

[64] If, then, there extended from the mirror's ⟨centre⟩, to any part of the form that occurs in this manner on the surface of the spherical convex mirror, a form similar to that occurring in that part, then that form would coincide with the form that has arrived from the visible point to that part, namely the reflection point. But if the point seen in this way were not on the perpendicular, and the form extended from it to the same place, i.e., the reflection point, then it would not occur in this place in the same [condition]
V 165b as that of a form arriving [there] from the centre. | And if the visible point were also on the perpendicular at a point other than the centre, and the form extended from it to the reflection point, then [that form] would not occur at the reflection point in the same [condition] as that of the form arriving there from the centre. For the centre is an imagined point, while the visible point has [a certain] magnitude, as has the form that occurs at the reflection point; and, therefore, every point on the form that occurs at the reflection point at the diameter's extremity will have the same position with respect to the centre as that of all [other] points in the form. But no point other than the centre is such that the parts of the form occurring at the reflection point will be similarly situated with respect to it, and the form that occurs at the reflection point therefore cannot be similar to any form that arrives there from any point other than the centre. That is why the form that sight perceives
V 166a in this manner can only be perceived on the line that joins | the centre of the eye and the mirror's centre. Thus, the image of the visible point on the eye's surface, which is [at] the intersection of this surface with the diameter drawn from the centre of the eye to the mirror's centre, i.e., the perpendicular drawn from the eye's centre to the mirror's surface, will be located on this diameter [only], because the form that occurs at the reflection point would not occur [there] in the condition it [actually] has when it arrives rectilinearly from the side facing the eye and sight perceives it directly and without reflection, unless it has come from the mirror's centre, or from a location on the cone whose vertex is the mirror's centre and whose base is that visible point on the eye's surface, that location [on the cone] being between the centre of the mirror and its surface, [and the form arriving at the reflection point] being as if it were a form extending from the centre.

[65] Now, let the [visible] point lie outside this diameter and let it be perceived by reflection in this mirror. The form that extends from it to the reflection point will then not
V 166b be the same in figure | and position as the form that extended from the same visible point to that location on the surface of the mirror when the visible point was located on the mirror's diameter passing through the reflection point. For the line along which the form of the point lying outside the diameter extends through the eye's centre to the reflection point will be inclined to the mirror's surface; further, this line is not an imagined line, but a body that has magnitude, since the visible point must have magnitude, and the parts of the extremity of this line will not be similarly positioned on the mirror's surface. And thus the parts of the form that occurs at the reflection point will not be similar to one another in shape or position or clarity. Yet the extremities of this form and all of its parts curve

towards the centre, despite the difference of their form[s], and, therefore, the form that occurs at the reflection point, if the form arrived [there] along an inclined line, would have a figure composed of a form arriving at that point from the centre | and a form that comes V 167a
from the visible point.

[66] Now, if a form proceeded from the mirror's centre, extending towards the circumference and occurring on the diameter drawn from that centre to the visible point in the manner [previously] specified – I mean, in such a way that all points on it are similarly situated with respect to the mirror's centre –, and if the form of the visible point extended along this same diameter and the two forms met each other, then they would produce a form mixed of both of them. But it is not possible for a form to proceed from the centre, in the manner described, and meet the form of the visible point so as to give rise to a form mixed of the two forms, unless these two forms meet on the diameter. The reason is that the figure of the form that occurs at the reflection point is a figure constituted from the form of the visible point with reference to the centre, and thus this figure can only be constituted on the diameter extending between the centre and the visible point. Therefore, the form produced at the reflection point is as if it were a form | that has arrived [there] V 167b
from the form located on the diameter joining the mirror's centre and the visible point, which [form] is mixed of the two forms mentioned [and] whose figure can be constituted only on this diameter. But every point on this diameter can be occupied by a form mixed of the two mentioned forms, and that form, namely the one at the reflection point, can only be seen along the line along which it is reflected to the eye, i.e., [the line] extending between the eye and the reflection point. Therefore, the location on the diameter that joins the centre and the visible point and which lies on the line along which the form is reflected to the eye is a location at which a form mixed of the two mentioned forms may exist; and this location is the image-point that lies on the diameter, i.e., the point of intersection between this diameter and the reflection line. Further, the form produced in the eye by the reflected form | will be as a form that has come to the eye from the location of the image. All natural V 168a
matters must be in accordance with their [own] principles; and the principles of natural matters are subtle, thoroughly hidden, and concealed to the senses. Thus, the reason why the image in this mirror lies on the diameter that joins the mirror's centre and the visible point is that the form that occurs at the reflection point, from which sight perceives the visible [object-]point, is as a form that has come from a point on this diameter, the like of which can only arrive at that [reflection] point from a visible object placed on this diameter at the same location as the image on the diameter.

[Cases of the remaining mirrors]

[67] In the very same manner, for spherical concave mirrors, the reason why the image is located on the diameter can be demonstrated. Again, in the same manner, for cylindrical mirrors, convex and concave, the [location] of the image on all the circles that occur in them can be explained. And again, in the same manner, an explanation can be given for the images produced by | reflection from [conic] sections occurring in cylindrical mirrors, V 168b
convex and concave, and in conical mirrors, convex and concave, so that the point, where the perpendicular drawn at the reflection point meets the perpendicular drawn from the visible point to the line tangent to the intersection between the mirror's surface and the reflection plane, will be as the centre in the circle's image.

[The universal reason]

[68] Thus, the universal reason why images lie on perpendiculars is that the forms occurring at the reflection point will have the figure of a form arriving rectilinearly from a visible object placed at the location of the image on the perpendicular; and an object of vision, from which a form having a figure identical to the figure of the form occurring at the reflection point may reach that same point, can only be located where the image is, namely, on the perpendicular alone.

V 169a **[69]** Thus, we have shown the reason why images are located on perpendiculars | drawn from visible objects to lines tangent to the intersections between the surfaces of mirrors and reflection planes; and this is the reason why the objects seen by reflection are perceived at the locations of their images. Having made that clear, we now need to describe the locations of the images in detail, and identify them for each of the mirrors previously described.

⟨⟨V 2: Third Inquiry. On the Locations of Images in the Seven Mirrors, in Six Propositions⟩⟩

⟨⟨V 2.III.1: General statement on the locations of images in each of the seven mirrors⟩⟩

[70] I say that, for every point seen in a plane mirror, if it lies outside the perpendicular drawn from the centre of the eye to the mirror's surface, the line along which the form of that point is reflected to the eye always meets the perpendicular drawn from that point to the surface of the mirror; and that the meeting-point, namely the location of the image, will lie behind the mirror, its distance from the mirror's surface being always equal to the distance of the visible point from that surface. [I say] also that sight will always perceive the form of the visible point in this mirror at this location, and that every point seen in this
V 169b mirror will have | only a single image.

[71] [I say] further that, for every point seen in a spherical convex mirror, if it lies outside the line joining the centre of the eye and the centre of the mirror, the line along which the form is reflected to the eye will always meet the line drawn from that point to the centre of the mirror, i.e., the perpendicular drawn from that visible point to the line tangent to the intersection between the mirror's surface and the reflection plane; and that the position of the meeting-point, namely the location of the image with respect to the mirror's surface, will vary with the position of the eye relative to that surface. Thus, the meeting-point may lie behind the mirror, or on the surface of the mirror, or in front of it. And [I say] that sight will always perceive all images in this mirror [as lying] behind it, regardless of their different locations, but will not discern their locations relative to the
V 170a mirror's surface; and that every point seen in this mirror will have only a single image. |

[72] [I say] further that, for every point seen in a cylindrical convex or in a conical convex mirror, if it lies outside the perpendicular drawn from the centre of the eye to the plane tangent to the surface of the mirror, the line along which the form is reflected to the eye will always meet the perpendicular drawn from that point to the line tangent to the intersection between the mirror's surface and the reflection plane; that the locations of the images in these two mirrors will also vary – some will lie behind the mirror, some on the surface of the mirror, and some in front of it; that sight will always perceive all images in these two mirrors behind the mirror; and that every point seen in each of these two mirrors will have only a single image.

[73] Furthermore [I say] that, of the lines along which the forms of visible points are reflected to the eye from a spherical concave mirror, some will meet the perpendiculars drawn from the visible points to the lines tangent to the intersections between the surface of the mirror | and the reflection planes, and some will be parallel to these perpendiculars. V 170b
Those [lines] that meet the perpendiculars will have different meeting-points: some of the meeting-points, or locations of the images, will lie behind the mirror, and others in front of the mirror. Of those [meeting-points] that lie in front of the mirror, some will be [located] between the eye and the mirror, some at the centre of the eye itself, and some behind the centre of the eye. Sight's perception of the forms of visible objects in this mirror will also vary: it will perceive some [forms] at the locations of the images, i.e., the meeting-points, and these it will perceive distinctly; others it will not perceive in meeting-points, and its perception of them will be indistinct. [I say] also that, of the visible objects perceived by sight in this mirror, some will have a single image, others will have two, three, or four images, but not more.

[74] Furthermore [I say] that, of the lines along which the forms are reflected to the eye from the cylindrical concave mirror, or from the conical concave mirror, | some V 171a
will meet the perpendiculars drawn from the visible points to the lines tangent to the intersections, and some will be parallel to the perpendiculars. Of those [lines] that meet the perpendiculars, some will have their meeting-points behind the mirror, while for others the meeting-points will be in front of the mirror, and of these, some will lie between the eye and the mirror, some at the centre of the eye, and some behind that centre. Sight's perception of the objects it perceives in these two mirrors will also vary: some visible objects it will perceive at the locations of their images, namely the meeting-points, and others it will not perceive in meeting-points. And [I say] that of the objects seen in each of these two mirrors, some will have only a single image, others will have two images, or three, or four, but not more. Let us now prove all | these notions [i.e., propositions] by V 171b
[geometrical] demonstrations.

⟪V 2.III.2: Demonstrations of the preceding propositions [for the plane mirror]⟫

[75] [Diagram V.1] Let point A be on a visible object, let point B be the centre of the eye, and let point G be on a plane mirror. Let A be outside the perpendicular drawn from the centre of the eye to the mirror's surface, and let the form of A be reflected to the eye B from G. We draw line AG, along which the form extends to the mirror's surface, and line GB, along which the form is reflected to the eye.

Lines AG and GB will be in the plane orthogonal to the mirror's surface, for it was shown that forms are reflected only in the plane orthogonal to the plane imagined to be tangent to the polished surface, and the imagined plane tangent to the surface of a plane mirror will coincide with the mirror's surface.

Let the intersection of this orthogonal plane with the mirror's surface be line DGE. We draw from G a perpendicular to line DGE in plane AGB that is orthogonal to the mirror's surface; let it be line GH. GH will then be perpendicular to the mirror's surface, because

V 172a it is perpendicular to the intersection between that surface | and the plane orthogonal to it; GH will lie between lines AG and BG, and angles AGH and BGH will be equal, because the lines along which the forms are reflected to the eye cannot be otherwise.

We imagine a perpendicular from A to line DGE, say AE, which will also be perpendicular to the mirror's surface, because it is perpendicular to the intersection. Perpendiculars AE and HG are therefore parallel. Line BG will be in their plane, since the three [lines] are in the orthogonal plane, and line BG will therefore meet the perpendicular AE, and the meeting-point will be behind line DGE.

We produce the two lines until they meet, say at point Z. Then Z will be the image-point.

Now ∠BGH = ∠AGH; therefore, by subtraction [from the right angle], ∠BGD =
V 172b ∠AGE; and therefore ∠ZGE, being equal to ∠BGD, will be equal to ∠AGE. |

And the angles at point E are right. Therefore, line ZE will be equal to line EA.

And [it will proceed] similarly for the image of any point seen in a plane mirror, if [the point] lies outside the perpendicular drawn from the centre of the eye to the mirror's surface.

⟪V 2.III.3: Determination of the position of the reflection point on a plane mirror⟫

[76] [Diagram V.1] To find the point of reflection, given the two points A and B, and the [plane] mirror DE, we drop from the visible point A the perpendicular AE to the mirror's surface and produce it rectilinearly through [the mirror]; we make EZ equal to EA, and join BZ, cutting line DE, say at point G. I say that point G is the reflection point.

Demonstration. [If] we join AG, it will be equal to GZ; and [if we] draw HG perpendicular to the mirror's surface, it will be parallel to the perpendicular AE.

Then the angles at G will be equal to the angles at points A and Z. But the angles at A and Z are equal, therefore the angles at G are equal to each other, and lines BG and GA are
V 173a thus reflected at equal angles. |

G is therefore reflection point. And that is what we wished to prove.

⟪V 2.III.4: Only one point on the perpendicular drawn from the centre of the eye to the surface of a plane mirror will be perceived by reflection⟫

[77] [Diagram V.2] Only one point on the perpendicular drawn from the centre of the eye to the surface of the plane mirror will be seen by reflection, namely the point on the surface of the eye through which this perpendicular passes; no other point on this perpendicular that is drawn to the plane mirror will be perceived by reflection in the mirror.

Let the centre of the eye be point A, and the plane mirror be BG. From A draw the perpendicular AG to the mirror's surface. Let D be a point on the surface of the eye, and E a point behind A; and let Z be a point between the eye and the mirror's surface.

Thus, the form of point E, having extended along the perpendicular EG, will not reach the mirror's surface, being prevented from passing through by the body of the eye. The form of point E will not therefore be reflected to the eye A from point G. I say that the form of point E will also not be reflected from any point other than G.

If possible, let [the form] be reflected from point B. We draw lines EB and BA, and
V 173b bisect ∠ABE by line BT. | Line BT will then be perpendicular to the mirror's surface, since

any two lines reflected from the surface of a plane mirror contain two equal angles with the perpendicular to the surface at the point of reflection in the reflection plane. Join BG. ∠TBG will therefore be right, but so is ∠AGB, which is impossible.

The form of point E will not therefore be reflected from any point on the mirror's surface other than G, and it has been shown that it cannot be reflected from G. Therefore, sight will not perceive point E in mirror GB.

And, similarly, sight will not perceive by reflection in mirror BG any point on the perpendicular AG behind point A.

[78] As for point Z: it can be shown, as it was shown for point E, that its form cannot be reflected from a point other than G.

But the form of it that extends along line ZD will be reflected from G and return | to point Z. V 174a

If Z is in an opaque body, then the form of it that is reflected from G will not reach the eye A.

If, however, point Z is in a transparent body, its reflected form will pass through the transparent body, and, having done so, will reach the eye mixed with the form of point Z that has extended rectilinearly from Z to the eye A. Sight will thus not discern the reflected form, and will not therefore perceive point Z by reflection in the plane mirror. Similarly, no point on line DG will be seen by reflection.

[79] As for the points on line AD inside the eye: their form[s] will not be reflected to the eye A from any point except G, as was shown for point E.

But their form[s] that extend along the perpendicular AG will be reflected along GA and reach the [centre of] the eye, since the interior of the eye is a transparent body. But these reflected form[s] will arrive [there] mixed with the rectilinear form[s] and will not therefore be discerned by sight. Moreover, as the body of the eye's interior is of a very delicate colour, its form is not discernible | to sight. Sight will not therefore perceive V 174b anything by reflection from [any part of] the eye's interior body that extends along line AD.

[80] As for point D on the surface of the eye: it too can be shown, as was shown for point E, that its form will not be reflected from the mirror's surface, except at G.

But the form of it that extends along the perpendicular AG will reach G and will be reflected along line GD and return to D, and pass through the body of the eye and reach the ⟨centre of⟩ the eye; sight will then perceive the reflected form [of D] mixed with its rectilinear form. Moreover, its image will lie on perpendicular AG, because AG is the line along which its form will be reflected to the eye. Moreover, the image [of D] will lie behind the mirror's surface, because it will be continuous with the images of the [points] that surround this point on the eye's surface, and the images of these surrounding [points] will lie behind the mirror. Further, the distance of this image from the mirror's surface will equal the distance of the visible point from that surface, because the distance of the surrounding [points] on the eye's surface from the mirror's surface will equal the distance | of the eye's surface from the mirror's surface. Sight will only perceive this point [by V 175a reflection] as mixed with the form of it that has reached the eye rectilinearly, because it is continuous with what surrounds it on the eye's surface.

Thus, when sight perceives ⟨the point on the surface of the eye⟩ in the mirror, it will perceive this point as [part of] the whole surface of the eye, and will not perceive it singly and discretely.

[81] Therefore, the image of every point perceived by reflection in a plane mirror will lie behind the mirror; and the distance of the image from the mirror's surface will equal the distance of the visible point itself from that surface. That being so, the image of every object perceived by reflection in a plane mirror will be behind the mirror, and the distance of the image from the mirror's surface will equal the distance of the visible object itself from that surface. And that is what we wished to prove.

⟪V 2.III.5: A visible point seen by one eye in a plane mirror will have only a single image and a single reflection point⟫

[82] [Diagram V.3] Having shown [what has just preceded], I say [that] every point that a single eye perceives by reflection in a plane mirror will have only a single image,
V 175b and the form of that point will be reflected to | that eye from a single point only on the mirror's surface. Similarly, every visible object that a single eye perceives by reflection from a plane mirror will have only a single image.

Let point A be the centre of the eye, point B in a visible object, and point G on a plane mirror. Let B lie outside the perpendicular drawn from the centre of the eye to the mirror's surface.

We draw line AB: it will not be perpendicular to the mirror's surface, because B is outside the perpendicular drawn from the eye A to that surface. And let the eye A perceive point B in mirror G.

I say: the form of point B will be reflected to point A from a single point only.

If possible, let the form of B be reflected from two points, say G and D.

Since the forms can be reflected only in planes orthogonal to the mirror's surface, the form of point B will be reflected to point A only in the orthogonal plane that contains
V 176a points A and B. But the plane surface that contains points | A and B will [also] contain the whole of line AB. Therefore, the form of point B can be reflected to point A only in the orthogonal plane that contains line AB.

Now if line AB is not perpendicular to the mirror's surface, then it will lie in only one of the planes orthogonal to that surface, for there exists no intersection between the intersecting planes orthogonal to a plane surface, unless it is perpendicular to that surface. For if the intersection were not perpendicular, and from a given point on it a line were drawn in each of the two orthogonal planes at right angles to the intersection, then the two [drawn] perpendiculars would be at right angles to the mirror's surface while meeting the given point, and this is impossible.

But line AB is not perpendicular to the mirror's surface; therefore, it is not an intersection between two planes that are orthogonal to the mirror's surface.

Now if the form of point B is reflected to A from the mirror's surface, then AB will be in
V 176b a plane at right angles to that surface. And [this being] so, and | AB cannot be in two of the planes orthogonal to the mirror's surface, then line AB will lie in only one of those planes.

But if AB lies in only one of the planes orthogonal to the mirror's surface, and the form of point B is reflected to the eye A from points G and D, then points A, G, and D all lie in the single plane orthogonal to the mirror's surface, which contains line AB, since reflection can take place only in a plane at right angles to the mirror's surface.

Now points G and D are on the surface of the mirror. Therefore, they lie on the intersection between the mirror's surface and the plane orthogonal to that surface, which contains line AB. Let the intersection that contains AB be line EGDZ. Draw lines AG, AD, GB, and DB.

The form of point B that extends along line BG will be reflected along line GA, and the form of B that extends along BD will be reflected along DA.

Therefore, ∠BGD = ∠AGE, | and ∠ADG = ∠BDZ. But ∠AGE > ∠ADG; then ∠BGD V 177a
> ∠ADG, and thus ∠BGD > ∠BDZ, which is impossible.

Therefore, from points A and B, lines will be reflected at equal angles from a single point only. The form of point B will therefore be reflected to the eye A from a single point only on the plane mirror. And thus the form of any point perceived by a single eye in a plane mirror will be reflected only from a single point on the mirror. And [this being] so, the image of that visible point will be a single point, because the line of reflection can only be a single line, and the perpendicular from the visible point to the mirror's surface will only be a single line.

[83] As for the point on the eye's surface through which the perpendicular drawn from the centre of the eye to the mirror's surface passes: we have shown that its form will be reflected from a single point, | namely the foot of the perpendicular, and that its image will V 177b
be continuous with the image of what surrounds that [point] on the eye's surface, and that its image can only be on this perpendicular. Now, only a single point on the perpendicular that intersects the image of the surrounding surface of the eye can be continuous with the image at that point [on that surface]. Therefore, the image of the point on the eye's surface, through which the perpendicular drawn from the centre of the eye to that surface, can only be a single point. Thus, every point that a single eye perceives by reflection in a plane mirror will have only a single image, and that being so, every visible object that a single eye perceives by reflection in a plane mirror will have only a single image. And that is what we wished to prove.

⟨⟨V 2.III.6: Perception of a visible object with both eyes by reflection from a plane mirror⟩⟩

[84] [Diagram V.4] When vision takes place with both eyes, a single object will be perceived as such with both eyes at once when reflected in the plane mirror. And this will be found to be the case when examined experimentally. For, let the observer look at the mirror with both eyes and see in it some visible object; | if he covers one eye he will see V 178a
the object with the other, and covering the other eye while the first is open he will also see the same object in the mirror, even though both eyes, the object, and the mirror are fixed in their places. Thus, when both eyes are opened while in that position, they both will perceive that object.

[85] This state of affairs is clearly manifest with reference to the form [image] of the [viewer's] face. Thus, someone looking in a mirror will, in most cases, see his face in it with both eyes. For if he covers one eye, he will in most cases see his face with the other eye; and if he covers the other eye and opens the first, he will also see his face in most cases. One who looks in a mirror will also see in it, with both eyes, many objects other than his face. I say, then, that if a single object is perceived with both eyes in a plane mirror, then the image of that object will be, for both eyes, a single, common image.

[86] Now, when a single object | is perceived in a plane mirror with both eyes, the two V 178b
forms of the object will not be reflected to both eyes from one [and the same place] on

the mirror. Rather, the form of the single object will be reflected to both eyes from two places on the mirror. For if the form of any point on the object is reflected to one eye from a [given] point on the mirror, then a form of that point [on the object] will not be reflected to the other eye from that [same] point [on the mirror], because the form of a single point on the object will extend to a single point on the mirror only along a single line, and a single line can only be reflected at equal angles along a single line; and thus the form of a single point on the single object will be reflected to both eyes from two points only on the mirror's surface. That being so, the form of a single object can only be reflected to both eyes from two different parts of the surface of the plane mirror.

[87] I say, nevertheless, that when the two forms of a single object are reflected to both eyes, the image [of the object] will be single with respect to both eyes.

Let point A be in a visible object, let the centre⟨s⟩ of the two eyes be points B and G, V 179a let the form of A be reflected to eye B from point T, and let the form | of A be reflected to eye G from point Z.

Now A will either be with both centres in a single plane at right angles to the mirror's surface [as in Cases 1 and 2], or in two such planes [as in Case 3].

If A lies with the centres of both eyes in a single plane at right angles to the mirror's surface, then [A] will either be located between the perpendiculars drawn from the two centres to the mirror's surface [Case 1], or to one side of [both] perpendiculars in the direction of one of the eyes [Case 2].

From A let us draw a perpendicular, say AD, to the mirror's surface, and produce it rectilinearly through [the surface]; and we draw lines BT, AT, GZ, and AZ.

Plane BTA is therefore orthogonal to the mirror's surface, and line AD is perpendicular to that surface. But no perpendicular can be drawn from a point on a plane orthogonal to the plane surface, except on the intersection between both planes. Point D is therefore on the intersection between plane BTA and the mirror's surface. Let the intersection be line DTK.

We produce BT to meet the extension of perpendicular AD behind line KD and behind V 179b the mirror, say at point E: | ED will then be equal to DA.

Again, point D is on the intersection between plane GZA and the surface of the mirror. Let this intersection be line DZH. Therefore, line GZ will also meet perpendicular AD behind line DH and behind the mirror, whether DH is continuous with KD, as in the first and second diagrams, or cuts line KD, as in the third diagram. And the line cut off by HZ from the perpendicular AD, namely that behind the mirror's surface, will be equal to AD.

But ED = AD, and [ED] lies behind the mirror. Therefore, GZ produced will meet perpendicular AD in point E. Let line GZ end at E.

V 180a Point E will be an image of point A for the eye B and also an image of A for eye G, | whether the plane[s] in which the two forms of point A are reflected to the eyes B and G are a single plane or two intersecting planes.

And the same holds for any point perceived with both eyes in a plane mirror, if the point is outside the perpendicular drawn from the centre of the eye to the mirror's surface.

[88] Furthermore, the point on the surface of each eye will be perceived with the other eye by means of a reflected line, and it will be perceived by that [other] eye as continuous with the image of what surrounds it on the eye's surface, and on the perpendicular that has been drawn from it to the mirror's surface. [The same point] will be perceived by the eye on which it is located, on the perpendicular passing through it to the mirror's surface, and will be perceived with this eye also as continuous with the image of what surrounds it on the eye's surface. Thus, the image of this point too will be single for both eyes. And, therefore, the image of every point perceived with both eyes in a plane mirror will be, for both eyes, a single, common point.

[89] That being the case, then every visible object perceived | with both eyes in a plane mirror will have only a single image; and the two cones drawn from the centres of the eyes, whose base is that image, are the cones through which the two forms of that object are reflected from the object to the eyes. Thus, every object perceived in a plane mirror will have a single image, whether [the object] is perceived with one eye or with both eyes. And that is what we wished to prove. V 180b

⟪V 2: Fourth Inquiry. On the Images of Spherical Convex Mirrors and Related Matters, in Eighteen Propositions⟫

⟪V 2.IV.1⟫

[90] [Diagram V.5] Let point A be in a visible object, let point B be the centre of the eye, and point G on a spherical convex mirror. Let A be outside the line that joins B and the centre of the mirror, i.e., the perpendicular to the plane tangent to the mirror's surface. ((Let the form of point A be reflected to the eye B from point G. Draw lines AG and GB. Plane AGB will be orthogonal to the plane tangent to the mirror's surface)) at G. Let line DGE be the intersection between plane AGB and the tangent plane, and let arc TGK be the intersection between the mirror's surface and plane AGB.

We imagine line HG perpendicular to DGE and in plane BGA: | it will then be perpendicular to the tangent plane, because it is perpendicular to the intersection, and ∠BGH will be equal to ∠AGH. V 181a

Now if the perpendicular to the tangent plane at the point of tangency be rectilinearly produced, it will meet the centre of the sphere. For the line drawn from this centre to the point of tangency will always be perpendicular to the tangent plane and will thus be continuous with the perpendicular drawn from that point. Otherwise, if rectilinearly produced, it would cut [the perpendicular]. But, if it did, two perpendiculars to the same plane would then have been drawn from a single point, which is impossible. Therefore, line HG, if rectilinearly extended, will reach the centre of the sphere.

Now line HG is in the plane of arc TGK; therefore, TGK is one of the circles whose planes pass through the centre of the mirror. And line DGE is tangent to this arc, because it lies in the plane tangent to the spherical surface.

Let N be the centre of the mirror, and let line HG end at point N. N will then be in plane ABG, and line AN, if drawn, will be | in plane ABG. V 181b

Now line BG cuts line DGE, which is tangent to the arc. Then BG produced will cut arc TGK and will cut line HGN; therefore, it will cut sector NGK, meeting line NEA, say at point Z.

Line AN is perpendicular to the mirror's surface; it is therefore perpendicular to the plane tangent to the mirror's surface at the point on this surface through which line AN passes, and perpendicular to the line tangent to the intersection between the mirror's surface and the reflection plane, i.e., arc GK. And line BG is the line along which the form of point A is reflected to the eye B.

Therefore, the line along which the form of point A is reflected to the eye B will meet the perpendicular from A to the line tangent to the intersection between the mirror's surface and the reflection plane.

And the same holds for every point perceived by the eye in this mirror, if [that point] lies outside the line joining the centre of the eye and the mirror's centre.

⟪V 2.IV.2⟫

V 182a **[91]** [Diagram V.5] Furthermore, $\angle NGZ$ is equal to $\angle AGH$, | because it is equal to $\angle BGH$; and angles EGN and EGH are right angles; therefore, by subtraction, $\angle AGE = \angle EGZ$.

Therefore, AE:EZ = AG:GZ.

Draw AH || GZ. Then $\angle AHG = \angle NGZ$; therefore, $\angle AHG = \angle AGH$, and thus line AH = line AG.

But GA:GZ = AE:EZ, thus AH:GZ = AE:EZ.

And AH:GZ = AN:NZ, therefore AE:EZ = AN:NZ.

[92] And [it will be] similar for every point perceived by the eye in this mirror, if it lies outside the line joining the centre of the eye and the mirror's centre: the line along which the form [of the point] is reflected to that eye will meet the perpendicular from that point to the line tangent to the intersection between the mirror's surface and the reflection plane,
V 182b [this perpendicular] being a diameter | of the mirror.

And the ratio of the line between the visible point and the mirror's centre, to the line between the mirror's centre and the meeting-point, i.e., the image-point, will be as the ratio between the segments of the line between the visible point and the image-point, as divided by the line tangent to the mirror's surface at the reflection point.

Let us call the point [E] that divides this line 'the extremity of the tangent'.

⟪V 2.IV.3⟫

[93] [Diagram V.5] Again, AN:NZ = AG:GZ, therefore AN:AG = NZ:ZG.

And NA > AG, since $\angle NGA$ is obtuse, therefore NZ > ZG.

Therefore, the line between the image-point and the mirror's centre will always be longer than that between the image-point and the reflection point.

⟪V 2.IV.4⟫

[94] [Diagram V.5] Again, since AN:NZ = AE:EZ, therefore NA:AE = ZN:ZE.

And NA > AE, therefore ZN > ZE.

Thus, the line between the image-point and the extremity of the tangent is always shorter than the radius; therefore, the line between the tangent's extremity and the
V 183a circumference | of the circle will be much shorter than the radius.[2]

2. Point Z is assumed to be inside the circle.

⟪V 2.IV.5⟫

[95] As for the perpendicular from the centre of the eye to the plane tangent to the mirror's surface, i.e., the line passing through the eye's centre and the mirror's centre: the eye will not perceive by reflection any of the points on [such a perpendicular], except that point on the eye's surface through which the perpendicular passes. This can be shown as was done with regard to the plane mirror: for if from any point on this line and from the eye's centre two lines are drawn to a point on the surface of this [spherical] mirror, then the diameter drawn from the centre of this mirror to that point on the mirror's surface will not cut the angle contained by those two lines. The eye will not perceive the form of any point on this line, which [form] has extended along this diameter, except the point on the eye's surface through which the diameter passes, as was shown in our discussion of the plane mirror. Sight will not, therefore, perceive by reflection any of the points on this line other than the point on the eye's surface through which | this line passes. The form V 183b of this point will extend along this line, which is perpendicular to the plane tangent to the mirror's surface, and will be reflected along this line itself; and the image of this point will be continuous with the image of what surrounds it on the surface of the eye.

⟪V 2.IV.6⟫

[96] [Diagram V.6] Having made that clear, I say that, of the images of visible objects perceived by reflection in spherical convex mirrors, some are located behind the mirror, some on the mirror's surface, and some in front of the mirror.

Let there be a spherical convex mirror on which the points A and B [are marked], let D be its centre, and G the centre of the eye.

Join GZD, and from line GD draw a plane that cuts the mirror's surface, producing in it a circle with D as centre, say circle ATB.

From G draw two lines tangent to the circle on either side of GD; let them be lines GA and GB. These two lines will lie on the surface of the cone tangent to the mirror's [spherical] surface, which [conical surface] will mark off the segment of the mirror's surface that faces the eye. Arc AZB will be all that faces the eye | of the circle ATB. V 184a

From G draw a line that cuts the circle, so that the segment of this line, cut off inside the circle, is equal to the radius of the circle;[3] let it be line GHT, and let HT be equal to the circle's radius.

Draw lines DT and DH, produce DH to M, and make ∠MHO = ∠GHM; then ∠MHO = ∠DHT. Now, ΔDHT is equilateral, since line HT is equal to the radius; therefore ∠DHT = ∠HDT, and thus ∠MHO = ∠HDT.

Therefore, HO || DT and every line drawn from a point on line HO to point D, even if HO is produced indefinitely, will cut line HT between points H and T.

On line HO take a point at random, say O, and join OD: it will cut HT, say at point C, and OD will cut the circle's circumference, say at Q.

Thus, if | through Q a plane is drawn tangent to the mirror's sphere, then line OQ will V 184b be perpendicular to it. And if O lies in a visible object, then the form of that point will extend along line OH and, reaching point H, will be reflected along line HG.

3. This elementary construction is left unexplained.

Now, line OQD is perpendicular to the plane tangent to the mirror's surface at Q, and therefore it is perpendicular to the line tangent to the intersection between the mirror's surface and reflection plane OHG. And C is the point of intersection of line OD with line GH, that is, the line of reflection; therefore, C is the image of point O, and it lies inside the sphere of the mirror.

And the same can be shown for every point on line HO, even if HO is indefinitely produced in the direction of O.

[97] [Diagram V.6] Further, draw line GYS below line HT; YS will be longer than the radius. Join DS and DY; then $\angle DYS < \angle YDS$.

((Produce DY to U, and make $\angle Uyt = \angle GYU$; then $\angle Uyt < \angle YDS$.)) Therefore, line
V 185a Yt will not meet line DS in the direction of t; and if from any point on Yt, even if | Yt is produced indefinitely in the direction of t, a straight line is joined to point D, then [this line] will cut YS between points Y and S. The point of intersection will be the image-point.

And thus, the image of any point on Yt, even if indefinitely produced, will be located on line YS between points Y and S. And, therefore, the images of all points on line Yt, even if indefinitely produced, will lie behind the mirror and within the sphere of the mirror.

And it can be similarly shown for every line reflected from a point on arc HZ that the images of all points on it will lie behind the mirror's surface.

[98] [Diagram V.6] Again, draw line GLK above line HT [where points L and K are on the circle]; then LK will be shorter than the circle's radius. Draw line DK; then $\angle DLK > \angle LDK$.

Extend DL to F and make $\angle FLE = \angle GLF$; then $\angle FLE > \angle LDK$. Therefore, line LE will meet line DK in the direction of E, say at point E.

V 185b Thus, if point | E is in a visible object, then the form of that point will extend along line EL, and reaching point L will be reflected along line LG, and sight will perceive it along LG. And line EK, being perpendicular to the plane tangent to the mirror's surface at K, will be perpendicular to the line tangent to the intersection between the mirror's surface and the reflection plane, that is, plane GLE. K will be the point of intersection of this perpendicular with line GLK, which is the line of reflection. Therefore, K is the image of point E, and it is located on the mirror's surface.

Produce LE in a straight line, take a point on it at random, say point d, and join dD. Produce GK in a straight line: it will cut dD outside the circle, say at point L_1. Then if point d is in a visible object, the form of that point will extend along line dL, and, reaching point L, will be reflected along line LG.

Now, line dD is perpendicular to the plane tangent ((to the mirror's surface at the point in which line dD cuts the circle's circumference, and perpendicular to the line tangent)) to
V 186a the intersection between the | mirror's surface and the reflection plane. L_1 is the point of intersection of this perpendicular with line GKL_1 which is the line of reflection for point d. Therefore, point L_1 is the image of point d, and it lies outside the mirror's surface.

And the image of any point on Ed, even if Ed is produced indefinitely in the direction of d, will be outside the mirror; and that can be shown, as has just been done.

Now every point [located] on line EL between points E and L will have its image behind the mirror. Because every line drawn from any point on EL to point D will cut line LK, the point of intersection will then be the image-point, and LK will be inside the mirror.

[99] [Diagram V.6] In the same way we [can] show, for every point on arc HB, that the line reaching it from G will cut the circle; that the segment cut off from this line inside the mirror will be shorter than the circle's radius; and that the extremity of every one of these lines can be an image of some point in a visible object.

And [it can be also be shown in the same way] that if | each of these lines is produced V 186b
in a straight line, then there will exist on the segment of it that lies outside the circle many points, each of which can be an image of some point in a visible object; and that every point on the segment cut off from each of these lines inside the mirror can be an image of some point on a visible object.

Thus, with regard to arc HZ, the images of all points on it from which forms are reflected to the eye at G will be behind the mirror; and, with regard to arc HB, [of] the images of the points whose forms are reflected from it, some will be behind the mirror, some on the mirror's surface, and some in front of the mirror; | and all the images located V 187a
on the mirror's surface, or outside the mirror, will be behind the diameter drawn from the mirror's centre to point B, because they will occur at or beyond the second of the points in which the line drawn from the centre of the eye cuts the circle's circumference.

And similarly for arc ZA: if a line is drawn to it from point G, similar to line GH, then the segment of the arc next to Z ⟪and lying between points Z and A⟫ will have all its images behind the mirror; while of the images of the segment of the arc next to point A, some will be behind the mirror, and some behind the diameter drawn from A.

[100] [Diagram V.6] Thus, if line GD is fixed in position, while line GH turns round it with ∠HGD unchanged, and a portion of the mirror's surface is thereby cut off by the circle drawn by GH on that surface, then the images of all the forms of visible objects reflected from that portion will be behind the mirror.

But, with regard to the portion of the mirror's surface that is cut off between this circle and the circle drawn by the line tangent to the sphere, of the images of the forms of | visible V 187b
objects reflected from it, some will be behind the mirror, and some in front of it.

[101] Thus, it is manifest from all that we have explained that the lines of reflection of the forms of visible objects perceived in spherical convex mirrors always meet the perpendiculars drawn from points in those objects at right angles to the lines tangent to the intersections between the surfaces of the mirrors and the reflection planes; and that of the meeting-points, i.e., the images, some are behind the mirror, some on the mirror's surface, and some in front of it.

And that is what we wished to prove.

⟪V 2.IV.7–8⟫

[102] And, along with what we have shown, it has been made clear that every point perceived in a spherical convex mirror will have its image on the diameter of the mirror that passes through that visible point and the mirror's centre. And from this it has become manifest that the images that may be perceived of all points that lie on one and the same diameter of the spherical convex mirror will lie on that same diameter.

[103] And it is manifest from this proposition also that every line | drawn from the V 188a
centre of the eye to the reflection point will cut the mirror's surface at two points, and that the reflection point is the first point at which the reflection line cuts the mirror's surface.

It remains for us to determine the locations of the images on each one of the diameters of this mirror.

⟪V 2.IV.9–16: Determination of the locations of images on any diameter of the spherical convex mirror⟫

⟪V 2.IV.9⟫

[104] [Diagram V.7] I say that, for any diameter of a spherical convex mirror, if a line drawn from the centre of the eye to a point on [the diameter] cuts the mirror's surface before reaching that point, so that the segment of that line between that point and the mirror's surface equals the segment cut off from the diameter between that point and the mirror's centre, then that point will mark, on the side of the centre, the limit of the images of all points on that diameter that can be perceived with that eye by reflection; I mean that there can be no images on the line [segment] between the point we have specified and the mirror's centre.

Let us prove this notion [i.e., proposition] also by demonstration.

Let circle ABG be in a spherical convex mirror with centre D; let E be the centre of
V 188b the eye, and | DH one of the diameters whose points can be perceived [by reflection]. Let line EBZ be drawn from the eye E to diameter DH, and let line BZ be [assumed] equal to line ZD. I say that there can be no image on line [segment] ZD for any of the points on diameter DH, even if DH is produced indefinitely.

Demonstration. We join DB and produce it in a straight line to T, and make ∠TBK = ∠EBT.

Since BZ = ZD, therefore ∠ZDB = ∠DBZ. But ∠TBK = ∠DBZ, since ∠TBK = ∠EBT; thus ∠TBK = ∠HDB, and BK will be parallel to DH and, therefore, BK will not meet DH.

Point Z will not therefore be an image of a point on diameter DH, even if DH is indefinitely extended; nor will Z be an image of a point outside the diameter DH, because
V 189a the image of a visible point can only be located on the diameter on which | that point lies.

Nor will there be an image on line ZE, except that of a point whose form is reflected from B. Therefore, no form of a point on diameter DH will be reflected from B, nor is Z a point on a diameter other than DH.

I say then that no point on line [segment] DZ will be an image of any of the points visible in mirror ABG.

On line [segment] DZ take a point at random, say O, and join EFO; join DF and continue it in a straight line to M, [and] make ∠MFN = ∠EFM.

Now ∠EOD > ∠EZD, and ∠FDO > ∠BDZ; thus, by subtraction, ∠DBZ > ∠DFO and [therefore] ∠BDZ > ∠DFO. Thus, ∠FDH is much larger than ∠DFO, and ∠MFN = ∠DFO, since ∠MFN = ∠EFM; therefore ∠HDF > ∠NFM. Thus line NF will meet line DH if extended rectilinearly in the direction of D, and point O will therefore not be an image
V 189b of a point | on the diameter DH.

Point O will not, therefore, be an image of one of the points visible in mirror ABG, and the same holds for any point on line DZ.

Point Z, then, is the point that marks, on the side of the centre, the limit of the images of all the visible points on diameter DH, even if DH is extended indefinitely.

And the same can be shown for every point corresponding to Z on every diameter of the mirror, whether the point lies inside the mirror, or outside the mirror's surface, or on the surface.

⟪V 2.IV.10⟫

[105] [Diagram V.7] From this demonstration it is manifest that if any line is drawn from the centre of the eye to a point on line [segment] DZ, then the line cut off from it between that point and the circle's circumference will be longer than the line cut off from DZ towards the centre, because ∠FDO and every angle corresponding to it will be greater than ∠DFO and every | angle corresponding to [the latter]. V 190a

Later on, [in Lemma V,] we will show how to draw a line like EBZ so that BZ = ZD.

⟪V 2.IV.11⟫

[106] [Diagram V.8] I say further, with reference to any diameter of a spherical convex mirror that cuts the portion of the mirror's surface that faces the eye, other than the diameter through the eye's centre, that every point on the segment of that diameter between the point marking the limit of the images on the side of the centre and the diameter's extremity can be an image of one of the visible points; and that the extremity of each one of these diameters will be ⟨the limit⟩, on the side of the visible objects, of all the images that can be on that diameter.

Let A be the centre of the mirror, and B the centre of the eye; let AGD be one of the diameters that cut through the portion of the mirror's surface facing the eye without passing through the eye's centre; and let point G on it be on the mirror's surface.

From diameter AGD let a plane be drawn through the eye's centre, and let it produce circle EGZ in the mirror's surface. From point B draw two tangential lines to the circle, say BE and BZ, and let them touch the circle | at points E and Z. Arc ZE will then be in the V 190b
portion of the mirror's surface that faces the eye.

Join AHB; line AB will then be the diameter through the centre of the eye. Let T be the point that marks the limit of the images on diameter AG, on the side of the centre.[4] I say that any point on line [segment] TG can be an image of one of the visible points; and that G is the limit of the images of this diameter on the side of the visible objects.

On line TG take a point at random, say K, and join BFT; then FT = AT. Draw line BK: it will cut arc FG, since BK will be below [i.e., closer to the centre than] the tangent BE and above [i.e., farther from the centre than] BT. Let BK cut arc FG in point M; join AM and continue it to L, and make ∠LMN = ∠LMB; join AB.

Then ∠AFB > ∠AMB, and so ∠AMK > ∠AFT. | V 191a

But ∠AFT = ∠TAF, and ∠TAF > ∠MAK, therefore ∠AMK > ∠MAK; thus ∠LMN > ∠MAK, since ∠LMN = ∠LMB, which is equal to ∠AMK.

And line MN will therefore meet line AG in the direction of N, say at D.

Then, if point D is in a visible object, the form of that point on the object will then extend along line DM, and, reaching M, will be reflected along line MB, whereby it will be perceived by the eye, and its image will be point K. And the same can be shown for every point on line [segment] TG.

Thus, any point on line [segment] TG can be an image of one of the visible points; and G, i.e., the extremity of the diameter, will be the limit, on the side of the visible objects, of all the images that can be on the diameter | AG. V 191b

---

4. I.e., FT = TA in Diagram V.8.

But no point on AGD outside the circumference of the circle can be an image of one of the visible points, and neither can point G, because it has been shown that the images outside the mirror's surface or on that surface are all behind the diameter drawn to point E. G will then be the point that marks the limit, on the side of the visible objects, of the images [located] on diameter AG.

And the same can be shown for every one of the mirror's diameters that cut arc ZE and for every arc corresponding to ZE on any circle that cuts the mirror.

Therefore, for any diameter of a spherical convex mirror that cuts the portion of the mirror's surface facing the eye, any point on the segment between the point that marks the limit of the images on the side of the centre and the diameter's extremity can be an image of one of the visible points; and the extremity of the diameter will mark the limit, on the side of the visible objects, of all the images on that diameter.

⟪V 2.IV.12⟫

[107] [Diagram V.9, see also Diagram V.9(2)] Furthermore, let there be a circle the plane of which passes through the centre of the mirror and the eye's centre, let A be the
V 192a centre of the eye and B the | mirror's centre, and let the circle be GDZE. Draw line ADBZ, and from A draw two lines touching the circle, say lines AG and AC: GC will be the arc that lies in the portion facing the eye.

Join BG and construct the right angle GBH. BH will then be parallel to AG, since ∠BGA is right. The diameter BH will therefore not meet AG, and every diameter drawn from B to a point on arc HG will meet AG.

We draw AME so as to cut the circle in such a way that segment ME will equal the radius. Join BM and BE, and carry BE through to meet line AG, say at point T.

Then, since ME equals the radius, it is equal to line BE; then E is the point that marks, on the side of the centre, the limit of the images on diameter BE. This can be shown as in the previous demonstration.

I say then that any point on line ET can be an image of one of the points on the visible objects.

On line ET we take a point at random, say Q. Join AQ: it will be situated below AG,
V 192b cutting the circle at two points, one | on arc DG, the other on arc GE.

Let N be the point on arc DG; N will then lie on arc MG between points M and G, because line AQ is between lines AE and AT.

We join BN and continue it to O. Then ∠AMB > ∠ANB, and therefore ∠QNB > ∠EMB. But ∠EMB = ∠MBE, and ∠MBE > ∠NBE, thus ∠QNB > ∠NBE.

We make ∠ONF = ∠QNB. Line NF will then meet line BQ, and will do so beyond point T.

For ∠ONF is acute; and every line drawn from a point on arc DG, and containing with the diameter drawn to that point an acute angle outside the circle, will cut line AG between points A and G; and NF will then cut AG between points A and G. And if [NF] cuts AG,
V 193a then it will not meet line BT between points B and T, for, if it did, | it would cut line GT at another point. Line NF will not therefore meet line BT between points B and T, and will therefore meet it beyond T, say at F.

But ∠FNO = ∠ANO, since ∠FNO = ∠QNB. Therefore, if point F is in a visible object, then its form will extend along line FN and will be reflected along line NA, and the eye will perceive it through AN.

But Q is the point of intersection between lines AN and FB, the latter being the mirror's diameter, therefore point Q is the image of point F.

And the same can be shown for every point on line [segment] ET. Therefore, any point on [segment] ET can be an image of one of the visible points. But all the images will be on the lines that extend from the eye to the reflection point, and all the lines that so extend will be below line AG; therefore, all the images will be below AG. And if all the images are below line AG, then T is the point that marks, on the side of the visible objects, the limit of the images | situated on line ET. V 193b

[108] [Diagram V.9(3)] From B we draw a diameter that cuts arc EG at random; let it be diameter BLS. ⟨⟨I say: where L is on the circumference, and S on the meeting point with AG produced.⟩⟩

Join AL: it will cut arc MG, since AL is between lines AE and AT; and point L is beyond arc MG. Let AL cut arc MG in point R. Line RL will then be shorter than line LB, because it is shorter than ME.

Therefore, the line drawn from A to diameter BL, such that the segment of it between BL and the circle's circumference equals what that line cuts off from BL ⟨on the side of B⟩,[5] will be [located] below line RL. For if this line were above RL, then RL would be longer than LB, as was shown earlier.

Therefore, the extremity of this line will be between points B and L, and that point [i.e., the extremity] will mark the limit of the image[s]; let that point be Y.

Then any point on line YS can be an image of one of the visible points; and this can be shown, as has been done in this proposition and in the preceding one.

Furthermore, S will be the point that | limits, on the side of visible objects, the images V 194a on line BS; and point Y will be inside the circle; and of the images of the visible points that are on diameter BL, some will be inside the mirror, some outside the circle, and one on the mirror's surface.

[109] [Diagram V.9(4)] We then draw from B a diameter that cuts arc HE at random; let it be diameter BKd. Join AK: it will cut arc DM, say at point W. Line WK will then be longer than line KB, because it is longer than ME.

And line Gd < dB, since ∠dGB is a right angle.

Therefore, the line drawn from A to Bd, such that the segment cut off from it between the circle's circumference and line Bd equals what is cut off from line Bd on the side of B, will be located above line WK and below line Gd; and the extremity of that line will be outside the circle, and that extremity will mark the limit of the images on the side of B.

Let the extremity of that line be point L_1. Then any | point on line L_1d can be an image V 194b of one of the points on the visible objects; and d will be the point that marks, on the side of the visible objects, the limit of all the images on that diameter.

Thus, the images on the diameters that cut arc HE will all be outside the mirror's surface; and of the images on diameter BE, one will lie on the mirror's surface, namely point E, and all the other images will be outside the mirror; and of the images on the diameters that cut EG, some will be outside the mirror, some inside the mirror, and one on the mirror's surface.

And the images on [all] these diameters are those that belong to one half of the circle.

---

5. I.e., JY = YB in Diagram V.9(3).

[110] [Diagram V.9] Now, if all of these diameters are produced in straight lines through the centre, they will all cut the circle on the other side, and it will be possible for the objects on that side, whose forms are reflected from arc DC, to have their images on these diameters. For, since each of them will contain with diameter AZ, on the other side
V 195a of the circle next to Z, an angle | greater than ∠HBZ, it will be possible for each to have other images located on it, on that side.

Similarly, it will be possible for the diameters that cut arc ZH to have images located on them next to arc CD, for ∠HBZ is acute. And, therefore, these diameters, being rectilinearly produced, will contain obtuse angles with diameter BZ, next to Z; and thus each of these angles will be greater than ∠ZBH; and, therefore, it will be possible for each of these diameters to have images located on it, on the other side of the circle.

[111] [Diagram V.9] But it is not possible for those diameters that cut arc ZH to have any images located on them next to arc ZH, because, being produced out of the circle in the direction of Z, they will meet neither line AG nor line AC, and so they will be
V 195b situated below lines AG and AC, | beyond points G and C; and nothing will be reflected to the eye from what lies below lines AG and AC, beyond points G and C, because the line drawn from them to a point on arc CG, from which alone the forms may be reflected, will cut the circle's circumference in another point before reaching arc CG. And, therefore, no form of a point below lines AG and AC, beyond points G and C, will reach the convex side of arc CG; and the eye will therefore not perceive in this mirror any of the points that lie below lines AG and AC, beyond points G and C. Therefore, on the diameters drawn to arc ZH, there will be no images next to arc ZH; and the same holds for the arc corresponding to arc ZH and continuous with it on the other side of the circle.

[112] [Diagram V.9] If thus we imagine line BH to turn about diameter AZ while ∠HBZ remains constant, then point H will draw on the surface of the mirror a circle that marks off on the sphere a segment between it and the circle ⟨⟨that marks off⟩⟩ the segment facing the eye; as a result, it will be possible for every diameter that is drawn from the centre of the sphere and which cuts that [middle] segment to have on it images outside the mirror. It will be possible for every diameter drawn from the centre of the sphere to cut
V 196a that [middle] segment to have on it images outside the mirror. Further, | each one of these diameters will meet the conical surface [with the eye as apex] that touches the mirror's surface, the meeting-point being that which marks, on the side of the visible objects, the limit of the images on that diameter.

[113] [Diagram V.9] If we also imagine line BE to turn while ∠EBZ remains constant, then point E will draw on the sphere's ⟨surface⟩ a circle that divides the segment between the first two circles into two parts. Of these two parts, the one next to the circle drawn by line BH will be such that the diameters drawn to it will have all their images located outside the mirror. The other part will be such that each diameter drawn to it will have its images: some outside the mirror, some inside the mirror, and only one image on the mirror's surface. As for any diameter that cuts the circle drawn by line BE, only the image [of a single point] on it will be on the mirror's surface. All other images [associated with it] will be located outside the mirror. And [finally], as regards the segment [of the mirror] that faces the eye, all images [associated with] the diameters drawn to it will be located inside
V 196b the mirror. We have thus completed our determination of all | that relates to the [locations of] images belonging to this mirror.

[114] All the images that belong to this mirror are perceived behind the mirror, regardless of their locations with respect to the mirror's surface. For the images of this mirror that lie outside the mirror or on the mirror's surface will be located behind the point of tangency on the mirror's surface. But nothing of the mirror's surface behind the point of tangency will be visible, because it will be in a part of the surface that is hidden from the eye. As for the images that lie outside the mirror, the eye will not perceive the locations below them on the mirror's surface, and the eye will not perceive the locations of the images on that surface. Further, since all images in this mirror are located behind the reflection point on the mirror's surface, the eye will, in all cases, perceive the image [as lying] behind the reflection point.

[115] If the image lies outside the mirror, ⟨the eye⟩ fails to perceive what is below it on the mirror's surface. If the image lies on the mirror's surface, the eye will not perceive its location on the mirror's surface. And, therefore, the eye only perceives all images in this mirror behind the mirror, regardless of their locations with respect to the mirror's surface, because these images are | always behind the reflection point, and because the V 197a
eye perceives neither the location on the mirror's surface that lies below the image that is situated outside the mirror nor the location of the image on the mirror's surface; and if it cannot perceive what lies below the image on the mirror's surface, then it cannot perceive that the image lies outside the mirror. Thus, if the eye cannot perceive the location of the image on the mirror's surface, then it cannot perceive the image as lying on the mirror's surface.

[116] Therefore, the images of all the objects that the eye perceives in the spherical convex mirror are exactly located: some behind the mirror, some on the mirror's surface, and some outside the mirror, but all are perceived to be behind the mirror. And that is what we wished to prove.

⟪V 2.IV.13⟫

[117] [Diagram V.10] All this having been clarified, I say [that] every point perceived by a single eye in a spherical convex mirror and located outside the diameter of the mirror passing through the centre of the eye will have only a single image, and its form will be reflected from the mirror's surface from only a single point.

Let point A be the centre of the eye, | B a point in a visible object, and G a point on a V 197b
spherical convex mirror. Let B be outside the diameter through the centre of the eye and the mirror's centre. Let the eye at A perceive point B by reflection from mirror G, the reflection point being point G.

We draw lines AG and GB. Then BG will be the line along which the form extends from B to G, and GA the line along which the form is reflected from G to eye A. Plane ABG will be orthogonal to the plane tangent to the mirror's surface at point G and will thus be one of the planes passing through the mirror's centre. Let the intersection between this plane and the mirror's surface facing the eye be arc DGE. Let the mirror's centre be point N. And join BLN.

It is clear that point A can be with point B in only one plane that is orthogonal to the mirror's surface or to a plane tangent to the mirror's surface.

For all planes orthogonal to the mirror's surface | and passing through point B will V 198a
intersect in line NB. But only one plane can be drawn from line NB through a single point

outside NB, and point A cannot be with point B ⟨in⟩ a plane orthogonal to the mirror's surface, unless that plane passes through line NB. Therefore, point A can be with point B in only one plane that is orthogonal to the mirror's surface.

Now, every plane that is orthogonal to a plane tangent to the mirror's surface and passing through the point of tangency will be orthogonal to the mirror's surface. For the line drawn from the point of tangency, perpendicular to the tangent plane, will be perpendicular to the mirror's surface and will pass through the mirror's centre; and this perpendicular will lie in the plane orthogonal to the tangent plane. But plane AGB is orthogonal to the mirror's surface, and point A lies with point B in plane AGB that is orthogonal to the mirror's surface. Point A cannot therefore lie with point B in a plane, other than that of AGB, that is orthogonal to a plane tangent to the mirror's surface.

Therefore, the form of point B can only be reflected to point A from arc DGLE on mirror G.

Now, it is clear that there can be no reflection between points B and A from point L. For line BL is perpendicular to the mirror's surface and to the plane tangent to the surface
V 198b at L, and the line drawn from A to | L will be inclined to the tangent plane.

[118] [Diagram V.10] Nor is it possible for the form of B to be reflected to eye A from arc LE, because for every line drawn from B to a point on arc LE, if from the extremity of this line, ⟪i.e., the point at which it meets arc LE,⟫ a line is drawn tangent to LE, then the exterior angle at E, contained by that line and the tangent line, will be obtuse, and the interior angle contained by the tangent line and the line drawn from A to the point of tangency will be acute, [this angle] being above the tangent line. Therefore, the form of point B will not be reflected to the eye A from arc LE.

[119] [Diagram V.10] I say further that the form of B cannot be reflected to eye A from arc LG.

If possible, let it be reflected from point Z on arc LG. Join AZ and continue it in a straight line, and let it meet line BN at point O. Draw AG to meet BN, say at point K. From points G and Z draw two lines tangentially to arc DGZE; it is clear that they will
V 199a meet line BL, since the tangent line drawn from G lies inside figure BGN. | For line BG cuts the circle, and the tangent line drawn from Z lies below this tangent [from G]. Let the tangent line from G meet line BL at F, and let the tangent line from Z meet BL at M. Point F will then be closer to B than M is, since line ZM, being produced in the direction of G, will meet GF.

Thus, BN:NK = BF:FK, and BN:NO = BM:MO, as was shown earlier, and BN:NK > BN:NO, NK being shorter than NO. Therefore, BF:FK > BM:MO; but BF < BM, and FK > MO, and that is impossible.

And the same can be shown for every point on arc GL.

V 199b **[120]** [Diagram V.10] Nor can | the form of point B be reflected to the eye A from arc GD.

If possible, let it be reflected from point T. Join AT and carry [the line] through to meet BN, say at point H. Join BT and draw the tangent line from T: it will meet BN, since it is inside figure BTH, say at point C.

Point C will be closer to B than F, because GF will meet TC in the direction of T. Thus, BN:NH = BC:CH, and BN:NK = BF:FK, and BN:NH > BN:NK. Therefore, BC:CH > BF:FK; but BC < BF, and CH > FK, and that is impossible.

Thus, the form of point B cannot be reflected to the eye A from arc DG.

Therefore, the form of point B cannot be reflected to the eye at A from anywhere on circle DGE, except at point G. But since the form of B cannot be reflected from any circle other than DGE, then the form of B cannot be reflected to the eye at A from anywhere on the surface of mirror DGE, except at one point. And if | it can be reflected from only one V 200a
point, then it can have only a single image. Therefore, in the spherical convex mirror DGE, point B will have only one image for the eye A.

And thus, for every point perceived by one and the same eye in a spherical convex mirror, if the point lies outside the diameter through the centre of the eye and the mirror's centre, the form of that point will be reflected to the same eye from a single point only, and it will have only a single image. And that is what we wished to prove.

⟨⟨V 2.IV.14⟩⟩

[121] [Diagram V.10] ((As for the point on the eye's surface and on the diameter through the centre of the eye and the mirror's centre, its image also will be a single point. This can be shown as in the case for the plane mirror. Thus, any point perceived by one and the same eye by reflection in a spherical convex mirror will have only a single image. And that is what we wished to prove.

⟨⟨V 2.IV.15⟩⟩

[122] And it is clear from what we have proved in this proposition that reflection at equal angles between two given points can only take place from a single point on the surfaces of spherical convex mirrors.))

⟨⟨V 2.IV.16⟩⟩

[123] [Diagram V.10] It can be proved, also by reference to this proposition [*or* diagram], that for any two points on a single diameter of a spherical convex mirror and on one side of the mirror's centre, the image of the point that is closer to the mirror's surface will be farther from the mirror's centre, and its reflection point will be farther from the eye's centre. And the image of the point that is farther from the mirror's surface will have its image | V 200b
closer to the mirror's centre, and its reflection point will be closer to the eye's centre.

Let C be a point assumed on diameter NB [between L and B]. I say that the image of C will be closer to the mirror's surface than point K is, and that the reflection point for C will be farther from the eye than point G is.

Demonstration. The form of point C cannot be reflected to the eye A from arc LE, for this can be shown as was done for point B.

Nor can its form be reflected from point G, for two forms cannot be reflected along a single line from a single point.

I say further that its form cannot be reflected from a point on arc GD.

For, if possible, let [the form of C] be reflected from point T. Join NT, CT, and AT.

Now, if diameter NT does not divide [i.e., bisect] ∠ATC, then the form of C will not be reflected to the eye A from point T.

So let diameter NT divide [i.e., bisect] ∠ATC. Produce diameter NT to S; join NG and continue it to Q.

Then ∠ATN will be greater than ∠AGN, and [therefore] ∠ATS will be smaller than
V 201a ∠AGQ, and [therefore] ∠CTS will be smaller than ∠BGQ. |

Therefore, [the sum of] angles TCN and TNC will be less than [the sum of] angles GBN and GNB, which is impossible, since ∠TCN > ∠NBG, for line CT cuts line BG, and ∠CNT > ∠BNG.

The form of point C will not, therefore, be reflected from arc GD or from arc LE, and can only be reflected therefore from arc GL, and between points G and L.

And it can be reflected from only one point, as was shown in this [present] proposition.

If, therefore, the reflection point ((lies between points G and L, then the image-point ⟨must⟩ be located between points K and L, and will thus be closer to the mirror's surface than point K is. Further, the reflection point will be)) farther from the eye at A than point G, since any line drawn from A to a point on arc GL will be longer than line AG.

Therefore, for any two points on a single diameter of a spherical convex mirror and located on the same side of the mirror's centre, if their forms are reflected to one and the same eye from this mirror, then the image of the point that is closer to the mirror's surface will be closer to that surface, and the reflection point for that form will be farther from the
V 201b centre of the eye, while the image of the point that is farther from | the mirror's surface will be farther from that surface, and its reflection point will be nearer to the centre of the eye.

⟨⟨V 2.IV.17: On determining the point of reflection⟩⟩

[Six Lemmas]

[124] To find the point of reflection if two points [related to each other as eye and object] are given and the spherical convex mirror is given.

We join the two points to the centre of the sphere with two straight lines, and draw the plane in which the two lines lie, thereby producing a circle on the surface of the sphere.

If the two lines are equal, bisect the angle contained by them at the [mirror's] centre, produce the bisecting diameter until it cuts the circle's circumference on the side of the two points, and join the point of intersection to the two points with two straight lines: these two lines will contain two equal angles with the diameter that is perpendicular to the plane tangent to the sphere [at the intersection-point]; the intersection-point, that is, the diameter's extremity, will be the point of reflection.

[125] If, however, the two lines that join the two [given] points with the centre of the sphere are unequal, then we put forward [six] lemmas, which we will use in this question and in others that will come up later. The lemmas are as follows.

⟨⟨Lemma I⟩⟩

V 202a **[126]** [Diagram V.11] Let | circle ABG, with diameter GB, be known; let BG be produced on the side of G; let line KE be given; and let point A be given on the circumference of the circle.

We wish to draw from point A a line, as AHD, so that the part of it that lies between the [produced] diameter and the circle, such as HD, is equal to line KE.

Now arcs AB and AG are either equal to each other or not; let them be equal [Diagram V.11]. Draw lines BA and AG, and make KZ × ZE = AG^2, then KZ > AG.

Make AT = KZ, and, with A as centre and with distance AT, draw an arc of a circle: it will always cut line GD, say at D. Join AD. Then AD = KZ.

And AD will always cut arc AG, since the tangent line drawn from A will be parallel to GB; for the line drawn from point A to the circle's centre will be perpendicular to line GB, because of the equality | of arcs AB and AG. Therefore, line AD will cut arc AG, say at point V 202b H. Join GH. Then $\angle AHG + \angle ABG = 2$ right angles. But $\angle ABG = \angle AGB$, therefore $\angle AHG = \angle AGD$, and $\triangle ADG \sim \triangle AHG$. Therefore, DA:AG = GA:AH, thus $DA \times AH = AG^2$.

But $KZ \times ZE = AG^2$, therefore $DA \times AH = KZ \times ZE$. And DA = KZ; therefore AH = ZE. Thus, by subtraction, line HD = line KE. And that is what we wished to do.

[127] [Diagram V.12] Now, let arcs BA and AG be unequal. We draw lines BA and AG, and draw GZ || BA.

Assume a known line at random; let it be TN. Make $\angle TNL = \angle DGA$ and $\angle TNM = \angle DGZ$; | produce line LN on the side of N to Q, and draw line MT parallel to NL [to V 203a intersect NM in M]. Draw TQ parallel to NM [to intersect LN extended in Q], and produce QT on the side of T ⟨⟨to O⟩⟩.

Then, through M, draw the hyperbola of which lines OQ and QL are asymptotes, as is shown in Proposition 4 in Book II of the *Conics* by Apollonius; let it be [conic] section SMW. We then make Y:TN = BG:KE.

In [conic] section SMW draw line MC equal to line Y; produce MC on both sides: it will meet lines LQ and QO, as is shown in Proposition 8 in Book II of the *Conics*, and let it meet them in points L and O. Lines OM and LC will then be equal, as is also shown in Proposition 8 of the said Book.

From point T draw line TF parallel to OL, and let it cut NM in point J: area LMTF will then be a parallelogram, and LM = FT. But LM = CO, so CO = TF. | V 203b

And MO = JT, because area JO is a parallelogram; then, by subtraction FJ = CM; and CM = Y, thus FJ = Y; and therefore FJ:TN = BG:KE.

On line GA and at point A construct $\angle GAZ = \angle NFT$. This line, namely AZ, will meet line GD, because the angles at points A and G are equal to the angles at points F and N; let it meet GD at D.

Now since ((angles AGD and ZDG are [respectively] equal to angles FNT and JNT, and $\angle GAD = \angle NFT$, then triangles AGZ, ZGD, and AGD will be [respectively] similar)) to triangles FNJ, JNT, and FNT, and, therefore, ZA:AG = JF:FN, and AG:GD = FN:NT, therefore AZ:GD = FJ:NT.

But FJ = Y, and Y:TN = BG:KE, therefore AZ:GD = BG:KE.

[128] [Diagram V.12] And since line AD meets BD outside the circle on the side of G, line DA will either touch the circle at point A [Case 1], or it will | cut arc AG [Case 2], or V 204a else cut arc AB [Case 3]:

For if arc AG is shorter than arc AB, then the tangent drawn from A will meet the diameter BG on the side of G, and the line drawn from A parallel to diameter BG will cut arc AB; and, therefore, the lines drawn from A to meet GD above the tangent will cut the part of arc AB that is cut off by the parallel line, and the lines drawn from point | A to meet V 204b GD below the tangent will cut arc AG.

[129] [Diagram V.12] Now, if arc AG is longer than arc AB, then every line drawn from A and meeting BG outside the circle on the side of G will always cut arc AG. For the tangent line drawn from A will meet diameter BG on the side of B; and the line drawn from A parallel to diameter BG will cut arc AG. From this it follows, provided arc AG is longer than arc AB, that all lines drawn from A so as to meet BG outside the circle on the side of G will cut arc AG. Thus, line AD will either touch the circle at A, as in the first diagram [Case 1], or cut arc AG, as in the second diagram [Case 2], or else cut arc AB, as in the third diagram [Case 3].

[130] [Diagram V.12, Case 1] [First,] let [AD] be tangent [to the circle]: then $\angle GAD$
V 205a $= \angle ABG$; but $\angle ZGD = \angle ABG$, then $\angle ZGD = \angle GAD$; and therefore $AD \times DZ = GD^2$. |

But $BD \times GD = AD^2$, because AD is a tangent, thus by subtraction, $AD \times AZ = BG \times GD$, therefore AZ:GD = BG:DA. But it was shown that AZ:GD = BG:KE, therefore BG:KE = BG:DA, thus line AD = line KE.

[131] [Diagram V.12, Case 2] Now, let line AD cut arc AG, say at H. We join GH. Then $\angle AHG + \angle ABG = 2$ right angles, and thus $\angle GHZ = \angle ABG$. But $\angle ZGD = \angle ABG$, therefore $\angle GHZ = \angle ZGD$, so $HD \times ZD = GD^2$.

But $AD \times DH = BD \times DG$. Therefore, by subtraction, $HD \times AZ = BG \times DG$.

Thus, AZ:GD = BG:HD. But [it was shown that] AZ:GD = BG:KE. Therefore, BG:HD = BG:KE, and thus line HD = line KE.

V 205b **[132]** [Diagram V.12, Case 3] Now, let line AD cut arc AB, say | at H. We join GH. Then $\angle GHA = \angle GBA$, and $\angle ZGD = \angle GBA$, therefore $\angle GHD = \angle DGZ$, so $HD \times DZ = GD^2$. But $HD \times DA = BD \times DG$, thus, by subtraction, $HD \times AZ = BG \times DG$. Therefore, AZ:GD = BG:HD. But AZ:GD = BG:KE. Therefore, BG:HD = BG:KE, and thus line HD = line KE.

Thus, it has been proved, in all [three] cases, how to draw from point A a line that meets diameter BG outside the circle on the side of G, so that the part of the line that lies between the circle and the diameter is equal to line KE. And that is what we wished to do.

⟨⟨Lemma II⟩⟩

[133] [Diagram V.13][6] Furthermore, let there be a circle ABG, with BG [given] as diameter and point A [given] on the circumference; and let line ZH be given. We wish to draw from A a line that cuts the diameter BG and extends to the circle so that the part of it
V 206a that lies between the circle and the diameter will be equal | to line ZH.

We draw lines AB and AG, and on line ZH at point H we construct $\angle ZHK = \angle ABG$ and $\angle ZHT = \angle AGB$. From Z draw line ZT || line HK, and line ZK || line TH; area TK will then be a parallelogram.

Through point T draw the hyperbolic section of which lines HK and KZ are the asymptotes; let it be [conic] section TC, and let the [conic] section opposite to it be WS. Produce lines HK and ZK on the side of K to L and Q.

With T as centre and [with] a distance equal to diameter BG, describe a circle, and let it meet section WS at point S. This circle will meet [conic] section WS if line BG is not shorter than the shortest line that can be drawn from [point] T to [conic] section WS.

As to what the shortest line is that can be drawn from T to section WS: this has been
V 206b shown in Propositions 34 and 61 in Book V of Apollonius's | *Conics*.

[134] [Diagram V.13] Thus, if the circle described about T with distance [i.e., radius] BG meets the section, it will either touch it at one point or cut it in two points.

6. The line TS' has been added in the diagram for clarification.

If [the circle] touches [the section], then ((only one line)) equal to BG can be drawn from point T to section WS. ((But if the circle cuts the section in two points, then only two lines equal to BG can be drawn from T to section WS.))

Thus, point S is either the point of tangency or one of the two points of intersection.

Draw line TS, then TS = BG, and TS will cut lines HK and KQ; let it cut HK in point F and KQ in point Q. From Z draw a line parallel to TS: it will cut lines HK and HT, since TS cuts these two lines; let that be line LZM.

Then ZM = TQ, since area MQ is a parallelogram.

Now, since CT and WS are opposite sections, and line TS cuts their asymptotes, then line TF = line QS, as is shown in Proposition 11 in Book II of the *Conics*. | V 207a

Now TF = ZL, ((since area LT is a parallelogram,)) therefore ZL = QS; and ZM = TQ, thus LM = TS.

But TS = BG, ((therefore LZM = BG.))

[135] [Diagram V.13] On BG and at point G on it we construct ∠BGD = ∠MLH:

then ∠MLH will be acute, because ∠LHM is right, being equal to [the sum of] angles ABG and AGB. Therefore, GD will fall inside the circle; let it cut the circle in point D. Join BD and AD, and let AD cut diameter BG in E.

Then ∠GDB is right, thus ∠GDB = ∠LHM; and ∠BDE = ∠BGA, which is equal to ∠ZHM; and ∠GBD = ∠LMH; therefore △BGD ~ △LMH and △DEB ~ △HZM.

Thus, GB:BD = LM:MH and BD:DE = MH:HZ, therefore GB:ED = LM:ZH.

But LM = BG, therefore DE = ZH.

Thus, from point A, line AED has been drawn so that | ED = ZH. And that is what we wished to do. V 207b

[136] But if two lines equal to BG go from point T to section WS, then there will go from point Z to lines KH and HT two lines equal to BG, such that they contain between them and HK two unequal angles.

Then, if two angles equal to those angles are constructed on line BG at point G, then two points will be produced on arc BG; and if two lines are joined between them and point A, there will be cut off from each of these two lines between arc BDG and diameter BG a line equal to ZH, this being shown by the demonstration we have [just] mentioned.

[137] But, should line | BG be equal to the shortest line that can be drawn from T to section WS, then only one line can be drawn from A to arc BDG so that the line [segment] between the arc and line BG will be equal to ZH. V 208a

If, however, BG is longer than the shortest line, then there will go from A to arc BDG two lines in each of which the [segment] between the arc and the diameter will be equal to line ZH.

And no more than two lines can be drawn from A to arc BDG so that the segment between the arc and diameter BG will be equal to ZH. For the circle about centre T cannot cut section WS in more than two points, since the centre of the circle lies outside the section.

And [finally,] if BG is shorter than the shortest line, then a line cannot be drawn from A to arc BDG so that the [segment] between the arc and the diameter is equal to ZH.

This ⟨construction⟩, therefore, is either impossible, or it can be carried out once, or twice, but not more. And that is | what we wished to prove. V 208b

⟨⟨Lemma III⟩⟩

[138] [Diagram V.14] Further, let triangle ABG have a right angle at B; let point D be given on line BG [as in Case 1, or on BG produced, as in Case 2]; and let the ratio of E to Z be known. We wish to draw from D a line, as DTK, so that TK:TG = E:Z.

We join DA and draw DM || BA, and on triangle DMG circumscribe circle DMG: MG will then be a diameter of the circle, because ∠MDG is right.

Make ∠DMN = ∠DAG; then line MN will cut ∠DMG, and thus [it will] cut arc DG in the first diagram [Case 1], or cut arc MG in the second diagram [Case 2]; let it cut [the arc] in point N.

Set AD:H = E:Z, and from N draw line NCL [to intersect the circle in C and diameter MG in L] so that CL = H, as was previously shown, [in Lemma I]. Join DC and produce it in a straight line: it will cut LM, say in point T; and join GC.

V 209a Then ∠GCD = ∠GMD, | therefore ∠GCD = ∠GAB, and thus ∠GCT = ∠TAK. But ∠CTG = ∠ATK.

Therefore, line CT, produced rectilinearly in the first diagram [Case 1], will meet AK at an angle equal to ∠TGC.

Produce line CT, and let it meet AK at point K. Then ΔAKT ~ ΔCGT in the two diagrams [Cases 1 and 2], thus AT:TC = KT:TG.

Also ∠DCN = ∠DMN, and ∠DMN = ∠DAT, thus ∠LCT = ∠DAT, therefore ΔLCT ~ ΔADT. Thus, AT:TC = AD:LC.

But LC = H, therefore AT:TC = AD:H. And AD:H = E:Z, thus AT:TC = E:Z.

V 209b But AT:TC = KT:TG, therefore KT:TG = E:Z. And that is what we wished to do. |

⟨⟨Lemma IV⟩⟩

[139] [Diagram V.15] Let circle AB with centre G be given, and let D and E be two given points.[7] We wish to draw from E and D two lines, as EA and DA, so that a tangent to the circle being drawn, as AH, will bisect angle EAD.

Join GD, GE, and ED, and produce EG in a straight line to [meet the circle at] B. Take a line at random, let it be MY, and divide it at S, so that YS:SM = EG:GD; bisect [line YM]
V 210a in N; and draw NO perpendicular to YM. |

Make ∠NMO = ½∠DGB, and from S draw line SQF, so that QF:FM = EG:GB [by Lemma III]. Make ∠EGA = ∠SFM, and join EA and QM, then ΔEAG ~ ΔQMF.

[140] [Diagram V.15] We make ∠EAZ = ∠QMS, then ∠ZAG = ∠SMO = ½∠DGB.

Produce AZ on the side of Z, and make AZ:ZK = MS:SY = DG:GE. Join EK and QY, and draw EL perpendicular [to AK]. Then the angles at points A, E, K, Z, and L will be equal to the angles at points M, Q, Y, S, and N, and, therefore, the triangles will be similar. Thus, AL = LK, and AE = EK, and KZ:ZA = YS:SM = EG:GD.

[141] [Diagram V.15] Draw AT || EK [to meet GE extended at T], then ∠TAZ = ∠ZAE,
V 210b | and EA:AT = EZ:ZT = KZ:ZA = EG:GD.

[142] [Diagram V.15] Now, make ∠GAW = ∠GAE, then ∠WAT is twice ∠GAZ, which [i.e., ∠GAZ] is equal to ∠FMN, which is equal to ½∠DGB. Thus, ∠WAT = ∠DGW, and therefore line WA will meet line GD, if AW meets GB; and the triangle cut off by line WA produced will be similar to triangle WAT.

---

7. The position of point A on the line is not given, but will be determined later.

I say then that WA meets GD at point D.

For EG:GD = EA:AT, and EA:AT = (EA:AW) × (WA:AT), thus EG:GD = (EA:AW) × (WA:AT). But EA:AW = EG:GW, because the angles at point A are equal, and WA:AT is as the ratio of WG to the line cut off by WA from GD. | V 211a

Therefore, EG:GD = (EG:GW) × (the ratio of GW to the line cut off by WA from GD), but EG:GD = (EG:GW) × (GW:GD). Thus, the line cut off by WA from GD is line GD, and line WA will thus end at point D, and therefore ∠TAD = ∠EGD.

[143] [Diagram V.15] Now, make the right angle GAH. Then ∠ZAH = ½∠EGD, because ∠ZAG = ½∠DGW, and therefore ∠ZAH = ½∠TAD.

But ∠ZAE = ½∠TAE, thus ∠EAH = ½∠EAD.

[144] [Diagram V.15] But if line AW || line GE, then ∠EGA = ∠GAE, therefore line AE = line EG.

The angle adjacent to ∠WAT is equal to ∠TGD, and the angle at the intersection of WA with GD is also equal to ∠TGD, because they are alternate angles, therefore line TA is equal to the line that is cut off by WA from GD.

But EA = EG, thus EA:AT is as the ratio of EG to the line cut off by WA from GD. | V 211b

And EA:AT = EG:GD, therefore the line cut off by WA from GD is line GD, thus ∠TAD = ∠EGD. But ∠ZAH = ½∠EGD, therefore ∠ZAH = ½∠TAD, and ∠ZAE = ½∠TAE.

Therefore, it is always the case that ∠EAH = ½∠EAD. And that is what we wished to prove.

⟨⟨Lemma V⟩⟩

[145] [Diagram V.16] Let circle AB with centre G be given, let diameter GB be given, and let point E outside the circle be given. We wish to draw from E a line, as EDZ [intersecting the circle in D and diameter GB in Z], such that DZ = ZG.

We join EG, and from E draw perpendicular ES to GB.

Make line KT = line ES, and on line TK construct the segment of a circle that admits an angle equal to ∠EGB, let it be segment TMK, and complete the circle. Bisect TK | at V 212a
L, and draw LM perpendicular to TK and produce it to N: line MN will then be a diameter of the circle.

Draw from K line KFC so that CF = ½GB [as in Lemma II], and join TF, then TF = FK. Draw CQ || FN, and QO || KL, then ∠CQO will be a right angle, and QF = FO, because TF = FK.

Then, since ∠CQO is right and line QF = FO, thus QF = FC, and FC = FO.

Construct ∠BGD = ∠KCQ. Draw line ED, and produce it to Z.

I say then that DZ = ZG.

[146] [Diagram V.16] Demonstration. From point D draw the perpendicular DY, and construct the right angle GDW; line DW will then meet GB, because ∠DGZ is acute, being equal to ∠OCQ; let the lines meet at point W.

Join TC, and from point Q draw the perpendicular QH. Draw TJ || CH, and produce | V 212b
HQ to meet [TJ], say at J. Draw the perpendicular TR, then TR = JH.

[147] [Diagram V.16] Then, since CF = ½GB, CO = GD. But TK = ES, thus TK:CO = ES:GD.

But GD:DY = GW:WD, and GW:WD = CO:OQ, therefore ES:DY = TK:QO = TF:FQ, thus ES:DY = TF:FQ.

But TF:FQ = JH:HQ, and JH = TR, therefore ES:DY = TR:QH. And GE:ES = CT:TR, since △GES ~ △CTR, therefore EG:DY = TC:QH.

V 213a And YD:DG = HQ:QC. |

Therefore, EG:GD = TC:CQ. But ∠EGD = ∠TCQ, thus the two triangles [EGD and TCQ] are similar, thus ∠GDZ = ∠CQF. ((But ∠DGZ = ∠QCF.))

Therefore, DZ:ZG = QF:FC. And QF = FC, thus ZD = ZG. And that is what we wished to prove.

⟪Lemma VI⟫

[148] [Diagram V.17] In triangle ABG, let angle B be right; produce AB on the side of B, and let point D be given on BG; and, further, let the ratio E:Z be known. We wish to draw from D a line, as TDK, such that TK:KG = E:Z.

Join AD, and make AD:H = E:Z.

Draw line DM || BA, thus ∠MDG will be right.

On triangle DMG circumscribe a circle with MG as diameter, and construct ∠DMC = ∠DAG. From C draw line CLN such that LN = H [as in Lemma II].

V 213b Join DKN and produce it on the side of D to T; and join GN. |

Then ∠DNG = ∠DMG = ∠BAG, and ∠NKG = ∠AKT, therefore line KT will meet line AB, say at T. Then △ATK ~ △NGK, thus TK:KG = AK:KN.

But ∠DNC = ∠DMC = ∠DAG, therefore △AKD ~ △NKL. Thus, AK:KN = AD:NL, which [ratio] is equal to E:Z. Therefore, TK:KG = E:Z. And that is what we wished to prove.

[149] [Diagram V.17] And it was shown earlier, [in Lemma II,] that there can be drawn from point C two lines in each of which the segment cut off between the circle and the diameter will be equal to the given line. Thus, if two such lines are drawn from C, then there will issue from D two lines in the given ratio; but [in this case] the two angles
V 214a produced | at G will be unequal, I mean ∠TGK and the one corresponding to it.

These lemmas having been proved, we now go back to proving what we have advanced them for.

[Finding the reflection point on the surface of a spherical convex mirror]

[150] [Diagram V.18] Let A and B be the given points whose point of reflection from the surface of the spherical convex mirror we wish to find. Let A be the centre of the eye, B the visible point, and G the centre of the sphere.

We join AG and BG; and let them be unequal. Draw the plane in which lines AG and BG lie, and let it produce in the sphere circle DW with radius GW.

Assume a straight line at random, as MN, and divide it at a point F, such that MF:FN = BG:GA. Bisect MN at O, and draw the perpendicular OC, and make ∠NCO = ½∠AGB. Produce line CO in the direction of O, and draw from F ((line FQ so that SQ:QN = BG:GW [as in Lemma VI]. Join SN. It has been shown that two lines can be drawn from F)) in this
V 214b ratio, which will produce at N two unequal angles. Let ∠SNC be the greater of the | two angles, assuming it is possible to produce in this triangle two lines in this ratio. I say that if ∠SNQ is obtuse, then it is possible for the form of point B to be reflected to the eye at A from the circumference of circle DW; and that, if this angle is not obtuse, then the form of B cannot be reflected to the eye A from the surface of this sphere.

Let ∠SNQ be obtuse. Make ∠BGD = ∠SQN, and join AD and BD. Then ΔBGD ~ ΔSQN, since BG:GD = SQ:QN. And the angles at points G and Q are equal, therefore ∠BDG is obtuse. I say then that the form of point B will be reflected to the eye A from point D.

[151] [Diagram V.18] Demonstration. We produce GD to E, and draw DZ tangent to the circle at D and continue it to T. GD will then be perpendicular to line ZT.

Make ∠GDK = ∠QNF [point K on BG], and produce DK in a straight line to Y. Then ∠DKG = ∠NFQ, which is acute. | V 215a

Draw the perpendicular BR [point on DY], make RY = RD, and join BY.

Draw DL || BY: it will meet line BG, say at point L. Then YK:KD = BK:KL = BY:DL. And BD = BY, thus BD:DL = BK:KL, therefore ∠LDK = ∠KDB.

We now make ∠LDH = ∠AGB, and produce HD in the direction of D: it will meet line GA, since angles LDH and LHD in ΔLDH are [respectively] equal to angles HGA and DHK; and, therefore, lines HD and GA will meet in the direction of A. Then the triangle marked out by line HD will be similar to ΔHDL.

And since the angles at G, D, B, K, and R are [respectively] equal to the angles at Q, N, S, F, and O, the triangles are therefore similar.

Thus, RK:KD = OF:FN, and YK:KD = MF:FN = BG:GA. | V 215b

Therefore, BD:DL = BG:GA.

Since ∠GDK = ∠QNF, and angle GDZ is right, therefore ∠ZDK = ∠NCO = ½∠AGB. But ∠LDH = ∠AGB, thus ∠ZDK = ½∠HDL. And ∠KDB = ½∠BDL, therefore, by subtraction, ∠BDZ = ½∠BDH. And since angles ZDG and ZDE are right, and ∠ZDB = ∠ZDH, thus ∠BDE = ∠HDG.

[152] We draw Ht || BD. Then ∠HtD = BDE, thus ∠HtD = ∠HDt, therefore HD = Ht. | V 216a

Thus, BD:DH = BD:Ht. But BD:Ht = BG:GH, therefore BD:DH = BG:GH.

Now BD:DL = (BD:DH) × (HD:DL), but [as before] BD:DL = BG:GA, therefore BG:GA = (BD:DH) × (HD:DL). And BD:DH = BG:GH, therefore BG:GA = (BG:GH) × (HD:DL). But [also] BG:GA = (BG:GH) × (GH:GA), therefore HD:DL = HG:((GA.

Now ΔHDL is similar to the triangle marked out by line HD as it meets line GA, therefore HD:DL is as the ratio of HG to)) the line that is cut off from line GA, thus HG:GA is the ratio of HG to the line cut off by HD from line GA.

Therefore, line HD will meet line GA at point A, so line HDA is a straight line, thus ∠ADE = ∠HDG. | V 216b

But ∠HDG = ∠BDE, therefore ∠ADE = ∠BDE. Thus, the form of point B will be reflected to the eye at A from point D. And that is what we wished to prove.

[153] [Diagram V.18] If ∠SNQ, namely the greater of the two angles produced at point N, is not obtuse, then I say that it is not possible for the form of point B to be reflected to the eye at A from the surface of the spherical convex mirror.

If possible, let it be so reflected. We draw the reflection plane to cut the sphere, thus producing in it a circle, say circle DW. The reflection point will be located on this circle; let it be point D.

We join AD, BD, and GD. Produce AD to H. Draw DZ tangent to the circle at D: it will bisect ∠BDH. ∠BDG will then be obtuse.

Make ∠HDL = ∠AGB. Then ΔHDL ~ ΔAGH, therefore HD:DL = HG:GA. But BD:DH = BG:GH, | thus (BD:DH) × (HD:DL) = (BG:GH) × (HG:GA), thus BD:DL = V 217a
BG:GA.

Make ∠ZDK = ½∠((BGA, then ∠ZDK = ∠NCO. Then, by subtraction, the complement ∠KDG = ∠CNO. But ∠ZDK = ½∠HDL, and ∠ZDB = ½∠))BDH, thus [by addition] ∠BDK = ½∠BDL, therefore BD:DL = BK:KL.

We produce DK in the direction of K and draw BY || DL: it will meet line DY [i.e., DK], say at point Y. Then YK:KD = BK:KL = BY:DL.

But BK:KL = BD:DL = BG:GA, thus BK:KL = BG:GA = MF:FN. And, therefore, BY = BD, since BD:DL = BK:KL.[8]

We bisect YD at point R. And join BR: it will be perpendicular to line DY. And produce
V 217b BR in the direction of R: it will meet line DG, | say at point J.

Then ΔDJR ~ ΔNCO, and DK:KR = NF:FO, and BKG:GD = SFQ:QN.

Thus, ∠BDG is equal to one of the two angles produced at point N.

And ∠BDG is obtuse, and, therefore, one of the two angles produced at N is obtuse.

Now ∠SNQ is the greater of the two angles produced at N, and [by assumption] it is not obtuse, which is impossible. And the impossibility results only from our assuming the reflection of the form of point B to the eye at A from the surface of the sphere DW. Therefore, the form of B cannot be reflected to the eye A from the surface of the spherical convex mirror if none of the two angles produced at N is obtuse. And that is what we wished to prove.

[154] [Diagram V.18] Nor is it possible for both of the angles produced at point N to be obtuse, because it would then follow that the form of point B would be reflected to eye
V 218a A from the circumference of circle DW | at two points.

For the two angles, namely ∠SNC and the one corresponding to it, would be unequal. As a result, if an angle equal to the other angle is constructed on line BG, then the line corresponding to GD would cut circle DW in an [additional] point other than D. And if a construction were made at that [other] point similar to what was made at D, then it would follow that the two lines drawn from A and B to that point would contain two equal angles with the diameter, and therefore, the form of B would be reflected to eye A from the circumference of circle DW at two points. But it was shown earlier that this is impossible. Therefore, if two lines can be drawn from point F in one and the same ratio, then the two angles produced at N in triangle SNC would not [both] be obtuse.

If, therefore, the greater of the two angles produced at N is obtuse, ((then the form of B will be reflected to A from a single point. If the greater of two such angles at T is not obtuse,)) then it is not possible for the form of B to be reflected to the eye at A.

From all that we have shown, it is now evident how to find the point of reflection; and
V 218b it has also been shown that this point can | only be a single point.

⟨⟨V 2.IV.18: Binocular perception of a single object in the spherical convex mirror⟩⟩

[155] Furthermore, a single object may be perceived with both eyes in the spherical convex mirror. When so perceived, the object will in most cases be perceived [to be] single. And if this situation is examined as we have shown for the plane mirror, it will be found to be as we described.

---

8. The manuscripts have BG:GA instead of BK:KL.

I say then that a single object perceived with both eyes in a spherical convex mirror will have a single image common to both eyes.

For if a single point is similarly situated with respect to both eyes, and both eyes are similarly positioned with respect to the mirror's centre, then the visible point is equally distant from the centres of the eyes, and these centres are equally distant from the mirror's centre. But, if one and the same point is equally distant from the centres of the eyes, and these centres are equally distant from the mirror's centre, then the two diameters drawn from the mirror's centre to the centres | of the eyes will be equal, and so will be the two V 219a angles contained by these diameters with the diameter drawn from the mirror's centre to the visible point, whether the eyes' centres lie with the diameter through the visible point in a single plane or in two intersecting planes.

So let the eyes' centres be in a single plane with the diameter through the visible point, and let this point be equally distant from the eyes' centres. Then the diameter through the visible point will lie between the two diameters through the centres of the eyes. The diameter through the visible point will then bisect the arc forming the intersection between the mirror's surface and the plane containing the three diameters, and this plane will be the reflection plane.

If, however, the eyes' centres lie with the diameter through the visible point in two planes, rather than in a single one, then these two planes will intersect at the diameter through the visible point, and the two arcs forming the two intersections | between these V 219b two planes and the mirror's surface will intersect at the point on the diameter through the visible point, which is the intersection of this diameter with the mirror's surface. These two arcs will then be equal; the two planes that produce these two arcs will be the planes of reflection for the two eyes; and the two points of reflection will then lie on these two arcs. Now, for either position, the diameter through the visible point is the perpendicular drawn from the visible point to the line tangent to the intersection, or [the perpendicular] to the two lines tangent to the two intersections between the mirror's surface and the reflection plane or the two reflection planes.

If, further, the visible point is equally distant from the eyes' centres, then, for either position, the distances of the points of reflection from the point of intersection between the diameter through the visible point and mirror's surface are equal. And the two arcs marked off by the reflection points next to the intersection-point | will be equal; and the V 220a arcs marked off by the reflection points next to the diameters through the eyes' centres will also be equal. Then the lines drawn from the eyes' centres to the reflection points will contain two equal angles with the diameters drawn from the mirror's centre to the reflection points. If, therefore, these two lines are rectilinearly produced to cut the arcs, they will meet the diameter through the visible point at one and the same point. And this point is the image-point for the two eyes.

Thus, if the visible point is similarly situated with respect to both eyes, and the eyes are similarly situated with respect to the mirror's centre, then the image of this point will be a single image with respect to both eyes.

[156] Moreover, the order of the parts of the image of an object perceived by reflection in a spherical convex mirror will be the same as the order of parts of the object itself.

For the images of the points on the object's surface ((will lie on diameters of the mirror. Now, let a line [segment] that lies on the surface of an object)) perceived in this | mirror V 220b cut the mirror's diameters; and imagine two lines drawn from the extremities of this line

to the mirror's centre. Then the triangle thus produced will contain the images of all the points on this line. For the image of any point on this line will lie on the diameter drawn from the mirror's centre to that point. And the diameters drawn to this line will all lie in the plane of the triangle whose base is this line and whose vertex is the mirror's centre. And, further, the order of [the points' images] in the forwards–backwards direction will be the same as the order of the points on this line, because their order will be the same as that of the diameters on which they lie. Now, only one perpendicular can be drawn from a single point to the mirror's surface, or to the plane tangent to the surface, or to the line tangent to the intersection, [and this perpendicular] is the diameter extending from that point to the mirror's centre. Therefore, the perpendiculars drawn from the points on the line [in question] are the diameters that lie in the plane of the triangle whose base is this line
V 221a and whose vertex is the mirror's centre. But it has been shown that, | for a single eye, the image of any point seen in this mirror is a single image. Thus, if the images of the points on a single line must lie on diameters of the mirror that all lie in a single plane, namely the plane of the triangle with this line as base and the mirror's centre as vertex; and if a single point can have only a single image; and if the order of the diameters on which the images lie is the same as that of the points on the line; then the order of the images of the points on the line is identical with the order of the points themselves that lie on the line, whether the eye lies in the plane of the said triangle or outside this plane. Thus, the order of parts of the object perceived in the spherical convex mirror is the same as the order of parts of the object itself.

[157] When the viewer perceives with both eyes a visible object by reflection in a
V 221b spherical convex mirror, and a point of this object | is similarly situated with respect to both eyes, then the image of this point will be a single point common to both eyes. If, on the surface of this object, a line extends in a straight line the diameter drawn from the mirror's centre to that point, then the image of that whole line will be a single image common to both eyes. For every point on this diameter will be similarly situated with respect to both eyes, since the two angles contained by the two diameters drawn from the eyes' centres to the mirror's centre with this diameter will be equal.

[158] And if, on the surface of the object, a line extends in the plane that cuts the two perpendicular planes that pass through the eyes' centres and the visible, similarly situated point in a single, common line bisecting the angle between these two planes, then the image of that line also will be a single image common to both eyes, because the diameter
V 222a drawn to any point on that line | will contain two equal angles with the diameters drawn from the eyes' centres. As for the points that lie on the surface of the object outside these two [latter] lines: each will not be equally distant from the two eyes, and the image of each of these points will be two points for the two eyes, and not one point. But the position of any two lines drawn from the eyes and reflected to one of those points will be similar in position with respect to both eyes and to both axes of the eyes, while their distances from the axes will be either the same or not excessively different. For the order of parts of the object's image will be the same as that of the parts of the object itself. And [this being] so, then the order of lines extending from each of the eyes' centres to parts of the image will be the same as that of the parts of the object itself. Therefore, the lines extending from the centres of the eyes to the two images of the single point that is displaced from the similarly
V 222b situated point will be | displaced from the similarly situated lines extending from the eyes' centres to the common image of the similarly situated point to one and the same side of

the eyes' axes and in the same direction as the displacement of the displaced point. For the image of the similarly situated point will lie on the axes common to both eyes. And if the displacement of these two lines from the axes is in a similar direction, and their distances from the axes are not excessively different, then these two lines will be similarly situated with respect to the eyes and to the eyes' axes, or the difference between their positions will not be appreciable.

[159] Now, these two lines are as the lines drawn from the eyes to a point of an object perceived rectilinearly, when perceived with both eyes, and when there meet on it [the object] the eyes' axes or two similarly situated lines with respect to the eyes. For, in the case of an object perceived rectilinearly with both eyes, while the eyes' axes meet at a point on it, there | will meet at every one of the other points on it two of the radial lines V 223a
drawn from the eyes. But not every two lines meeting at a point of the object perceived rectilinearly are perfectly similar in position, but rather their position[s] will be in a similar direction with respect to the eyes' axes, while they may or may not be equally distant from the axes. For the distances of these two lines from the axes are measured by the two angles contained between the axes and these two lines. But the axes do not contain equal angles with every pair of these lines. Rather, the two angles contained between these two lines and the axes may differ. But a single point of an object perceived rectilinearly with both eyes is not seen as two points [i.e., double] when the axes or two similarly situated rays meet at a point on the object; rather it is the case that every point of an object of this description is seen as a single point, despite the difference | between the distances of the V 223b
lines meeting at it from the [respective] axes. The reason is that the difference between the distances from the axes is small, and a small difference has no sensible effect. This notion we have clearly explained in Book III,[9] in the course of our explanation of the notions examined by means of the board.

[160] Likewise, the lines drawn from the eyes' centres to the two images of the point perceived by reflection with both eyes are either similarly situated with respect to both eyes, or the difference between their positions is not such as to produce a sensible effect. Thus, the two lines drawn to any two points marking the images of a point perceived by reflection in the spherical convex mirror with respect to both eyes will either be similarly situated with respect to both eyes, or the difference between their positions will be not such as to produce a sensible effect. That being so, the two images of an object of this description will be for the two eyes two overlapping images, | so that some parts of these V 224a
two images will be common to them, while the other parts will be such that every two points of them forming the two images of a single point will lie on[10] a single diameter of the mirror and displaced to one and the same side from the common points whose position is perfectly similar.[11] The two lines extending from the eyes' centres to those two points will be similarly situated with respect to the eyes and to the eyes' axes, or the difference between their positions will be not such as to produce a sensible effect. The two reflected forms that extend along these two lines to the eyes will thus occur at two similarly situated

9. Chapter 2, paragraphs [25–85], esp. [41] and [49]. See Sabra, *The Optics of Ibn al-Haytham. Books I–III*, I, pp. 240 and 242.

10. From here there is a lacuna in MS K until paragraph [279] at V 297a.

11. That is, the same in direction and distance.

points of the eyes, as far as appears to sense. These two forms will then occur at a single point in the common nerve. And the forms of the two overlapping images will occur as a single form in the common nerve where the sensation takes place. Thus, the form perceived by the eye at the place of the two overlapping images will be a single form.
V 224b | And this form is the image of the object perceived by reflection. Therefore, the object perceived with both eyes by reflection in the spherical convex mirror will have a similar, single image for both eyes, if a point of the object is similarly situated with respect to both eyes.

[161] Also, if the object perceived with both eyes in the spherical convex mirror does not have a perfectly similar position with respect to both eyes, then the eyes' axes may meet at a point of its image. For if a point of the object has a perfectly similar situation with respect to both eyes, then the image of this point will lie on the common axis; and the two radial lines reflected to this point will be the eyes' axes; and the axes will meet at the image-point. But not every point at which the axes of the two eyes meet lies on the common axis; rather, the axes of the eyes may always meet at any one of the points
V 225a that surround the common axis, | provided the point is not excessively remote from the common axis; and most objects seen rectilinearly are of this description. This notion was explained in Book III.[12]

[162] It was also shown in Book III that when the viewer faces a visible object, the two axes of the eyes will meet at a point of that object. And when the viewer fixes the gaze of both eyes together on the object, then the eyes' axes will meet at a point of that object, whether the object lies on the common axis or outside it. Now, the image, i.e., the form of the object perceived by reflection, faces the eye, at the extremities of the rectilinearly extending radial lines. Thus, when the viewer contemplates this image, nay, if he [just] looks at it so as to perceive it distinctly, the axes will [be found to] meet at a point of this form, and these axes will have been those reflected from the mirror's surface
V 225b to the point of the object | whose image is that point, whether the object lies on or outside the common axis. But if the two axes meet at a point of the object, then the two forms of that point will occur in two similarly situated points in the eyes, and they will occur at a single point in the common nerve in which the faculty of sense perceives the form of the object. Thus, all the points that surround this point will be equal in condition to the point on the common axis. And the lines drawn from the eyes' centres to the points surrounding this point will be similar in position to the lines drawn from those centres to the points surrounding the point on the common axis. That being so, then the image of such an
V 226a object, when perceived with both eyes, will also have a single form in the | common nerve. And if the form of the image perceived with both eyes occurs in the common nerve as a single form, the image perceived at once with both eyes, will be single. Therefore, every object perceived by reflection in spherical convex mirrors will always have a single image, whether it is perceived with one eye or with both eyes at once. And that is what we wished to prove.

---

12. For the references to Book III here and in paragraph [162], see Chapter 2, paragraphs [1–10]; and Sabra, *The Optics of Ibn al-Haytham. Books I–III*, I, pp. 228–32.

⟪V 2: Fifth Inquiry. On the Images of Cylindrical Convex Mirrors, Consisting of an Introduction and Seven Propositions⟫

⟪Introduction⟫

[163] With regard to the right-cylindrical convex mirror: the intersections of its surface with the planes in which the forms are reflected differ, as was shown earlier. Some of these [intersections] are straight lines, some are circles, and some are [conic] sections. As for the points whose forms are reflected from this mirror in the planes that intersect [with the mirror's surface] in straight lines: their images are the same as those of the plane mirror, I mean that the lines along which the forms [of these points] are reflected | will meet the V 226b perpendiculars drawn from them, and all their images will be located behind the mirror, at distances from the mirror's surface equal to the distances of the visible points from that surface. These intersections are the intersections between the mirror's surface and the plane passing through the mirror's axis and the centre of vision. As for the points whose forms are reflected in the planes whose intersections are circles, their images will be the same as those for the circles on the spherical convex mirrors; I mean that their reflection lines will meet the perpendiculars [i.e., diameters] drawn from [those points], and of their images some will lie behind the mirror, some in front of the mirror, and some on the surface of the mirror. As for the points whose forms are reflected from the mirror's surface in planes whose intersections are [conic] sections, I say that for these points also, the lines of reflection of their forms will meet the perpendiculars [i.e., diameters] drawn from them, and that of the images of these points some will lie | behind the mirror, some in front of the V 227a mirror, and some on the surface of the mirror. Let us show the locations of these images also by demonstration.

⟪V 2.V.1⟫

[164] [Diagram V.19] Let line ABG be one of the [conic] sections in this mirror. Let point D, the centre of the eye, be in the plane of this [conic] section, and let point E, on a visible object, be also in the plane of this [conic] section. Let the form of point E be reflected to the eye at D from point B.

We draw lines DB and BE. Let BT be the perpendicular to the plane tangent to the mirror's surface at the reflection point; and let EZ be the perpendicular from E to a line tangent to this [conic] section, such as line LZM. Line ZM will thus fall inside figure BZE, which is contained by lines BE and EZ and the periphery of the [conic] section. Line ZM will thus cut line BE, say at point M.

Let line NBO be the intersection between the plane | of the [conic] section and the V 227b plane tangent to the mirror's surface at point B. Line TB will then be perpendicular to NBO, ⟨and line⟩ BO will fall inside figure BMZ, which is contained by lines BM and MZ and the periphery of the [conic] section. Thus, line BO will cut line ZM and line ZE; let it [BO] cut ZM in O and ZE in C. Points O and C will then lie outside [conic] section ABG; but points B and Z are on the periphery of the [conic] section, and therefore point Z will be behind line BC with respect to point T, and B will be behind ZM with respect to E.

Join BZ. Then BOZ will be a triangle, and angles TBZ and BZE will be obtuse. If, therefore, lines TB and EZ are rectilinearly produced, they will meet behind the [conic] section ABG; let them meet at K.

Now line DB cuts ∠KBE, because it cuts lines KB and EB [at B]; so DB cuts EK, say at point H. H will then be the image of point E.

V 228a But line BO cuts | ∠EBZ; so it cuts EZ between points E and Z. Point C will thus be located between E and Z.

Therefore, since ∠EBT = ∠DBT, thus ∠EBT = ∠KBH. The angles CBT and CBK are right, thus, by subtraction, ∠EBC = ∠HBC. Therefore, EB:BH = EC:CH.

Draw HF || EB; then ∠EBT = ∠HFB, therefore ∠HBF = ∠HFB, and line HF = line HB, thus EB:BH = EB:FH. And FH || EB, therefore EK:KH = EB:BH. But EB:BH = EC:CH, thus EK:KH = EC:CH. ⟪And KH will be longer than HB⟫, which is equal to HF,
V 228b since | ⟪∠HFK, which is equal to ∠EBK⟫, is obtuse.

⟪V 2.V.2⟫

[165] [Diagram V.19] And, similarly, for every image that lies in one of the planes of the [conic] sections that occur in this mirror, the ratio of the perpendicular [i.e., EK] drawn from the visible point to the line tangent to the [conic] section between the visible point and the point at which this perpendicular meets the perpendicular to the plane tangent to the mirror's surface at the reflection point, to the line [i.e., KH] between the meeting-point and the image-point, will be as the ratio [i.e., EC:CH] between the segments of the line between the visible point and the image-point, which segments are set apart by the line tangent to the mirror's surface at the reflection point.

⟪V 2.V.3⟫

[166] [Diagram V.19] And the line [KH] between the meeting-point and the image-point will be longer than that [HB] between the image-point and the reflection point, that is, if the visible point lies outside the perpendicular drawn from the centre of the eye to the plane tangent to the mirror's surface. And that is what we wished to prove.

⟪V 2.V.4⟫

[167] And it is clear from this proof that every two perpendiculars on two lines tangent
V 229a to the periphery of one of | these [conic] sections at two points will meet behind the segment of the periphery on which these two points lie. As for the perpendicular from the eye's centre to the plane tangent to the mirror's surface, the eye will not perceive any of the points on it, except that point alone on the eye's surface through which this perpendicular passes. For the eye does not perceive the form of every point on this perpendicular, as has been previously shown ⟪for the spherical mirror⟫. And ⟨the forms [of any point on this perpendicular]⟩ that extend along [a line] other than this perpendicular will not be reflected ⟨to⟩ the eye. For if two lines are drawn from any point on this perpendicular and from the eye's centre to a point on the mirror's surface, the perpendicular to this point will not divide the angle contained by those two lines. Because, if it did, then it would meet that perpendicular that passes through the eye's centre between ⟨this⟩ point and the

eye's centre. And this perpendicular contains a | right angle with the mirror's axis, while the other perpendicular from the eye's centre also contains a right angle with the axis, thus giving rise to two right angles in a triangle, which is impossible. The eye does not, therefore, perceive any point on this perpendicular other than the one on the eye's surface through which this perpendicular passes. V 229b

⟪V 2.V.5⟫

[168] [Diagram V.20] That having been made clear, I say that images produced by these [conic] sections are [located as follows]: some are behind the mirror, some on the surface of the mirror, and some in front of the mirror.

Let ABG be [points] on a [conic] section in this mirror; and let AD be one of the lines drawn in the plane of this [conic] section perpendicularly to the plane tangent to the cylinder's surface. Draw from point A the intersection between the plane of this [conic] section and the tangent plane, say EAZ. Then EAZ will be tangent to the [conic] section, and ∠DAZ will be right.

Assume on the periphery of the [conic] section a point at random other than | that at V 230a
which a tangent line is parallel to DA, which [latter point] can only be a single point on one side of the [conic] section; and let G be the assumed point. From G draw GZ parallel to DA. Then GZ will meet AZ, say at Z, and will cut the [conic] section.

From G we draw a line tangent to the [conic] section, as has been shown in Book II of Apollonius's *Conics*:[13] ⟨this tangent line will then⟩ cut GZ, since GZ cuts the [conic] section. And since ⟨the tangent⟩ cuts GZ, it will then meet DA outside the [conic] section, say at H. Join GA and produce it in a straight line to T.

From G draw GK perpendicular to GH and produce it in the direction of G: it will meet line AD, as was shown in the preceding diagram, say at D. On GK assume a point at random, say point K, and join | KA. V 230b

[169] [Figure V.20(1)] Now, ∠KAH is either equal to, or greater or smaller than ∠TAH; and, first, let them be equal.

If ∠KAH = ∠TAH, then the form of point K will be reflected to point T from point A, and its image will be point G, which is [located] on the mirror's surface.

Extend AZ rectilinearly to meet KG, say at point t. On Kt assume a point at random between points K and t; let it be S. Make ∠HAM = ∠SAH. Produce MA in the direction of A: it will cut tG, say at point O. Then the form of point S will be reflected to point M from A, and its image will be point O, which is [situated] in front of the mirror's surface.

Produce GK in the direction of K and assume on it a point at random, say F; join FA.
Make | ∠HAN = ∠FAH. Produce NA in the direction of A: it will cut GD, say at Q. The V 231a
form of point F will therefore be reflected to point N from point A, and its image will be point Q, which is [located] behind the mirror.

[170] [Figure V.20(2)] If ∠KAH < ∠TAH, we make ∠HAS = ∠TAH, which will take this case back to the previous one. Then point G will be the image of point S for the eye at T.

Join FA.[14] Make ∠HAN = ∠FAH. Then point Q will be the image of point F for the eye at N.

---

13. Actually Book I, proposition 35.

14. Point F is on DS behind S.

If we cut off from ∠SAt an angle equal to ∠TAM, there will be produced on line St a point [in Diagram V.20(2)][15] whose image will be point O.

[171] [Diagram V.20(3)] If ∠KAH > ∠HAT, we make ∠HAM = ∠KAH, and produce line MA in the direction of A to cut the periphery of the [conic] section between points G
V 231b and A, say in point B; and it will cut GK, say in point O. Thus, if from B we draw | a line parallel to AD, it will cut the [conic] section and the tangent line AZ. If we then draw from B a line tangent to the [conic] section, it will cut line AH, say at point L. ∠BAD will be acute since the angle at A, ⟨i.e., ∠ZAD⟩, is right.

Now line BL cuts the tangent line GH, containing with it an acute angle adjacent to point H, ⟨because ∠BLH is obtuse⟩, ⟨⟨being greater than ∠BAL which is obtuse, because ∠ZAL is right⟩⟩. Thus, if the perpendicular ⟨to B⟩ is rectilinearly produced it will contain with GH an acute angle adjacent to point G.

Now KA cuts GH, say in point W. ∠KWG will then be acute, since ∠KGH is right. The perpendicular from B will therefore meet KA, say at point C. Let us produce the perpendicular in the direction of B: it will meet line AD, as was shown in the [previous] diagram, say in point J.

Then the form of C will be reflected to M from A, and its image will be point B, which
V 232a [lies] on the mirror's surface. And the form of point K will be reflected to M from | A, and its image will be point O, which is in front of the mirror.

We produce BC rectilinearly in the direction of C, and mark on it a point at random, say point Y. We join YA, and make ∠HAN = ∠YAH, and produce NA in the direction of A, it will cut BJ, say at X. Then the form of point Y will be reflected to N from A, and its image will be point X, which [lies] behind the mirror.

[172] It is thus clear from what we have shown that the images produced by [conic] sections in right-cylindrical convex mirrors will be [located as follows]: some behind the mirror, some on the mirror's surface, and some in front of the mirror. And that is what we wished to prove. Moreover, the images of these mirrors that lie in the plane passing through the whole axis will all be behind the mirror; as for the images that lie in a plane
V 232b parallel | to the mirror's base and passing through the eye's centre, i.e., a plane whose intersection [with the mirror's surface] is a circle, some of them will [lie] behind the mirror, some on the mirror's surface, and some in front of the mirror, as was shown for the images [produced] by circles.

[173] All images of these mirrors are perceived by the eye [to be] behind the mirror, because they are all behind the reflection point on the mirror's surface. As for those [images that lie] on the mirror's surface, the eye cannot perceive their position on that surface; and as for those [that lie] outside the mirror, the eye cannot perceive the points below them on the mirror's surface, because these positions on the mirror's surface that lie below the images or which are the same as the images' positions all lie in the portion of the mirror's surface that is concealed from the eye. The eye will therefore perceive the reflection points behind which the images are [located], but not the positions of the images that are [located] on the mirror's surface, nor the position[s] [on the surface] that lie below the images [located] outside the mirror. Thus, the images of right-cylindrical
V 233a convex mirrors are exactly [located]: | some behind the mirror, some on the mirror's

15. Indicated by the label L_1, added by the translator.

surface, and some in front of the mirror, but, to sense, all are [perceived to be] behind the mirror.

⟪V 2.V.6⟫

[174] [Diagram V.20+] All that having been proved, I say that every point perceived by reflection in the right-cylindrical convex mirror can only have a single image, and the form of [such a point] can only be reflected to one and the same eye from a single point on the surface of a given mirror.

Let A be the centre of the eye; let point B be in a visible object, and point G in a right-cylindrical convex mirror. Let the eye A perceive point B by reflection in mirror G, and let G be the reflection point; and we join AB.

[First position]

[175] If both the centre of the eye and point B lie in a plane passing through the whole axis of the mirror, then line AB will lie in the plane passing through the whole axis; and point G will be on the straight line that is the intersection between the surface of the mirror and the plane passing through the whole axis. | Then the form of point B will only be V 233b
reflected to the eye A in the plane passing through the whole axis. For any plane in which the form of a point is reflected to the eye at A is one in which all [the following] points lie: point A, point B, the reflection point, and the point on the axis reached by the perpendicular drawn from the reflection point. The perpendicular drawn from the reflection point will meet line AB and the axis, and therefore it lies in the plane passing through the whole axis, since both of its extremities, of which one is on line AB and the other on the axis, are located in the plane passing through the axis. Therefore, if line AB lies in a plane passing through the whole axis, the form of point B can then only be reflected to the eye A in the plane passing through the whole axis. And [this being] so, the ⟨reflection⟩ can then only take place [at a point] on the straight line forming the intersection between the surface of the mirror | and the plane through the whole axis. Now, this straight line is a unique line, V 234a
because the plane through the whole axis and through line AB can only cut the portion of the mirror's surface facing the eye in a unique line; and, therefore, the form of point B can be reflected to the eye A from only one of the lines that extend along the length of the mirror. And, as was shown for the plane mirror, the reflection of lines [to one another] between two points, and at equal angles, from a straight line, can only take place at a single point. Therefore, if point B and eye A lie together in a plane passing through the whole axis of a ⟨right-⟩cylindrical ⟨convex⟩ mirror, the form of [that point] will be reflected to eye A from only one point.

[Second position]

[176] If, however, line AB lies in a plane ⟪parallel to the mirror's base⟫, then the form of point B will be reflected to the eye A only in the plane parallel to the base. For if the form [of B] were reflected in a plane other than this [parallel] plane, then the perpendicular drawn from the reflection point | would meet the mirror's axis, and would be perpendicular V 234b
to it, because it would be perpendicular to the line extending along the length of the mirror

and passing through the reflection point. Now, this perpendicular meets line AB, because it bisects the angle of reflection. Thus, if, from the point at which this perpendicular meets line AB, a line is drawn to the centre of the circle whose plane passes through the eye's centre, then it would contain a right angle with the axis, because the axis is perpendicular to the plane of the circle and passes through its centre, and the perpendicular that meets line AB also contains a right angle with the axis, and thus two angles in a triangle would be right, which is impossible. Therefore, if line AB lies with eye A in a plane parallel to the mirror's base, the form of point B will not be reflected to eye A in a plane other than the plane parallel to the base containing line AB. And if the form of point B is not reflected to eye A except in the plane parallel to the base, then the form will not be reflected except
V 235a from the circumference of the circle produced by this plane | in the cylinder's surface. Now, only a single plane can be drawn from the centre of the eye parallel to the base of the cylinder. Thus, if point B lies with eye A in a plane parallel to the mirror's base, then the form of [B] can be reflected to eye A only from the circumference of the circle produced by this plane in the cylinder's surface. If, therefore, point B has been reflected to eye A from point G, then point G is located on the circumference produced by this plane in the cylinder's surface. And it was previously shown that reflection between two points, at equal angles, from the convex [side] of a circle can only take place from a single point. If, therefore, point B lies with eye A in a plane parallel to the mirror's base, then the form of [B] can only be reflected to eye A from only a single point.

[Third position]

[177] [Diagram V.20+] If the reflection of form B to eye A takes place in one of the planes whose intersections are [conic] sections, then point G will be located on the
V 235b periphery of the [conic] section produced by that | plane in the mirror's surface.

I say then that the form of point B cannot be reflected to eye A from a point other than point G.

If possible, let the form also be reflected from point D. Point D cannot then be located on the periphery of the [conic] section passing through point G, because no form is reflected from the periphery of one [conic] section on the surface of the cylindrical convex mirror, except from a single point ⟨⟨as was shown in Chapter 5 of Book IV⟩⟩. If, therefore, the form of point B has been reflected from point G, there is then no other point on the periphery of the [conic] section from which the form of G can be reflected to eye A.

[178] [Diagram V.20+] Let point D, then, lie in a reflection plane other than the plane in which point G is located; this other plane will not pass through the whole axis, nor will it be parallel to the mirror's base. For if it passed through the whole axis or parallel to the base, then it would not be possible for the form of point B to be reflected to eye A at point
V 236a G, which lies neither in the plane through | the whole axis nor in the plane parallel to the base. And thus the plane containing point D neither passes through the whole axis nor is it parallel to the base. Therefore, the intersection of this plane, in which point D lies, with the surface of the mirror, will also be a [conic] section.

[179] [Diagram V.20+] Let us draw from the centre of the eye a plane parallel to the mirror's base: it will produce a circle in the mirror; let it be circle EZ.

From point G draw a line extending along the mirror's length: it will meet the circumference of mirror EZ, say at point Z. Let TK be the axis of the mirror. Let GT be the perpendicular drawn from point G to the tangent plane, and continue it in a straight line: it will meet line AB, say at point N.

Now the centre of circle EZ lies on the axis TK; let the centre be point K. Join ZK. Draw from point B a perpendicular to the plane of the circle, I mean the plane parallel to the mirror's base, which contains the circle; let [the perpendicular] be BH. Join HZ, ZA, and AH; these three lines will lie in the plane of the circle. | V 236b

From point A draw a line in the plane of triangle ABG parallel to perpendicular GT; let it be AM. Produce line BG in a straight line on the side of G: it will meet line AM, say at point M. Join MZ.

Then, since BH is perpendicular to the plane of the circle, it will be parallel to line GZ, for GZ is perpendicular to that plane, because it is parallel to the axis; then lines BH and GZ are in one plane, and line BGM is in their plane. Then points G, Z, and M are in the plane containing lines HB and BM.

But lines TK and GZ are parallel, so they are in one plane; and lines GT and ZK are in the plane of these two lines; and GT and ZK are [respectively] in the planes of the parallel circles passing through points G and Z. Therefore, lines GT and ZK are parallel to each other.

Now line GT is parallel to AM, therefore ZK is parallel to AM, and they are therefore in one plane. And AZ is in the plane of these two lines; line AM is thus in the plane containing lines ZK and ZA. And the plane containing ZK and ZA is the plane of | circle V 237a EZ, ⟪therefore AM is in the plane of circle EZ⟫. And points A and B lie in the plane of circle EZ, and points Z and H are in this plane, ⟪therefore the three points H, Z, and M are in the plane of circle EZ⟫. The three points are thus in the plane containing lines HB and ZM, and points H, Z, and M lie on the intersection of the plane of the circle with the plane orthogonal to it that contains lines BH and ZM. Therefore, HZM is a straight line, and triangle HMA is in one plane and is in the plane of the circle.

Line ZK is in the plane of triangle HMA, and ZK cuts side MH; thus, if ZK is produced in a straight line, it will also cut side AH, say in point L.

Now since AM is parallel to NG, angles GAM and GMA are therefore [respectively] equal to angles AGN and BGN. And these [latter] angles are equal; therefore angles GAM and GMA are equal, and thus lines AG and MG are equal.

Now line GZ is perpendicular to the plane of circle EZ, therefore each of angles AZG and MZG is right. Therefore the square of AG is equal to [the sum of] the squares of AZ and | ZG, and the square of MG ⟪is equal to the [sum of the] squares of MZ and ZG⟫. And V 237b lines ⟪AG and MG⟫ are equal, therefore lines AZ and MZ are equal, thus angles ZAM and ZMA are equal. But AM || ZK [= ZL], therefore angles AZL and HZL are equal; and line LZK is a diameter for circle ZE, therefore lines AZ and HZ are reflected [to each other] at equal angles from the circumference of circle EZ.

[Third position, first assumption]

[180] Now, point D either lies on line GZ or outside it; let it first be on line GZ.

We draw from D a perpendicular to the tangent plane, namely a diameter of the circle passing through point D: it will meet axis TK at right angles to it; let it be perpendicular DO. Then perpendicular DO is parallel to perpendiculars GT and ZK, and thus parallel to line AM; [the two perpendiculars] are then in one plane.

We join ⟨⟨BD and DA⟩⟩. Line OD is then in plane BDA, because it is the reflection plane; and it will cut line AB, say in point F. Thus, lines AM and BD are in one plane that contains line BF.

V 238a Now line BD lies in the plane that contains | lines DF and BF; thus, if line BD is produced in a straight line, it will meet line AM, but not at point M: because if it did, it would then contain an area with the straight line BGM, which is impossible. Thus, [BD] will meet [AM] at a point other than point M, say at point Q. We join QZ. Then it is shown, as was shown for line MZH, that line QZH is continuous as a straight line.

But point Q is outside line MZH, because line AM can meet line MZH only at point M; and if point Q lies outside line MZH, then line QZ will cut line MZH; and, thus, line QZ will cut line MZH, with which it is continuous in a straight line, which is impossible.

Therefore, point D does not lie on line GZ.

[Third position, second assumption]

[181] Let point D lie outside line GZ.

We draw from point D a line extending along the length of the cylinder: it will meet the circumference of circle EZ at a point other than Z, say at point C. We join CA and CH.

V 238b It is clear, as was shown for point Z, that lines | HC and AC are reflected [to each other] from the circumference of circle EZC at equal angles; and thus, between two [given] points, lines would be reflected from two points on the circumference of circle EZC at equal angles, which is impossible.

Thus, the form of point B is not reflected to eye A from a point outside line GZ, nor from a point on line GZ other than point G, nor from the circumference of a circle, nor from a point in the plane passing through the whole axis. Therefore, the form of point B cannot be reflected to the eye at A, when reflected from point G, except from G.

Now point G was [assumed] to be with line AB in the plane passing through the whole axis, or in the plane parallel to the base, or with line AB in one of the planes whose intersections are [conic] sections. And if the form of point B can be reflected to the eye A from only one point, then it can only have a single image.

And the same can be shown for every point perceived by the eye in this mirror, if [the point] lies outside the perpendicular drawn from the centre of the eye at right angles to the
V 239a | plane tangent to the surface of the mirror.

[182] The point on the ⟨surface⟩ of the eye through which the perpendicular drawn from the centre of the eye passes, namely the only point on ⟨this⟩ perpendicular that is perceived by sight, will have only a single image; for the form of the surface of the eye, surrounding that point, can only cut the perpendicular drawn from the centre of the eye ⟨and⟩ passing through that point at a single point [of intersection].

Therefore, every point perceived by the eye in a right-cylindrical convex mirror can have only one image. And that is what we wished to prove.

⟨⟨V 2.V.7⟩⟩

[183] [Diagram V.20+] As to how we find the point of reflection when two points are given, that can be done by the reverse of the method we followed in the demonstration of this [foregoing] proposition.

That is, we draw from one of the two points, say A, a plane parallel to the cylinder's base: it will produce a circle in the cylinder; let it be circle | EZ. And we draw from the other point, let it be point B, a perpendicular to the extended plane of the circle; let [the perpendicular] be BH. V 239b

We then find the point on the circumference of circle EZ from which the form of point H is reflected to point A, in the way shown for the spherical convex mirror; let it be point Z.

We join HZ and AZ and draw perpendicular ZK and continue it to L. Draw from point A line AM parallel to this perpendicular, [and] produce HZ in a straight line until it meets this line, say at point M; lines AZ and ZM are then equal. Draw from point Z a line extending along the length of the cylinder; let it be ZG. This line is then parallel to the axis of the cylinder, and it is parallel therefore to line BH. And line HZM passes through them; therefore, it lies in their plane, and therefore, point M lies in this plane.

Join MB: it will cut ZG, say in point G. Join AG. And draw line GT perpendicular to the axis, and continue it to N. Lines AG and GM are then equal, and angles AGN and BGN will therefore be | equal. V 240a

Therefore, the form of point B will be reflected to point A from the surface of the cylindrical convex mirror at point G. And that is what we wished to prove.

⟪V 2: Sixth Inquiry. On the Images of Right-Conical Convex Mirrors, Consisting of an Introduction and Three Propositions⟫

⟪Introduction⟫

[184] With regard to the right-conical convex mirror, its sections also differ: some are straight lines, and some are [conic] sections. No forms of visible objects can be reflected in the planes of its circles ⟪as was shown in Book IV⟫, because these planes are not orthogonal to the planes tangent to the cone. With regard to the planes in which the forms are reflected from this mirror, namely those whose sections are straight lines: their images will be like the images of the plane mirror, in that their reflection lines will meet their perpendiculars, and the meeting-points, i.e., the images, will lie behind the mirror. As for the sections of this mirror that are [conic] sections: their conditions are like those of the sections in the cylindrical convex mirrors – their reflection lines will meet their perpendiculars, and their images will be [located as follows]: some behind the mirror, some in front of the mirror, and | some on the surface of the mirror. V 240b

[185] To show this for the [conic] sections in this mirror is the same as showing it for the [conic] sections of the cylinder. Thus, it will be clear, as it was for the [conic] sections of the cylinder, that every line perpendicular to a line tangent to the [conic] section will meet the perpendicular drawn from the reflection point, and will meet the line drawn from the centre of the eye to the reflection point. With regard to these [conic] sections, if we follow ⟨the method⟩ we followed for the [conic] sections of the cylinder, it becomes clear that the images of these [conic] sections are [located as follows]: some behind the mirror, some in front of the mirror, and some on the surface of the mirror.

[186] With regard to the perpendicular drawn from the centre of the eye at right angles to the plane tangent to the surface of the mirror: the eye will not perceive any point on it,

except the point on the surface of the eye through which that perpendicular passes; and ⟨it⟩ will perceive this point from the point's form that extends along this perpendicular. ⟨No other point on this perpendicular⟩ will be perceived by the eye, as has been shown
V 241a previously. And no form of any of these points will be reflected from any point on | the mirror's surface other than the point at the foot of this perpendicular. For this perpendicular meets the axis of the cone, and therefore it lies in the plane passing through the whole axis. ⟨And, for any other point⟩ on this perpendicular, if two lines are drawn – one from [that point], and another from the centre of the eye – to a point on the surface of the mirror outside the plane passing through the whole axis, then the perpendicular drawn from that point will not divide the angle contained by those two lines, because, if it did, it would meet the perpendicular drawn from the eye's centre, and thus that perpendicular would lie in the plane passing through the whole axis. And if that perpendicular were in that plane, then the reflection point, which is assumed to be located outside that plane, would lie in it, which is impossible. That perpendicular, therefore, will not divide the angle contained by the two lines drawn from the centre of the eye and from the visible point on the perpendicular. Thus, if two lines are drawn from that point on the perpendicular and from the eye's centre to a point on the mirror's surface, among those points that lie
V 241b in the plane | passing through the whole axis, then that point will be on the straight line that is the intersection between the mirror's surface and the plane through the whole axis. Further, the perpendicular drawn from that point will be parallel to the perpendicular from the centre of the eye, because each of them will contain a right angle with the straight line forming the intersection. And if that perpendicular is parallel to the perpendicular drawn from the centre of the eye, then that perpendicular will not divide the angle contained by the two lines drawn to that point. And if the perpendicular does not divide this angle, the form will not be reflected from that point. The eye will therefore not perceive any point on the line or perpendicular drawn from the centre of the eye at right angles to the plane tangent to this mirror's surface other than that point alone on the surface of the eye that is among the [possible] objects of vision whose images have been shown in detail. And the eye will perceive all these images [as lying] behind the mirror, because all of them will
V 242a be [located] behind the reflection point | on the mirror's surface with respect to the eye['s location], as in the case of the images of the cylindrical convex and spherical convex mirrors.

⟨⟨V 2.VI.1⟩⟩

[187] [Diagram V.21] I say, further, that any point that the eye perceives by reflection in the right-conical convex mirror can ⟨have⟩ only one image, and its form can be reflected from only one point.

Let A be the centre of the eye, B a point in a visible object, and G a point on a right-conical convex mirror. Let the eye A perceive point B in mirror G, and let point G be the point of reflection. I say that the form of point B cannot be reflected to the eye A from a point other than G.

We draw line AB, and from G draw a plane parallel to the base of the cone: it will produce a circle in the cone; let it be circle DGZ. Let E be the vertex of the cone and EH its axis, and let HGQ be the perpendicular drawn from point G at right angles with the plane tangent to the cone.

We join EG. From | point A draw a line parallel to line EG, let it be AN: it will meet the plane of the circle, say at point N. And from point B draw a line parallel to line EG, let it be BM: it will meet the plane of the circle, say at point M. And join MG and NG. V 242b

Let T be the centre of the circle. Join TG, and produce it to K. From point A draw line AL || GH; and from point N we draw a line parallel to line GT; ⟨let it be NF⟩. The plane containing lines AN and NF will then be parallel to plane EGH ⟪I say, because GT also lies in plane EGH⟫.

We produce BG in a straight line in the direction of G: it will meet line AL, say at point L. And we produce MG in the direction of G: it will meet line NF, say at point F. And we join LF.

Then, since line BM || EG, they are both in the same plane, and lines BG and MG are in the same plane with them. Therefore, points L and F are in plane EGBM. Now, this plane cuts plane EGH, with EG as their intersection; therefore, the plane containing lines BM and EG cuts the plane containing lines | AL and NF, and FL will be the intersection between the two planes. Line FL will therefore be parallel to line EG, and thus parallel to ⟨line⟩ AN. V 243a

From point G we draw the intersection between circle DGZ and the plane tangent to the cone and passing through line EG, to which HG is perpendicular; let it be GS. This line will then cut line NF and be perpendicular to it, because it cuts line GT, which is parallel to it [NF]. And it [GS] is perpendicular to it [NF], because it is tangent to the circle; let it cut it in point S.

Also from point G we draw the intersection between the reflection plane and the plane tangent to the cone and passing through line GE; let [the intersection] be GC. This line will cut AL and be perpendicular to it, because it cuts line GH and is perpendicular to it; let [AL] be cut in point C. We join CS: it will be the intersection between the tangent plane and the plane containing lines AL and NF.

Since points C and S are in the plane tangent to the cone and passing through line GE, | and both are in the plane containing lines AL and NF, and [since] the tangent plane cuts plane EGH, which is parallel to the plane containing lines AL and NF, line CS will thus be parallel to line EG. Therefore, line CS is parallel to each of lines AN and FL, which are parallel to each other, and the three lines will be in the same plane. V 243b

Therefore, AC:CL = NS:SF. And therefore, since ∠AGQ = ∠BGQ, and GQ || AL, ∠GAL = ∠GLA, and thus line AG = line GL.

Now line GC is perpendicular to line AL, thus line AC = line CL.

And AC:CL = NS:SF, therefore line NS = line SF.

Line GS has been shown to be perpendicular to line NF;

thus line NG = line GF, therefore ∠GNF = ∠GFN and ∠MGK = ∠NGK;

thus lines MG and GN are reflected [to each other] at equal angles.

[188] [Diagram V.21] We may similarly prove [the same result] if points A and B are on either side of the plane of circle DGZ, since the method | of construction is one and the same. But if one of the two points [A, B] lies in the plane of circle DGZ, then that point can be considered as one of points M and N. We then draw from that point two lines, one parallel to line GH and the other parallel to line GK, and the proof will be completed as previously shown. V 244a

[189] [Diagram V.21] If, however, the form of point B is reflected to the eye A from a point other than point G, then that point will either be on line EG or outside it.

If that point lies on line EG, then the perpendicular drawn from the reflection point will be parallel to line AL, because it is parallel to line GH. Therefore, line AL lies in the reflection plane, and the two reflection planes will intersect in line AL. And yet they meet at point B, which lies outside line AL – because point B lies on line BM, which is parallel to FL – and that is impossible.

Therefore, the other point cannot be on line EG.

[190] [Diagram V.21] But if [that other point] lies outside line EG, let it be as point W.
V 244b We then join EW and extend it to | meet the circumference of circle DGZ, say at point t. Through point W we draw a plane parallel to the base of the cone: it will produce a circle in the cone; let it be circle W, with U as centre.

This plane will cut lines BM and AN, because it cuts line EG, which is parallel to them. Point E will be outside this plane, because this plane cuts the cone. Let the plane passing through point W and parallel to the cone's base cut lines BM and AN in points X and R. We join [lines] EX and ER and produce them in straight lines; they will meet the plane of the circle DGZ.

Now line EX is in the plane of lines BM and EG: ⟨⟨thus it will meet plane DGZ at the intersection between the plane of lines BM and EG⟩⟩ and the plane of circle DGZ. But the intersection between these two planes is line GM; therefore line EX will meet line GM, say at point J. Similarly, it is clear that line ER will meet line GN; let it meet it in point Y.

We draw lines XW, RW, UWd, Jt, Yt, and Ttf. It is then clear, as was shown for circle
V 245a DGZ, that ∠RWd = ∠XWd. |

The planes of the two circles are parallel, and they are cut by plane EJt; therefore line XW || line Jt. Similarly, plane EYt cuts these two planes.

Therefore, line RW || line Yt. And, similarly, line dWU || ftT. Therefore, angles Jtf and Ytf are [respectively] equal to angles XWd and RWd, which are equal to each other; thus ∠Ytf = ∠Jtf. But ∠YGK = ∠JGK, therefore lines have been reflected between points J and Y at equal angles from two points on the circumference of circle DGZ, which is impossible.

Thus, the form of point B cannot be reflected to the eye A from the surface of cone DEZ, except from one point only, and [point B] will therefore have only one image.
V 245b Therefore, any point perceived by the eye in the conical convex mirror | will have only one image. And that is what we wished to prove.

⟨⟨V 2.VI.2⟩⟩

[191] [Diagram V.22(a)] If the two points are given, and the conical [convex] mirror is given, we may find the point of reflection as we now describe.

Let the two points be A and B; let the conical convex mirror be GDZ, with axis GT. We draw through point G a plane parallel to the cone's base; let it be the plane containing lines MG and GN. Then, points A and B either lie in this plane; or they lie towards the cone, outside the plane; or lie behind the plane; or [separately] on either side of the plane; or one of them lies in the plane and the other on one or other side ⟨⟨of it⟩⟩.

And first, let the two points lie on the side towards the cone [with respect to plane MGN]. Let us draw through one of the points, say A, a plane parallel to the cone's base: it will produce a circle in the cone; let it be circle DEZ, with T as centre. Join GB and
V 246a produce it | in a straight line: it will meet the plane of circle DEZ, say at point H. Thus,

points A and H are in the plane of circle DEZ and outside the circle's circumference, because points A and B are outside the cone's surface, and thus line GBH will be outside the cone's surface.

We then find the point on the circumference of circle DEZ from which the form of point H is reflected to point A, as we showed earlier; let it be point E.

We join AE, GE, HE, and TE, and produce TE to Y. Draw the plane containing lines GT and GE: it will cut line AB, say at point F. From point F draw a perpendicular to line GE; let it be FSX. Thus, let line SX be perpendicular to the plane tangent to the cone and passing through line GS.

From point A, draw line AL || SX, and line AJ || ET; then plane JAL || plane GET.

Join AS and BS, and produce BS in the direction of S: it will meet line AL, say at point L. Join | EH in the direction of E: it will meet line AJ, say at point J. Join JL; then line JL V 246b
|| EG, these being the two intersections [of planes AJL and TGE with plane HGE].

Now draw from point E the intersection between the circle's plane and the plane tangent to the cone and passing through line EG; let it be EO [with point O also on line AJ]. EO will then be perpendicular to line AJ, because it is perpendicular to line ET.

We also draw the intersection between the reflection plane and the plane tangent to the cone; let it be SC [with point C also on line AL]. It will be perpendicular to line AL, because it is perpendicular to line SX.

Join CO. It will be parallel to line EG, because the plane tangent to the cone cuts planes GET and JAL which are parallel to each other. Then line OC || line LJ. But [line] AJ || line ET, and the angles at point E are equal. Therefore, the angles at points A and J are equal, and therefore line AE = line JE.

Now EO is perpendicular to AJ, therefore AO = OJ, | but OC || JL. Therefore, AC = V 247a
CL, and SC is perpendicular to CL; thus line AS = line LS. Therefore, the angles at points A and L are equal, and thus ∠ASF = ∠BSF. Therefore, the form of point B will be reflected to point A on the surface of cone GDZ from point S. And that is what we wished to prove.

[192] [Diagram V.22(b)] Now, if the two ⟨given⟩ points lie in plane MGN, let them be M and N. We join MG and GN, bisect angle MGN by line GW [with point W also on line MN], and draw the plane | containing lines TG and GW: it [the plane] will cut the surface V 247b
of the cone, producing in it a straight line, say GE. ⟪And from W we draw WE at right angles to GE; then EW is perpendicular to the plane tangent to the cone at GE.⟫

We draw from point E a plane parallel to the cone's base: it will produce a circle in it; let it be circle DEZ. From points M and N draw two lines parallel to line EG; they will meet the plane of circle DEZ, say at points H and A. Join AE and HE.

Now since lines EG and MH are parallel, they are in one plane; and this plane cuts the plane of the circle and plane MGN. Therefore, lines MG and HE are parallel.

And, similarly, it is clear that lines GN and EA are parallel.

We draw the intersection between plane WGT and the plane of the circle; let it be TEY. It will be parallel to GW, and [also] a diameter of the circle, because the centre of the circle is the only [point] on the cone's axis that lies in the plane of the circle.

Now, since lines MG, WG, and NG are [respectively] parallel to lines HE, TE, and AE, the angles AEY and HEY are [respectively] equal to angles | NGW and MGW. And these V 248a
[last] two angles are equal, so angles AEY and HEY are equal. Thus, if we draw from point A two lines parallel to lines ET and EW, and follow the preceding method, it will be clear that angles MEW and NEW are equal.

And point M will therefore be reflected to point N from point E on the surface of the cone. And that is what we wished to prove.

[193] [Diagram V.22(c)] If the two ⟨given⟩ points lie behind plane MGN, then let them be points J' and g.

Join J'g. We draw the cone opposite to ⟨⟨cone GDZ⟩⟩. Draw through one of points J' or g, ⟨⟨let it be point g⟩⟩, a plane parallel to the base of the cone: it will produce a circle on the surface of the opposite cone; let it be circle tU.

That plane will or will not pass through point J'. If it does, then we use J'.

But if it does not, then we draw GJ' ⟨⟨which will necessarily meet the plane of circle tU, say at Q; then points g and Q will be in the plane of circle tU, and lie outside the circle's circumference. Then we find the point⟩⟩ from which the form of either g or Q is reflected to the other from the concave [side] of circle tU; let it be point R. Join QR and
V 248b gR. We then draw | from R the diameter of circle tU: it will cut line Qg; let it be diameter Rd. Then ∠QRd = ∠gRd. Later on, when we discuss concave mirrors, we shall show how we find that point. In our explanation there we will not use anything that we have used in the present proof.

We join RG and produce it in a straight line: it will extend along the convex surface of the cone GDZ; let it be line RGE.

Draw the plane containing lines RE and GT: it will cut line gJ', say in point L_1. And drop from point L_1 a perpendicular to GE; let it be L_1EK. Then EK will be perpendicular to the plane tangent to the cone. And we join J'E, gE.

[194] I say then that ∠J'EL_1 = ∠gEL_1.

Demonstration. We draw through point E a plane parallel to the cone's base: it will produce a circle on the surface of the cone; let it be circle DEZ ⟨⟨with centre T⟩⟩. ⟨⟨Join ET, and continue it to Y. And draw from g a line parallel to line RGE: it will meet the plane of circle DEZ, say at A. And draw from J' a line parallel to RGE, [and] let it meet the plane of
V 249a the circle at H; and produce it in the direction of J' until it meets RQ, say at L_2.⟩⟩ |

Join AE and HE. Then lines gA and RE are in one plane that cuts the plane of circles tU and DEZ [in lines gR and AE]. Then line gR || line AE, and L_2R || HE, and also Rd || TEY; therefore angles AEY and HEY are [respectively] equal to angles gRd and L_2Rd, which are equal to each other; therefore ∠AEY = ∠HEY.

Thus, if we draw from point A two lines parallel to lines ET and EK, and draw lines HE and EQ to meet them, and complete the construction as before, then it will be clear that ∠gEL_1 = ∠J'EL_1. Therefore, the form of point J' will be reflected to point g from point E. And that is what we wished to prove.

[195] [Diagram V.22(d)] If one of the two ⟨given⟩ points lies in plane MGN, and the other lies [outside the plane] towards the cone, let M be the point in plane MGN, and A the point that lies towards the cone.

We draw through point A a plane parallel to the cone's basis; let it produce in the [cone's surface] the circle DEZ, with centre T. From point M, draw a perpendicular to the
V 249b plane of circle DEZ; | let it be ⟨perpendicular⟩ MH.

Join HT, and draw from point A line AEK so that EK = KT, as we showed in the lemmas. Join TE, and produce it to Y. Draw line GE and [line] AM. Draw the plane containing lines EG and GT: it will cut line MA; let it meet [MA] at point F.

Draw from F a perpendicular to line GE; let it be FOC. Then line ⟨⟨OC is perpendicular to the plane tangent to the cone⟩⟩ and passing through the whole of GO. Join AO and MO.

I say that ∠AOF = ∠MOF.

Demonstration. We draw from point ⟨⟨H⟩⟩ in the plane of circle DEZ a line parallel and equal to line TE, say HB. Join MB: it will be equal and parallel to line GE, since MH is equal and parallel to axis GT. Join BE.

Since HB is equal and parallel to TE, thus BE || HT, therefore ∠BEY = ∠HTE. But ∠ETK = ∠KET, since EK = KT, and ∠KET = ∠AEY, | thus, ∠AEY = ∠BEY. V 250a

Now lines GE and MB are parallel, and they are in one plane; and lines MO and BE are in one plane with them. Thus, if we draw from point A a line parallel to line TE and follow the method we followed in the preceding diagram, it will be clear that ∠AOF = ∠MOF. Therefore, the form of point M will be reflected to point A from point O on the surface of the cone. And that is what we wished to prove.

[196] [Diagram V.22(d)] If one of the two ⟨given⟩ points lies in plane MGN and the other lies behind this plane, we draw the cone opposed to the cone forming the mirror. Then we find the reflection point on this cone, I mean the opposed [cone], then transfer the reflection point to the cone forming the mirror, as we did in the diagram preceding this one.

[197] [Diagram V.23] If the two ⟨given⟩ points lie on either side of plane MGN, | let A V 250b
and L be the two points; and let A be located towards the cone, and L behind plane MGN.

We join AL, and draw through point A a plane parallel to the cone's base; let it produce in the cone circle DEZ ⟨⟨with centre T⟩⟩. Join LG, and produce it in a straight line: it will meet the plane of circle DEZ, say at point Q. Then points A and Q will be in the plane of circle DEZ.

We then find on the circumference of the circle the point that, if two lines are drawn to it from A and Q, the line drawn tangentially to the circle from that point will then bisect the angle contained by the two lines, as we showed in Lemma IV; let the point be E.

Join AE and QE, and draw SE tangent to the circle; then ∠AES = ∠QES.

Join GE and draw QE, in ⟨the direction of⟩ E, to B; lines LQ and QB are then in one plane, and line GE is in it. We draw from L a line parallel to line GE: it will meet line QEB, say at point B. And let the centre of | circle DEZ be point T. Join TE and produce it to Y: V 251a
perpendicular SE is then perpendicular to line YET.

[198] Now, ∠AES = ∠QES, therefore ∠AEY = ∠QET; and ∠QET = ∠BEY, thus ∠AEY = BEY. We draw the plane that contains lines TG and GE: it will cut line AL, say at point W. We draw from point W a perpendicular to line GE; let it be perpendicular WOd. And we draw lines AO and LO.

Thus, if we draw from point A a line parallel to line Od and a line parallel to line ET, it will be clear, as was shown in the preceding diagram [Diagram V.22(d)], that ∠AOW = ∠LOW. And, therefore, the form of point ⟨L⟩ will be reflected to point A from point O on the surface of the cone GDZ. And that is what we wished to prove.

It is thus clear from what we have shown how to find the reflection point on the surface of the conical convex mirror if the ⟨single⟩ eye-point and the visible point are given, for all positions of these two points. And all that [holds as stated] when vision | takes place by V 251b
means of one eye.

⟨⟨V 2.VI.3⟩⟩

[199] If perception of a single object takes place by means of both eyes, in this [conical convex] mirror and in the cylindrical convex mirror, then the object's image will be like that [produced] in the spherical convex mirror. I mean that the two forms, which are the two images of the single object with respect to the two eyes, will overlap. For the form perceived in each of these two mirrors, when perceived ⟨with one eye⟩ by the viewer, will be perceived distinctly. ⟨But when the form is perceived with both eyes⟩, the axes of both eyes will meet in a single point of the form; and when the axes of the two eyes meet in one point of the form, that form will be perceived with both eyes as a single form, as was shown earlier.

[200] Now, the point on the surface of one eye, through which the perpendicular from [the centre] of that eye passes, will be perceived by the eye [to be] on the perpendicular drawn from its centre; and ⟨perceived⟩ as continuous with the image of what surrounds [that point] on the eye-surface, while the other eye will perceive [the same point] by means
V 252a of a reflected line, and on the perpendicular | itself that is drawn from the centre of the eye in which [the point] lies. This eye will also perceive [that point] by means of the reflected line as continuous with the image of what surrounds [that point] on the surface of the eye in which [the point] lies. This reflected line will then meet the perpendicular drawn from the centre of the eye in which [the point] lies, at the point that is continuous with the image of what surrounds it on the surface of that eye. Therefore, the image of this point, when perceived by both eyes, will be a single point in which the perpendicular drawn from the centre of the eye in which [the point] lies meets the reflection line along which [the point's] form is reflected to the other eye. And the image of every point perceived in right-conical convex mirrors and right-cylindrical convex mirror[s] can only be a single point. And the form of any object perceived in each of these two [types of] mirror can only be a single form, whether [the object] is perceived ⟨with one eye or⟩ with two at the same time. And that is what we wished to prove.

⟨⟨V 2: Seventh Inquiry. On the Images of Spherical Concave Mirrors, Consisting of an Introduction and Thirty-Three Propositions⟩⟩

⟨⟨Introduction⟩⟩

[201] As for spherical concave mirrors: all of their sections are circles, like the sections
V 252b of the spherical convex mirrors. But | reflection in the convex [mirror] will be from the convex side of the arc, while in the concave it will be from the arc's concave side. Most of the lines along which the light is reflected in this mirror will meet the perpendiculars drawn from the visible points at right angles with the lines tangent to the intersections between the surfaces of these mirrors and the planes of reflection, while some of [these lines] will not meet the perpendiculars. Of those that meet the perpendiculars, some will have their meeting-points, i.e., the images, behind the mirror, and some in front of the mirror; and, of the meeting-points that lie in front of the mirror, some will be between the eye and the mirror, some on the centre of the eye, and some behind the centre of the eye.

⟨⟨V 2.VII.1⟩⟩

[202] [Diagram V.24] Let A be the centre of the eye, and BG ⟨⟨an arc⟩⟩ on a spherical concave mirror with its centre at D. And let us draw a plane passing through ⟨the centre of the mirror and thus cutting it in a circle⟩; let it be circle CBG. The plane of this circle will then be one of the planes in which the forms will be reflected from the surface of this mirror, since this circle is orthogonal to | the plane tangent to the spherical mirror at a point V 253a on the circumference of the circle.

We join AD and continue it to B and to Y, ⟪and let Y be placed relative to D in the direction of A⟫. Draw a line from point A to the circle's circumference, ensuring that it is longer than AD, which is possible; and let it be AE. From point D, draw a line parallel to AE, say DH, continue it rectilinearly in the direction of D to point F, and join DE.

Make ∠DET = ∠AED. Line ET will then be inside the circle, since ∠DET is acute, being equal to the acute ∠AED. Line ET will then cut line DH, say in point T. And if ET is produced in a straight line, it will cut line DB, because ∠ADE > ∠DET, since line EA is longer than line AD, and thus ∠ADE > ∠AED, and, therefore, greater than ∠DET.

Let us produce ET in a straight line, and let it cut DB in point Z. Draw line AN between lines AE and DH. This line, if rectilinearly produced, will then meet line DH ⟪behind the mirror in L. We join DN and make ∠DNM = ∠DNA. Line NM will then cut DH⟫, say | V 253b in point M.

We draw line AG between lines AE and AY: this line will then meet line DH in the direction of F; let O be the meeting point. We join DG and we make ∠DGK = ∠DGA. This line ⟨GK⟩ will then meet line DH, if arc GH is shorter than the arc cut off from the circle by line DO; let K be the meeting point. ⟪I say: he means that the meeting takes place inside the circle.⟫

We draw line AC so as to cut line FD, and we join DC and make ∠DCQ = ∠DCA; let AC cut line FD in point S. Line CQ will then meet line DH if CS > SD, because ∠SDC > ∠DCS and thus greater than ∠DCQ; let line CQ meet line DH at point Q.

[203] [Diagram V.24] Thus, if point T lies in a visible object, its form | will extend V 254a along line TE and be reflected along line EA, because ∠TED = ∠DEA. Line DE will be perpendicular to the plane tangent to the circle at point E; and line DT is the perpendicular drawn from T at right angles to the line tangent to the circle, i.e., the intersection between the plane of the circle and the reflection plane, since DTH is a diameter of the circle; and line AE is parallel to line DH, and thus does not meet line DH. Therefore, the line along which the form of point T is reflected to the eye A will not meet the perpendicular drawn from point T at right angles with the line tangent to the circle.

Now, if point M lies in a visible object, its form will extend along line MN, and will be reflected along line NA, since DN will be perpendicular to the plane tangent to the sphere at point N. NA will meet DM, which is the perpendicular drawn from point M at right angles with the tangent line; and the meeting-point is point L, which is behind the mirror.

And the form of point Z will extend along line EZ, and be reflected along line EA. Line DZ | is the perpendicular from point Z drawn at right angles with the line tangent to the V 254b circle's circumference; and line EA meets line DZ in point A, which is the centre of the eye, and which lies in front of the mirror.

And the form of point K will extend along line KG, and be reflected along line GA. Line DK is the perpendicular from point K drawn at right angles to the line tangent to the circle; and line GA meets line DK, and the meeting-point is O, which lies in front of the mirror, and behind the centre of the eye.

And the ⟨form of⟩ point Q will extend along line ⟨QC, and be reflected along line CA. QD is the perpendicular from point Q drawn at right angles to the plane tangent to the circle⟩; and line CA ⟨will meet line QD, and the meeting-point⟩ is point S, which is between the ⟨centre of the⟩ eye and the mirror's surface.

[204] It is thus shown by this exemplification that, of the objects perceived by sight in the spherical concave mirror, some are such that the lines of their reflection do not meet their perpendiculars, and some have their lines of reflection meeting their perpendiculars; and that the meeting-points of these lines, which are the images [produced by] this mirror, are located [as follows]: some behind the mirror, and some in front of the mirror. And of
V 255a those that lie in front of the mirror, some are at the centre of the eye, some | behind the eye, and some between the eye and the surface of the mirror.

[205] Moreover, the eye will perceive all of these images as facing it, because the reflected forms only come to the eye from the sides facing it, and because sight only perceives the visible objects along the directions of the radial lines along which the forms are reflected to the eye, a notion that was made clear earlier. Therefore, in all cases, sight perceives the form of the object as being in front of itself.

⟪V 2.VII.2⟫

[206] If the image of the visible object lies behind the mirror, or between the eye ⟨and the mirror's surface, sight will perceive it distinctly. If sight perceives the objects whose images are at the centre of the eye⟩ or behind the centre, and whose reflection lines are parallel to their perpendiculars, it will perceive their form[s] joined to the forms of the images that it perceives in their [own] locations.

⟪V 2.VII.3⟫

[207] With regard to the points whose forms are reflected along lines parallel to their
V 255b perpendiculars: assuming they lie upon a visible object, some of the points | surrounding [one of them] will have their images behind the mirror, and some will have their images in front of the mirror. As a result, it may happen that the ⟪reflected⟫ form of an object of this description will intersect the surface of the mirror. Consequently, one portion of the form will lie behind the mirror and one portion in front of it, and the intersection will be at the extremity of the parallel line, namely the reflection point. The reflection point will then be the image-point.

⟪V 2.VII.4⟫

[208] As for the points whose images lie behind the eye or on the centre of the eye: the eye will perceive them facing it, and perceive each of these points on the radial line on which the image of that point lies, but ⟨the eye's⟩ perception of it will be indistinct. If the point adjoins others whose images face the eye, and the eye perceives these images, it will
V 256a perceive them | joined to the forms of the adjoining points. But the points whose images lie behind the eye or on the eye's centre, when perceived by the eye, will not be perceived distinctly; rather, their forms will be confused. If [the forms] contain subtle features or minute parts, these will not be distinguished to the eye, ⟨nor will the eye⟩ perceive them accurately. If these images adjoin images facing the eye, the eye will distinctly perceive the parts of the form [constituted by] the images that face the eye, and indistinctly perceive

the parts whose images lie behind the eye or on the centre of the eye. If these images are examined experimentally, the experimenter will find them to be as we have described, that is, indistinct [as we now will show].

⟨⟨Experimental examination⟩⟩

[209] To examine these images, the experimenter should set up a thin straight rod on the surface of the spherical concave mirror in a straight line with one of the mirror's diameters, and attach it to the surface of the mirror with a piece of wax. Let the rod be less than half the mirror's diameter in length. Let him position his eye in line with the rod, then move his eye away from | this straight line [to a point] a little closer to the mirror's surface, V 256b then look into the mirror at the place of it whose distance from the base of the rod exceeds the distance of his eye from the diameter of the mirror on the same straight line as the rod: he will see the form of the rod to be uncertain, and the shape of the form will not be truly distinct to him. Small and distinct dots, or small distinguishable parts that may be on the rod, would not be perceived by the eye. At this position of the eye in relation to the rod, ⟨⟨the images of the dots over it⟩⟩ will be located behind the eye.

⟨⟨V 2.VII.5⟩⟩

[210] If the eye is positioned on the very diameter that is in line with the straight rod, it will perceive of the rod only a certain colour. If the rod is extremely thin, the eye will also fail to identify its colour. It is clear from this experimental examination that objects perceived by sight in this mirror will not be perceived distinctly when their images lie behind the centre of the eye.

⟨⟨V 2.VII.6⟩⟩

[211] As for the images that lie behind the mirror: if they adjoin images located in front of the mirror, then the form of the | visible object will partially lie behind the mirror V 257a and partially in front of it. The mirror's surface will cut this form. In this situation, if the extremity of the object is in touch with the mirror's surface, then the image will be arc-shaped, with one extremity adjoining the base of the object [appearing to be] in touch with the mirror's surface, and the other extremity [appearing to be] on the mirror's surface. The place [of the latter extremity] on the surface of the mirror will be that of the intersection of the form with the mirror's surface. When the image is arc-shaped in this manner, the eye will not perceive its curvature, especially if the object is fairly large, but rather perceive [the object's] colour spread upon the mirror's surface without perceiving its shape, nor will the form have a shape similar to that of the object. If, while in this condition, the viewer moves his eye, the arc-shaped image will appear to him behind the mirror. If this situation is examined in the manner we have described, it will be found to be as we said.

⟨⟨V 2.VII.7⟩⟩

[212] Thus, what the eye distinctly perceives in these [spherical concave] mirrors is the visible object all of whose images are located behind the mirror, or all of whose images

V 257b are between the eye and the mirror. Otherwise, the object will not be | distinctly perceived by the eye.

⟪V 2.VII.8⟫

[213] As for the perpendicular drawn from the centre of the eye at right angles with the plane tangent to the mirror's surface: some of the points on [this perpendicular] are perceived by the eye in this mirror, and some are not. For this perpendicular passes through the centre of the eye, and thus if the centre of the eye is at the centre of the mirror, then any line drawn from the centre of the eye to the surface of the mirror will be perpendicular to the tangent line. In this situation, the eye will not perceive any visible objects lying outside the perpendiculars, nor any of the objects on the perpendiculars other than the points on the surface of the eye that lie only on these perpendiculars. For, in this situation, any point whose form extends to a point on the surface of the mirror is such that the line drawn from the centre of the eye to that point on the mirror's surface will be perpendicular to the tangent plane, and therefore that form will not be reflected on that perpendicular; and, therefore, the eye will not perceive that form. And the points on the surface of the mirror that lie on the perpendiculars are such that their form[s], which extend along the perpendiculars to the mirror's surface, can be reflected only along the perpendiculars themselves.

V 258a **[214]** Thus, if the centre | of the eye coincides with the centre of the spherical concave mirror, the eye will not in this situation perceive anything other than the portion of the eye's surface that is cut off by the cone whose vertex is the centre of the eye and whose base is the mirror's surface. Nor will [the eye] perceive the shape of this portion either, but will perceive only the colour of the eyeball's darkness all over the surface of the mirror, and thus perceive only a darkness in the eye, without discerning what this darkness is.

[215] If the centre of the eye lies outside the mirror's centre, then, if it lies between the mirror's centre and the surface of the mirror, the eye will not perceive any of the points on that radius of the mirror where the centre of the eye is located. For if two lines are drawn from the [i.e., any] point on this radius and the centre of the eye to a point on the mirror's surface, then the line drawn from the mirror's centre to that point [on the mirror's surface] will lie outside the angle contained by those two lines. And ⟨this line lies in the plane of the two lines⟩ and will therefore not divide that angle. And that line [from the centre to the point on the mirror's surface] is the perpendicular drawn from that point on the mirror's surface, at right angles to the plane tangent to the mirror at that point. Therefore, since
V 258b [that line] does not divide that angle, | the form of that point [on this radius] will not be reflected to the eye. And, therefore, the eye will not perceive any of the points on that radius of the mirror where the centre of the eye is located.

[216] As for the remaining points on this diameter when the centre of the mirror is [located] between them and the centre of the eye: they can be perceived by the eye. Also, if the centre of the eye is farther from the mirror's surface than the mirror's centre is, and it lies outside the sphere of the mirror, then the eye will not perceive any of the points on that diameter, other than those on that radius when the centre of the mirror is [located] between them and the centre of the eye.

[217] Thus, [in regard to] the perpendicular drawn from the centre of the eye at right angles to the plane tangent to the mirror's surface, [which perpendicular] is the diameter of the mirror passing through the eye's centre: the eye will not perceive any points on it that lie with the eye's centre on one and the same side with respect to the mirror's centre. But it is possible [for the eye] to perceive on this diameter the points [in the case] when the centre of the mirror lies between them and the eye's centre.

⟪V 2.VII.9⟫

[218] [Diagram V.25] I say, further, that, for every point perceived by sight in a spherical concave mirror, if the form of that point is not | reflected along a line parallel to the diameter V 259a
of the mirror that passes through that point, then the ratio of the line [BE] between the visible point and the centre of the mirror to the line [EH] between the centre of the mirror and the image-point is as the ratio of the line [BT] between the visible point and the extremity of the tangent [line] drawn from the reflection point to the line [TH] between the extremity of the tangent [line] and the image-point, whatever the position of the image-point.

Let A be the centre of the eye, B the visible point, and GD the spherical concave mirror. We draw the plane passing through the centre of the eye, the centre of the mirror, and the visible point; let it produce in the mirror circle GD, with centre, and let G be the reflection point.

We draw lines AG, BG, and EG. And we join BE; if it is not parallel to line AG, it will meet AG, but not at point G, because B lies outside line GE. If line BE meets line AG, it will either meet it at a point behind point G, as in the first diagram [Case 1], or between points G and A, as in the second diagram [Case 2], or at point A itself, | as in the V 259b
third diagram [Case 3], or behind point A, as in the fourth diagram [Case 4]. Let H be the meeting point.

[219] [Diagram V.25] We draw from point G a line tangent to the circle, say line GZ. Then GZ either meets BE, or will be parallel to it. First, let the meeting-point be point T.

We draw from B a line parallel to line AG: it will meet the tangent line, say at point K. And draw from H a line parallel to BG: it will meet diameter EG, say at point L.

Now since ZG is tangent to the circle, EG is therefore perpendicular to GZ; and $\angle EGB = \angle EGA$; therefore $\angle AGZ = \angle BGK$. And $\angle AGZ = \angle BKG$, since line AG $\parallel$ line BK, therefore $\angle BGK = \angle BKG$, [and] thus BK = BG.

And thus, since HL $\parallel$ BG, $\angle BGE = \angle HLE$. And $\angle BGE = \angle AGE$, | therefore $\angle HLG$ V 260a
$= \angle AGE$.

Now, it is either the case that $\angle AGE = \angle HGL$, as in the first diagram [Case 1], or $\angle AGE$ is itself $\angle HGL$, as in the remaining three diagrams.

But $\angle HLG = \angle HGL$, therefore HL = HG. And BK = BG, thus BG:HL = BK:HG. But BG:HL = BE:EH, and BK:HG = BT:TH, therefore BE:EH = BT:TH.

[220] [Diagram V.25] The same [holds], if the tangent line in the second diagram [Case 2] meets BE towards point B, because this would be the same as in the third diagram [Case 3] if point A lies behind point | H. V 260b

And the same [holds again] if the centre of the eye is point B in the third diagram [Case 3], or lies on line BG in the fourth diagram [Case 4], and the visible object is point A in the third diagram [Case 3] and also in the fourth diagram [Case 4].

If tangent GZ is parallel to BE [Case 5], then BE meets AG, next to point A, because angles BEG and ZGE are [both] right. Thus, line BE meets line AG at point A or between points A and G, or behind point A. Let [BE] meet it at point H, as in the fifth diagram; point H is then the image.

If GZ is parallel to BE, then BE = EH. But BG = GH, since ∠BGE = ∠HE, and EG is perpendicular to BH. Therefore, point G is as point T.

Thus, in all five positions the ratio of [line] BE, between the visible point and the mirror's centre, to line EH, between the centre and the image-point, is as the ratio of BT, between the visible point and the extremity of the tangent line, to TH, between the
V 261a extremity of the tangent and the image-point. | And that is what we wished to prove.

⟪V 2.VII.10⟫

[221] [Diagram V.26] Again, let A be the centre of the eye, and B a point in a visible object, and let the mirror be GD, with E as centre. We draw the plane through points A, B, and E: it will produce in the mirror an arc of a circle. And complete the circle; let it be circle GTDK. Join AE and continue it until it cuts the circle; let it cut the circle at points G and D.

Now, point B lies either on this line or outside it.

If point B lies on line AG, and the centre of the circle lies between this point and point A, then it is possible for the form of point B to be reflected to eye A from either one of the semicircles GTD or GKD. For if two lines are drawn from any point on the circumference of the circle other than points G and D, one [line] from a point on EG, the other from point A, then [the two lines] will contain an angle, and the line drawn from point E to that point will divide the angle.

Therefore, if point B lies on diameter AEG, and point E lies between it and point
V 261b A, | then it is possible for the form of point B to be reflected to eye A from each of the semicircles cut off by diameter AEG.

⟪V 2.VII.11⟫

[222] [Diagram V.26] If point B lies outside this diameter, then, first, let eye A lie inside circle GTDK. We join BE and continue it to cut the circle on both sides, say in points T and K. I say then that the form of point B cannot be reflected to eye A from either arc GT or arc KD, and that it is possible for the form of B to be reflected to eye A from arc KG and arc TD.

We join AB and continue it in both directions to cut the circle, say in points Z and L. Assume a point at random on arc ZT, say point Q. Join AQ and BQ.

Line BQ will lie between lines BT and BZ; and, if QB is produced in a straight line, it will cut line E, say at O. Then, points O and A will be on the same side with respect to point E. And thus, if a line is drawn from E to Q, it will not cut angle AQB. Therefore,
V 262a the form | of point B cannot be reflected to eye A from point Q. And the same [holds] for every point on arc ZT.

And the form of B, if it extends along line BZ, cannot be reflected along line BZ, since BZ does not pass through the centre.

Let us mark a point at random on arc ZG, say point F. ⟨And join BF and AF; then AF will cut BE at a point⟩, say Y. Then points B and Y will be on one side with regard to point E. And thus if a line is drawn from E to F, it will not cut angle AFB.

Therefore, the form of point B will not be reflected to eye A from point F. And the same [holds] for every point on arc GZ.

Now, line AE is perpendicular to the plane tangent to the surface of the mirror at point G. Then the form of B, if extended to G, will not be reflected along line GA. And the same [holds] for point T.

Therefore, the form of point B cannot be reflected to eye A from arc GT.

A similar proof shows that the form of point B cannot be reflected to eye A from arc KD.

[223] [Diagram V.26] We assume ⟨on⟩ arc KG a point at random, say H. We join AH and BH. | V 262b

Then line AH cuts line EK, say at point C. Points B and C will then be on the same diameter, with point E between them. And thus if a line is drawn from E to H, it will cut angle BHA. And the same [holds] for every point on arc GK between points G and K. Therefore, it is possible for the form of point B to be reflected to eye A from arc GK.

Now, on arc TD we mark a point at random, say M. ⟨⟨Join AM and BM: a four-sided figure BMAE is produced. If EM is joined⟩⟩, it will cut angle AMB. And the same [is true] for every point on arc TD between points T and D.

Therefore, it is not impossible for the form of point B to be reflected to the eye A from any part of arcs KG and DT on the surface of this mirror. But it is not possible for the form of point B to be reflected to eye A from any part of arcs GT and KD.

We can show the same [to be true] if point B lies on the circumference of the circle, such as K, or outside [the circle] and on diameter ET; and also if the eye A lies on the circumference of the circle, as point D, whether B is inside | or outside the circle, or on V 263a
the circumference.

If the eye lies on line EG, then arcs GT and KD will be [the arcs] from which it is possible for the form of point B to be reflected to the eye A; and arcs GK and TD will be [the arcs] from which it is not possible for the form of point B to be reflected to the eye A.

[224] [Diagram V.27(a,b)] Furthermore, let the eye be outside the plane continuous with the surface of the mirror, as in the second diagram [Diagram V.27(a,b)].

Let arc HGZ be in the segment of the mirror's surface that may face the eye [Diagram V.27(a)]. Join AH and AZ: they will be tangents to the circle. And line ADEG will lie between them, and if BE is produced in a straight line ⟨⟨in both directions⟩⟩, then ⟨one extremity of it will cut⟩ the semicircle on which D, H, and G lie, and the other extremity will cut the semicircle on which D, Z, and G lie; let [BE] cut DHG | in point K, and let it V 263b
cut DZG in point T.

[225] [Diagram V.27(a)] Then, a line drawn from A to any point on arc KG will cut line EK; and point E will lie between the meeting-point and point B. Thus, if a line is drawn from B to the point on arc KG, it will contain an angle with the line drawn from A, and this angle will be cut by the line drawn to it[s angular point] from point E.

It is, therefore, not impossible for the form of point B to be reflected to the eye A from arc KG, whether point K lies between points G and H, or it is the same as point H, or it lies between points H and D.

Further, if line KET cuts arc ZG, then, for any point on arc ZT, if two lines are drawn to it from points A and B, they will contain an angle, and will produce a four-sided figure of which the sides are these two lines, and AE and EB. Thus, if a line is drawn from E to [the vertex of] that angle, it will cut the angle. Therefore, it is not impossible for the form of point B to be reflected to the eye A from arc ZT.

As for arc GT, any line drawn to it from point A will cut ET; then the meeting-point
V 264a | and point B will be on the same side with respect to point E. And if the meeting-point is B, then the form of point B will not be reflected on that line, because it does not pass through the centre. If the meeting-point is other than point B, then the line drawn from B to that point will contain an angle with the line drawn to it from point A, and thus the line drawn from E to that point will not cut that angle. Therefore, the form of point B cannot be reflected to the eye A from arc GT.

And the same [holds] for arc HK in the mirror ⟪he means [the arc] behind the point of tangency⟫ [Diagram V.27(b)], because every line drawn from point B to arc HK will cut line AE, and the point of intersection will lie ⟪with A⟫ on the same side with respect to point E. Therefore, it is not possible for the form of point B to be reflected to the eye A from arc KH.

Similarly, [the same] can be shown if point B lies on the circumference of the circle, as point T, or outside the circle and on diameter ET.

⟪**Result**⟫

[226] Let the visible point lie outside the diameter of the mirror that passes through
V 264b the centre of the eye. If a line is drawn from that visible point | to the mirror's centre and cuts the circle's diameter that passes through the mirror's centre and the eye's centre, then it will, together with the diameter through the eye's centre, divide the circumference of the circle into four segments, two of which are opposed to each other and from which the form of that point may be reflected to that eye, while it will be impossible for the form of that point to be reflected to that eye from the two other segments, whether the centre of the eye and the visible point are both located within the mirror's surface, or both lie outside the mirror's surface, or both on the mirror's surface, or one of them within and the other ⟨either⟩ outside or on the surface of the mirror, or one of them on the surface of the mirror and the other outside it. As for the two segments from which it is possible for the form of that point to be reflected, these are [the segments] one of which is the base of the sector on whose sides the eye's centre and the visible point are located, while the other segment is
V 265a the base of the sector opposed | to this sector. And it will be possible for the form of that point to be reflected to that eye from any part of these two segments on the surface of the mirror. But reflection of the form of that point to that eye will not be possible from [any part of] the two other segments.

⟪**Refinement**⟫

[227] In the first diagram [Diagram V.26] we draw ⟪from A⟫ a line parallel to line BE: it will always cut the circumference of the circle, say in point S. Therefore, if point B is reflected to eye A from a point on arc TS, between points T and S, then the image will lie behind the mirror. If the reflection takes place at S, then the form will be at the reflection point, namely S. If the reflection is from arc SD, then the image will be behind the eye's centre, and the form will be between point A and arc SD. If the reflection is from arc GK, then the image will be between the eye and the mirror, on line EK. And similarly [this will be the case] for the image, if the eye A lies on the mirror's surface, as point D.

⟨⟨**Another refinement**⟩⟩

[228] If, however, the eye | lies outside the mirror's surface, as in the second diagram [Diagram V.27(a,b)], then, if either one of angles KEG or TEG is smaller than angle GAH or angle GAZ, which is equal to it [i.e., angle GAH], ⟨⟨and let the line drawn through TEKL be like this⟩⟩ [Diagram V.27(a)], then the line drawn from point [A], parallel to line BK, cuts the mirror's surface, as AM, and the image will be the same as in the first diagram [Diagram V.26]. ⟨And⟩ if the reflection is from arc TM, the image will be behind the mirror; and if it is from point M, the form will be at point M. ⟨⟨And if [the reflection] is from arc MZ, the image will be behind the eye's centre, and the form between the eye and arc MZ.⟩⟩ And if the reflection is from arc HG, the image will be between the eye and the mirror, because the line drawn from A to any point on arc HG cuts line LE, which is the mirror's diameter through ⟨⟨B⟩⟩. ⟨⟨I say [Diagram V.27(b)]: and if neither one of angles KEG or TEG is smaller than angle GAH or GAZ [which is equal to it], then if the reflection is from arc TZ, all the images will be behind the mirror; and if it is from arc HG, the images will lie between the eye and the mirror.⟩⟩ V 265b

⟨⟨**Notice**⟩⟩

[229] As to how we find the locations of the images in this mirror, [it should be remembered that] all the images of the visible objects perceived in the spherical concave mirrors lie on the mirrors' diameters, as was shown in [our] experimental examination that ⟨the image of⟩ any point perceived by sight in this mirror will lie on the perpendicular drawn from that point at right angles with the line tangent to the circle; and it is clear that ⟨such a⟩ perpendicular | must be a diameter of the mirror. V 266a

⟨⟨**V 2.VII.12**⟩⟩

[230] [Diagram V.28] I say then that if any one of the diameters of this mirror is produced in a straight line in both directions, then it is possible for every point on it to be the image of some visible point. Let us also prove this notion [i.e., proposition].

Let circle ABG be a ⟨section⟩ in a spherical concave mirror ⟨passing through its centre⟩. Let us draw in it a diameter at random, say diameter AG.

And let the mirror's centre be point D. Produce diameter GA in both directions to points Z and H. Draw DER at right angles to this diameter; let point E be the centre of the eye, and let it be inside the circle.

Assume on line AZ a point at random, say Z, and join EZ: this line will then cut the circle, say in point T. Join DT, and make angle DTL equal to angle ETD. Then point Z will be the image of point L. And, similarly, every point assumed on line AZ, even extended indefinitely, will be an image of some point on line AD.

Now let us assume on line AD a point at random; let it be point L. Join | EL and produce it in a straight line in the direction of E: it will cut the circle, say in point B. Join DB, and make ∠DBK = ∠DBL. Then line BK will meet line DG, ⟨because angles⟩ DBK and DBL are acute. Let BK meet line DG in point K. ⟨Then point L will be⟩ the image of point K, and, similarly, every point on line AD will be the image of one of the points on line DG. V 266b

Moreover, we produce ED to M. Then the form of any point on line GH, even if extended, will be reflected to eye E from arc AM. The images of [that point] will lie on line AD, because if any point on line GH, even if extended indefinitely, is joined to a point on arc AM, then that ⟨joining⟩ line will cut line DM.

Then, if a line is drawn to join that point [on arc AM] to point D, and an angle is constructed on that line equal to the angle produced at that point, then the second line will
V 267a cut line AD. If it is produced in a straight line, | then it is possible for an eye to be located on it, and the same [holds] for any point on line DG. Therefore, it is possible for any point on line DAZ to be the image of some visible point. And the same [holds] for lines AZ and GH, even if they are extended indefinitely.

Therefore, it is possible for every point on diameter AG, even if indefinitely extended in both directions, to be the image of some visible point. And, therefore, the images of the spherical concave mirrors do not have limited locations on their diameters.

⟨⟨Précis⟩⟩

[231] All of that having been proved, I say that of the visible points perceived in these [spherical concave] mirrors, some have a single image; some have two images; some have three images; some have four images; but none will have more than that. The reflection of the forms of the points that have four images takes place from four points, and their forms are four points. Those that have three images are reflected from three points, and their
V 267b forms are three points. Those that have two images are reflected from two points, | and their forms are two points. And, of those that have a single image, some have their image reflected from a single point, and their form is a single point; and some have their form reflected from the circumference of one of the circles that occur in the mirror, and the form of this point will be the form of the circle's circumference.

⟨⟨V 2.VII.13⟩⟩

[232] [Diagram V.29(a,b)] Let ABG be a spherical concave mirror, with centre D; and let E be the centre of the eye. We join ED and continue it ⟨in both directions⟩ in a straight line to cut the surface of the mirror, say in points A and Z. Let point H be in a visible object.

First let [H] be on diameter ED; and let point D lie between this point and point E, which is the centre of the eye.

We draw from diameter AZ a plane that cuts the mirror at random: it will produce in
V 268a the mirror a circle, say | circle ABZG.

Now, lines HD and DE may be equal or unequal. Let them first be equal, as in the first diagram [Diagram V.29(a)]. Draw line DG at right angles with diameter ADZ and continue it in both directions to points G and B; and draw lines EG, GH, EB, and HB.

Then ∠HGD = ∠EGD and ∠EBD = ∠HBD.

If we then imagine diameter ADZ to be fixed, while the semicircle AGZ rotates about it, then point G will draw in the mirror's surface one of the great circles that may occur in the sphere of the mirror; and, as a result of its rotation, point G will coincide with point B. Then, between every point on this circle and points H and E, there will be a triangle equal to triangle HGE; and, therefore, the line drawn from point D to that point will bisect the triangle.

This proof shows that the form of point H will be reflected to the eye E from every point on the circumference of the circle traced out by point G on the surface of the mirror. And | the images of all the points that can be imagined on this circle will be point E, which V 268b
is the centre of the eye.

[233] [Diagram V.29(a)] It is clear, moreover, that if lines HD and DE are equal, then the form of point H will only be reflected to the eye E from the circumference of circle AZG at points G and B. The reason is that if two lines are drawn from points H and E to any point on the circumference of circle AZG, or on the circumference of any circle that may occur in this mirror, other than the two points corresponding to B and G, then the two lines will be unequal. The ratio of these two lines to each other will not be the same as the ratio of lines HD and ED to each other. And, therefore, the line drawn from point D to that point on the circumference of the circle, to which those two lines are drawn, will not bisect the angle contained by the two lines. Therefore, the form of point H will not be reflected to the eye E from mirror AZG, except from the circumference traced by point G as it moves round diameter AZ.

[234] [Diagram V.29(b)] Now, let lines HD and DE be unequal, | as in the second V 269a
diagram [Diagram V.29(b)]; and let ED > DH.

⟪We produce EH to K,⟫ and make the product EK × KH equal to KD^2; and, with K as centre and radius KD, draw ⟨a circle⟩: it will cut circle AGZ at two points, say B and G. And draw lines EB, EG, HB, HG, DB, DG, KB, and KG.

Then KB = KG = KD. Thus, EK:KG = [EK:KD = KD:KH =] GK:KH, therefore △EKG ~ △HKG, and thus EG:GH = EK:KG = EK:KD. But EK:KD = KD:KH = [(EK–KD):(KD–KH) =] ED:DH, ⟨therefore ED:DH⟩ = EG:GH; thus ∠EGD = ∠HGD.

Therefore, the form of point H will be reflected to the eye E from point G.

Now △EBH = △EGH, and the angles at point B are equal, and | ∠BDH = ∠GDH. V 269b

Thus, if diameter ADZ is fixed and the semicircle AGZ rotates [about it], then point G will trace out a circle in the mirror's surface, and, by turning, point G will coincide with point B, and the circle produced in the mirror's surface will be one of the small circles [whose planes] do not pass through the centre of the sphere. And between every point on the circumference of the produced circle and points E and H, there will be a triangle similar to triangle EGH.

And therefore, ⟨the form of point H will be reflected from every point on the circumference of the produced circle to point E, and the image of that point will be point⟩ E, which is the centre of the eye.

And if line ED is shorter than line DH, the reflection will take place on the side of point Z; and this can be shown as was done in the case of point ⟨K⟩.

And the form of point ⟨H⟩ will not be reflected to the eye E from a point on the circumference of circle AZG other than points B and G, because any two lines drawn from points E and H to a point on the circumference of circle AZG will be different from the two lines EG and GH [as follows].

[235] [Diagram V.29(b)] Furthermore, if the point is located on arc GAB, then the line drawn from point E to that point will be longer than line EG, and the line ⟨passing⟩ through point H ⟨to that point will be shorter than line HG; and if the point is⟩ | on arc BZG, then V 270a
the two lines will be in reverse: I mean that the line drawn from point E will be shorter than line EG, and the one drawn from point H will be longer than line GH. And therefore, the ratio between those two lines will differ from that between lines EG and GH. But

EG:GH = ED:DH. Therefore, the ratio between them will differ from the ratio between lines ED and DH, and thus the line drawn from point D to that point will not bisect the angle contained by these two lines, and point H will be reflected to the eye E by the mirror ABG only ⟨from⟩ the circumference of a single circle.

⟪Result⟫

[236] Thus, for every point located with the eye on a single diameter of a spherical concave mirror, while the centre of the mirror lies between [that point] and the centre of the eye, the form of that point will be reflected to that eye from the circumference of a circle on the mirror's surface; and the centre of the eye will be as a pole for that circle; and the form of that point will not be reflected from the surface of that mirror, except from that
V 270b circle. | The image of such a point, as described, will be according to [i.e., produced by] the whole circle from which the form [of that] point is reflected, and will be a single point, the same as the centre of the eye.

⟪V 2.VII.14⟫

[237] However, the point located on the diameter passing through the centre of the eye and the centre of the mirror is an imagined point, because the diameter through the eye's centre and the mirror's centre, with which the demonstration is concerned, is an imagined line. Thus, the point on this diameter, whose form is reflected from the circumference of a circle, is an imagined point. And, therefore, that point, by itself, would not be perceived by the eye, nor would its form be perceived in the mirror. But if it lies on the surface of a visible object, then the images of all points surrounding that point will all lie outside the diameter passing through the eye's centre. As a result, when the eye perceives that surface in this mirror, and the images of that surface or of one portion of it lie behind the eye's
V 271a centre, the eye will have an indistinct perception of the form of | that surface, or of that portion of it whose images lie behind the eye's centre; and [the eye] will also perceive the shape of that surface [or portion] differently from the [actual] shape of the surface. Now, if the eye perceives that surface indistinctly, and fails to perceive its shape as it is, then the location of that point will not be distinguished by the eye, nor will the point itself be perceived by the eye, because it is an imagined point. And also the part of that surface on which that point lies – [namely] the smallest part of it from which the form can be reflected from the mirror, and which is taken to be a point when its form is reflected from the circumference of the circle – the form [of that smallest part] will not be distinguished by the eye when the form is uncertain and [when] its shape differs from the shape of the object's surface. [The reason is] that the portion of the surface is visually distinguished in the form in the mirror by means of its colour and the location [of that part] in the form. If the form has a shape different from the object's shape, [then] the eye would distinguish
V 271b the shape of the object's parts in the form | only by means of their colours. And if the part is extremely small, and it is reflected from the circumference of a circle, then the reflected form will be extremely subtle. And if the form is uncertain and has a colour that is extremely subtle, then the eye will not perceive that extremely subtle part, even if it differs in colour from the remainder of the form, because if the eye does not distinctly perceive the object, it will not distinguish the object's parts that are extremely subtle.

[238] If the point whose form is reflected to the eye from the circumference of a circle lies in a surface whose images are [located] behind the centre of the eye, then the eye will not perceive the form of that point. If the images of a part of that surface or of the whole [surface] are [located] in front of the eye and between the eye and the mirror, or if the images of that part lie behind the mirror's surface, then the form of this point will be a limit of that form, and [the point's] form will therefore be indistinct, and the form of what adjoins this point on that object will also be indistinct. [The reason is] that the image of this point is [at] the centre of the eye, and the part adjoining this | point on the surface of the object, V 272a
whose image lies in front of the eye, will be touching the eye's centre. And the form of any thing whose image touches the centre of the eye or lies close to it cannot be distinctly perceived by the eye. And when the eye fails to perceive the form of this part distinctly, then the eye will not distinguish the form of the point, namely the limit of this part whose image is the eye's centre, on account of the small size of [this point] and its being confused with the form of the adjoining surface whose images adjoin the eye's centre inside the body of the eyeball. Therefore, the form of the points that are located on the diameter of the mirror through the eye's centre cannot be distinguished by the eye, nor can sight make a judgement regarding its perception. If this notion is examined experimentally by attaching a long object of a fair size to the surface of the mirror at right angles with it, and the form of that object is observed in the mirror, the form will be found as we have described.

[239] [Diagram V.29(b)] Now, if points E and H lie on a diameter of the mirror, and H lies inside the mirror, then $EA:AH > ED:DH$. [It is assumed that $ED>HD$ and that the two circles GDB and GAB intersect at points G and B.]

For $\angle ADG$ will | always be acute, because GB is a chord in circle GDB. Thus, the line V 272b
drawn from point G tangent to circle GAB will meet diameter EDA on the side of A, and the ratio of the line [segment] between point E and the extremity of the tangent line, to the line [segment] between the extremity of the tangent line and point H, will be as the ratio of ED to DH. For this was explained in the proposition with the five diagrams, in which we determined the ratios of the segments of the diameter on which the image-point lies, in the third diagram, and it is the reverse of the ratio shown in the third diagram [Diagram V.25(3)].

But $EA:AH$ is greater than the ratio of the line between point E and the tangent's extremity to the line between the tangent's extremity and point H. [Call X the extremity of the tangent, then $EX>EA$. Thus, $EH:AH>EH:XH$, therefore $(EH+AH):AH>(EH+XH):XH$.] Therefore $EA:AH > ED:DH$.

Thus, if points E and H lie on the mirror's diameter, and H lies inside the mirror, and $EA:AH$ is not greater than $ED:DH$, then, two lines will not be reflected [to each other] from points E and H at equal angles.

[240] As for the points that lie outside the diameter | passing through the eye's centre: V 273a
these will be the ones discerned by the eye, while of ⟨those⟩ points [outside the diameter], some will have a single image, some will have two images, some will have three, and some will have four ⟨but none will have more than four⟩.

⟨⟨V 2.VII.15⟩⟩

[241] [Diagram V.30] Let the mirror be ABG. We draw the plane passing through the centre of the eye, the mirror's centre, and the visible point: it will produce a circle in the mirror's surface, say circle ABG, with D as centre; let the eye's centre be point T, and let point H be in a visible object.

We join TD and HD and produce these lines until they cut the ⟨circle's⟩ circumference. Let HD meet the circumference in points A and G, and let TD meet it in points B and K. And first let line TD = line HD.

Bisect ∠TDH by line DE and extend DE in the other direction to meet the circle, say at point Z. Join TZ and HZ. Then ∠TZD = ∠HZD, and, therefore, the form of point H will be reflected to the eye T from point Z.

V 273b **[242]** [Diagram V.30] And I say that the form of point | H will only be reflected from arc AZG to the eye T at point Z alone. It has been shown that [the form of H] cannot be reflected from arc GK.

As for the reflection, if possible, from [arc] AK, let [the form] be reflected from another point, say O.

We join TH: it will be bisected by line ZD, say at point L. We draw lines TO, HO, and DO. Produce DO in the direction of D: it will cut HT at a point between points T and L, say in point M.

Then, since TO is closer to diameter TK than line TZ, TO is therefore longer than TZ. And HO is farther from diameter HD than HZ, therefore it is shorter than line HZ. But TZ = HZ, because TL = LH, and ZL is perpendicular to TH. ⟨⟨Therefore TO > OH, and TO:OH = TM:MH because of the equality of the angles at O; thus TM > MH, and that is impossible.⟩⟩

And [a similar conclusion] can be proved if point O lies between points A and Z. Therefore, the form of point H will be reflected from arc AZG to the eye T at point Z only.

And [the same] can be proved if points H and T are located inside the circle or on its
V 274a circumference. |

[243] [Diagram V.30] We draw from point E a line tangent to the circle: it will meet lines DT and DH, because each of the angles TDE and HDE is acute. Let the tangent line meet line DT at F, and meet line DH at point Q.

Then, if points T and H lie behind points F and Q, the form of either one of them will be reflected to the other from the circumference of circle ABG only at point Z. [The reason is] that the lines drawn from points T and H to point E will both lie outside the circle; no part of these two lines will lie inside the concave [surface] of the mirror. Therefore, the form of neither one of those two points will be reflected to the other from point E on the concave [surface] of the mirror.

Now, for every point on arc BEG other than E, if two lines are drawn to it from points H and T, then one of [the lines] will lie wholly outside the mirror, and thus the whole of this line will fall outside the concave surface of the mirror.

Therefore, the form of point H will not be reflected to the eye T from arc BEG if points
V 274b T and H lie outside the concave surface of | the mirror behind the tangent line. And the same can be shown if points T and H are in the location of points F and Q.

[244] Therefore, if the form of neither of the points T and H can be at all reflected to the other from arc BEG, and neither one can be reflected from arc AZK except from a single point, then the form[s] of both of the points T and H can be reflected to each other from the circumference of circle ABG from only a single point. [The reason is] that the form of either one of points T and H can be reflected to each other only from arcs AK ⟨⟨and BG⟩⟩; and the forms of points T and H can be reflected from the surface of this mirror only in the plane of circle ABG, since only a single plane can pass through points T, H, and D.

⟪**Result**⟫

[245] Therefore, for every two points equally distant from the centre of a spherical concave mirror, [when] both located outside the mirror's surface, but not on a | single V 275a
diameter, and both [are] behind the line tangent to the circle in the plane through those two points and the mirror's centre [and touching] the extremity of the diameter bisecting the angle contained by the diameters through those two points, or [when both points are] at the extremity of the tangent line: the form of either one of [the two points] can be reflected to the other from the surface of that mirror only from a single point.

⟪**V 2.VII.16**⟫

[246] [Diagram V.31] Furthermore, let us redraw the circle and the three diameters, and drop from E two perpendiculars to lines DB and DG; let them be perpendiculars ET and EH. Then ∠TED = ∠HED. Then if one of points T and H is the eye's centre and the other in a visible object, the form of that visible point will be reflected from two points. I say then that the form of such a point will be reflected from only two points, namely the extremities of ZE.

As for [the reflection] from arc AZG, it has been shown that [the form] can be reflected only from point Z.

As for [the reflection] from arc ABG, I say that the [point's] form cannot be reflected from arc AB, so let its form be reflected | from either arc BE or arc EG; let the reflection V 275b
point be point O, ⟪and let it be on ⟨arc⟩ BE⟫. We draw lines HO, ⟨TO⟩, and DO; then ∠HOD = ∠TOD.

We draw a circle on diameter DE: it will pass through points T and H, because the angles at T and H are right. And this circle will be tangent to circle ABG, because both are constructed on a single diameter.

Then, circle DTE will meet circle ABG only at point E. And arc TE will thus cut line DO, say in point Y. We draw lines HY and TY. Then ∠HYD = ∠TYD, because arc TD = arc HD. Thus, ∠HYO = ∠TYO; and [by assumption] ∠TOY = ∠HOY; and line OY is common to both triangles; thus line HO = line TO, but that is impossible, because TO < TE, and HO > HE, and TE = EH.

The same [impossibility] can be shown if the reflection is from arc EG.

And, therefore, the form of neither of the points T and H can be reflected to the other from the circumference ABG, except from points E and Z only. | V 276a

Therefore, the form of either of the points T and H can be reflected to the other [point] from mirror ABG from two [reflection] points only.

[247] [Diagram V.31] Further, we assume on lines TD and DH two points equally distant from point D; let them be points M and N. We join ME and NE. Then ∠MED = ∠NED; and therefore the forms of points M and N will be reflected, to each other, from points E and Z.

I say then that the forms [of M and N] cannot be reflected to each other from any points other than E and Z.

For, [with] angles DME and DNE being both obtuse, we construct the right angle DMC. Join NC, then ∠CND also is a right angle. On diameter DC we draw a circle, which will pass through points M and N; let the circle be DMC.

Then, if the form of either one of points M and N were reflected to the other from [any point] other than points E and Z, ⟨that reflection⟩ would then take place only from arc BG. But, for any point on arc BG, the line drawn to it from point D will cut circle DMC on one of arcs MC and CN, thus giving rise to the impossibility that arose in circle
V 276b | DTEH.

Therefore, the forms of points M and N will not be reflected to each other from the circumference of circle ABG, except from points E and Z; and, therefore, the forms of points M and N cannot be reflected to each other from mirror ABG, except from points E and Z.

⟪**Result**⟫

[248] [Diagram V.31] Thus, for any point [N] as distant from the ⟨centre⟩ of the mirror [D] as is the centre of the eye [M], if the ⟨line⟩ [EN] drawn from [that point] to the extremity [E] of the diameter [ED], bisecting the angle between the line [ND] joining the point with the mirror's centre and the line [MD] joining the eye's centre with the mirror's centre, contains, with the line [ND] joining the point and the mirror's centre, an angle [END] next to the centre not smaller than a right angle, then the form of that point [N] will be reflected [to M] from two points and from two points only [E and Z].

⟪**V 2.VII.17**⟫

[249] [Diagram V.32] Let us redraw the circle and the three diameters ADG, BDK,
V 277a and EDZ. Let | ∠BDE = ∠GDE. From point E we draw a line tangent to the circle: it will meet the lines [extending] the two diameters. On diameter BD assume a point T behind the foot of the perpendicular ⟨drawn⟩ from point E to diameter BD but closer [to D] than point B. Join ET; ∠ETD is then acute. Cut off DH = DT and join HE: ∠DHE is then acute and equal to ⟨∠DTE⟩.

⟪Construct the right angle DTO. And ∠TDE is acute, being half ∠BDG. Therefore, line TO meets DE outside the circle, say at O. Join HO: then ∠DHO is equal to⟫ the right angle DTO. About diameter DO draw a circle passing through points T and H; let it be circle DTO.

This circle will cut arcs BE and EG. [The reason is] that line BTD cuts this circle in points T and D, such that point B will be outside this circle. And, if so, and point T lies on the circle's circumference, then the circle cuts arc BE. We can similarly show that the circle cuts arc EG. Let ⟨circle DTO⟩ cut arc BE at point L and arc EG at point M.

V 277b We draw lines TL, DL, HL, TM, DM, HM, TZ, and ZH. Then | ∠TLD = ∠HLD, ∠TMD = ∠HMD, ∠TED = ∠HED, and ∠TZD = ∠HZD. Thus, the forms of points T and H are reflected to each other from four points. And, therefore, if either point T or H lies in a visible object, and the eye is at the other point, the form of that point will be reflected to the eye from four points, and it will have four images.

[250] [Diagram V.32] And the form of point T or of point H will not be reflected to each other except from those four points.

As for arc[s] BA and KG, ⟨⟨it is impossible, and as for arc AK⟩⟩ it is clear that the forms of those two points cannot be reflected to each other except from point Z.

I say further that there will be no reflection from arc BEG at [any points] other than L, E, and M. If possible, let the reflection take place from point F.

We draw lines TF, DF, and HF; produce DF ⟨to Q⟩; and join TQ and HQ; then ∠TFD = ∠HFD, ⟨⟨and also ∠TFQ = ∠HFQ⟩⟩. Now, also ∠TQF = ∠HQF, because | arc TD = arc V 278a
DH, and thus line HF = line TF, which is impossible, because line HF is longer than line HE, and line TF is shorter than line TE, and TE = line EH.

[251] [Diagram V.32] And if point F lies on arc BL, then point Q will be on arc TL, and this would imply the impossibility implied for points F and Q. And the same [implication] would follow for every point on arcs BE and EG.

Therefore, the forms of points H and T can be reflected to each other from the circumference of circle ABG at only four points.

And the forms of those two points cannot be reflected to each other in a plane other than the plane of the circle, because points T, H, and D can only be in a single plane. Therefore, the forms of those two points can be reflected to each other from the surface of mirror ABG at only four points.

[252] [Diagram V.32] For every point [T] that is as distant from the centre of the spherical concave mirror as is the centre of the eye [H], and that lies ⟨⟨inside⟩⟩ that mirror, such that the line [TE] drawn from that point [T] to the extremity [E] of the diameter bisecting the angle [TDH] between | ⟨the line joining the eye with the mirror's centre and⟩ V 278b
the line joining that point with the mirror's centre contains, with the diameter [DT] joining that point with the mirror's centre, an acute angle next to the mirror's ⟨⟨centre⟩⟩, the form of that point will ⟨not⟩ be reflected to that eye from more than four points.

And, if points T and H lie on the circumference of the circle, as points B and G, or outside the circumference and before the tangent line, then their forms will be reflected to each other from point E, [the reason being that] the lines drawn from the two [points] to point E cut arcs BE and EG, and fall inside the circle, and contain with diameter DE two equal, acute angles inside the concave [surface] of the mirror. And, therefore, their forms will be reflected to each other from the point ⟨of tangency⟩ E, provided that the mirror's surface does not so extend as to cut ⟨one of⟩ the two lines | of reflection before reaching V 279a
point E. [Any] two points of this description will have their forms reflected to each other from point Z. Thus, two such points will have their forms reflected to each other from two points. And, therefore, [their forms] will be reflected from two points only, because if the two points lie on the circle's circumference, as points B and G [Diagram V.32], or outside the circumference and below the tangent line, then circle TOHD does not cut arc BEG.

If the reflection from arc BEG is assumed to take place at a point other than E, then the line drawn from D to that point will cut circle TOHD outside circle BEG, and this will give rise to the impossibility that occurred for point Q.

Thus, if the two points equally distant from the mirror's centre lie on the circle's circumference, or outside the circumference and before the tangent line, then their forms will be reflected to each other from two points on the surface of the mirror, and only from two points.

⟪Results of the Three Propositions [VII.15–17]⟫

V 279b As for the points | that lie outside the circle's circumference and behind the tangent line, and those that lie on the tangent line that are equally distant from the mirror's centre: their form[s] will be reflected from only a single point.

As for the points that lie outside the circle and before the tangent line, and those lying on the circle's circumference: their forms will be reflected from only two points.

As for the points that lie between the foot of the perpendicular and the mirror's centre: their forms will be reflected from only two points.

As for the points that lie between the foot of the perpendicular and the circumference of the circle: their forms will be reflected from only four points.

Therefore, for every two points that lie at equal distances from the centre of the spherical concave mirror and that do not lie on a single diameter [of the mirror]: if one of them lies in a visible object and the other at the centre of the eye, then the form of that visible point will be reflected to that eye from the surface of that mirror, either from a single point, or from two points, or from four points. The form of [that point] will not be reflected from more than four points.

⟪V 2.VII.18⟫

[253] [Diagram V.33(a,b)] Furthermore, let us redraw the circle and the three diameters.
V 280a Let DE bisect ∠BDG, | and let TD and HD be different. I say that the forms of points T and H will in all cases be reflected to each other from arc AZK, [regardless of] whether T and H are [both] located outside or inside the circle, or one of them outside and the other inside the circle, or one of them on the circumference of the circle and the other either outside the circle or inside it.

[254] [Diagram V.33(b)] Let us assume a straight line at random, say KL, and divide it into two parts so that the ratio of one to another will be as that of lines HD and DT to one another; and let M be the dividing point.

Let KM > ML, and let HD correspond to KM and TD to ML. We could also make line TD the longer [of the two].

We bisect KL at N, and draw from N a perpendicular to KL; let it be perpendicular NF. Make ∠NLC = ½∠TDA; then ∠NLC will be acute. And line LC will meet line NF, say at
V 280b point F. Produce NF in a straight line to | Q. From point M draw line MC ⟪to meet NQ at Q⟫ so that QC:CL = HD:DA.

I say first that no line [such as] MC can be drawn in this ratio from point M except line MCQ alone.

If possible, let another [such] line be drawn, say MSO, and let OS:SL = QC:CL. From the extremity of whichever of these two lines is closer to point F, and I mean lines QC and OS, draw a line parallel to the other [line]. Let the closer of the two lines be SO, and draw SY || CQ. Point Y will lie between points O and Q.

Line QC will then be longer than line SY ⟪because they are in the ratio of CF:SF⟫.

But QC:CL = OS:SL; and, therefore, [since OS:SL = QC:CL > SY:CL > SY:SL], OS > SY, ⟪but SY > SO⟫, which is impossible. Therefore, there can be no more than one line from point M that cuts [off] lines QC and CL in the assumed ratio.

[255] [Diagram V.33(a)] That having been shown, let us draw from point D a line that contains with line DA an angle equal to angle MCL, say line DY. | V 281a

And make ∠DYg = ∠CLM, which is equal to ⟨⟨½∠TDA⟩⟩, and let ⟨line Yg⟩ meet line DA at point g. Produce Yg in a straight line to point t. Then ∠Hgt will be acute, since it is equal to the acute ∠CMN.

From H draw ⟨⟨a perpendicular to gt⟩⟩; ⟨let it be⟩ perpendicular HJ. Make Jt = JY and join Ht. From Y draw a line parallel to Ht, say YW: this line will then meet line Hg, say at W. Produce YD in the direction of D until it meets perpendicular HJ, say at point L_1.

[256] [Diagram V.33(a,b)] Now, since the angles in ΔDYg are equal to the angles in ΔCML, these two triangles are therefore similar. And since ∠L_1JY is right, therefore ΔL_1JY ~ ΔLFN. Also ΔHL_1D ~ ΔQFC, because ∠DL_1H = ∠CFQ and ∠HDL_1 = ∠QCF.

Therefore, HD:DL_1 = QC:CF. Now, QC:CL ⟨⟨= HD:DY(=DA), and, therefore FC:CL = L_1D:DY. Thus, FL:LC = L_1Y:DY, and CL⟩⟩:LM = DY:Yg, | because the two triangles are V 281b
similar; therefore FL:LM = L_1Y:Yg. Now, NL:NF = JY:JL_1, because the two triangles are similar, and thus NL:LM = JY:Yg; and therefore KL:LM = tY:Yg, and, by separation [of ratios], KM:ML = tg:gY.

But KM:ML = HD:DT, therefore tg:gY = HD:DT.

Now ∠WYg = ∠Htg, and ∠Htg = HYg, therefore HY:YW = Hg:gW = HD:DT; thus HY:YW = HD:DT.

[257] [Diagram V.33(a,b)] Now, ∠WYD > ∠DYH, since ∠WYg = ∠gYH.

Let us then make ∠DYd = ∠HYD, and join Td.

Since ∠HYg = ½∠HYW, and ∠HYD = ½∠HYd, then the remaining ∠DYg = ½∠dYW; | and ∠DYg = ½∠TDA, because it is equal to ∠NLF, thus ∠dYW = ∠TDA. V 282a

Now HY:YW ⟨⟨I mean HD:DT⟩⟩ = (HY:Yd) ⟨⟨I mean (HD:Dd)⟩⟩ × (dY:YW), and HD:DT = (HD:Dd) × (dD:dT), therefore dD:dT = dY:YW.

But ∠dDT = ∠dYW, and therefore ΔdDT ~ ΔdYW; thus ∠DdT = ∠YdW, therefore, line TdY is continuously straight.

And ∠HYD = ∠DYT, therefore, the forms of points H and T will be reflected to each other from point Y, and similarly for any two points on diameters BD and DG.

Further, if lines DB and DG are produced as far as possible, provided the two points are not equally distant from point D, then the forms of those two points will in all | cases V 282b
be reflected to each other from a point on arc AZK.

⟨⟨V 2.VII.19⟩⟩

[258] [Diagram V.34] Moreover, I say that it is possible for the forms of an unlimited [number] of points on one of diameters DB or DG to be reflected from every point on arc AZK other than Z to an unlimited [number] of the points on the other diameter; and that the distances from centre D of any two points whose forms are reflected to each other will be unequal.

Let us redraw the circle and the three diameters, and assume on arc AZK a point at random, say H, and draw DH. On diameters BD and DG assume two points at random that are equally distant from D; let them be points L and M, [and let them both be located] inside or outside the circle, or on its circumference, as we like.

We join LM: it will be bisected | by line ZD, say in point F. V 283a

Produce HD in the direction of D: it will meet line LM between points F and L, say at point N. Draw lines LH and MH.

Lines HD and DL will then be equal to lines HD and DM.

But $\angle HDL > \angle HDM$, thus line LH > line HM. And line LN < line NM, thus $\angle LHN < \angle NHM$, since the line that bisects $\angle LHM$ falls between points N and M.

Make $\angle DHT = \angle LHD$, which is smaller [than $\angle NHM$]. Then the forms of points L and T will be reflected to each other from point H; and the distances LD and DT are unequal.

[259] Similarly, if we assume on diameters DB and DG two points other than M and L, and make them equally distant from point D, it is clear that the form of one of these two points will be reflected from point H to a point on the other diameter other than points L and T.

It is thus clear from this demonstration that, from point H, there will be reflected the
V 283b forms of an unlimited [number] | of points on [either] one of diameters DB and DG to an unlimited [number] of points on the other diameter; and the distances of those points from centre D will be unequal.

And the same can be shown ⟨for every point⟩ on arc AZK.

Thus, from every point on arc AZK other than point ⟨Z⟩, there will be reflected the forms of an unlimited [number] of points on [either] one of diameters DB and DG to an unlimited [number] of corresponding points on the other diameter, and the distances of those points from centre D will be unequal.

And, for every two points on diameters DB and DG, that are equally distant from centre D, and regardless of how long lines BD and DG are extended, the forms of those two points will be reflected to each other from point Z.

And, for every point on [either] one of diameters DB and DG, however far extended in length, the form of that point will be reflected from every point on arc AZK to a point on the other diameter.

And the form of that point [on one of the two diameters] will be reflected ⟪from Z⟫ to the point on the other diameter that is equally distant from centre D as that point is from
V 284a centre D. And | the reflection of the form of that point will take place from every one of the remaining points on arc AZK to a point on the other diameter, and the distances of all of these [latter points] from centre D will be unequal to the distance of that point from centre D.

⟪V 2.VII.20⟫

[260] ⟪Diagram V.35⟫ That having been shown, I say that, for every two points located ⟨on BD and DG⟩ at unequal distances from the centre D, if the form of either point is reflected to the other from arc AZK, then ⟨the reflection will not⟩ take place, except from one point only.

Let us redraw the circle and the three diameters. Let points L and T not be equally distant from centre D, and let the form of either point be reflected to the other from point H. I say then that the forms of points L and T will not be reflected one to the other from arc AZK, except from point H alone.

If possible, let [the form] be reflected from a point other than H, say Q. We draw lines
V 284b | HL, HD, HT, QL, QD, and QT. Join TL. Produce lines HD and QD in the direction of D to meet line TL, say at points N and C.

Then line HL > line QL and line HT < line QT, therefore LH:HT > LQ:QT.

But LQ:QT = LC:CT, and LH:HT = LN:NT, thus LN:NT > LC:CT, which is impossible.

In like manner we may show, for every two points located at unequal distances on diameters DB and DG, however these are extended in the directions of B and G, that their forms will always be reflected to each other from a point on arc AZK, but only from a single point on arc AZK. | V 285a

And the same can be shown if point ⟨Q⟩ is assumed, anywhere, on arc ZA.

And, similarly for every two points located on two diameters of the spherical concave mirror, at unequal distances from the mirror's centre: their forms will be always reflected to each other from a point on the semicircle whose plane passes through those diameters, namely [the semicircle] opposite the one in which those two points lie; and the forms of those two points will not be reflected to each other from more than a single point on that semicircle.

⟪V 2.VII.21⟫

[261] [Diagram V.36] Furthermore, let point A be the centre of the eye, [and] point B the centre of the [spherical concave] mirror. We join AB and produce it in a straight line to meet the circle, say at points G and D. From points G and D we draw a plane that cuts the mirror, say in circle GE.

And, first, let the eye be located inside the mirror. Draw from the centre of the eye a straight line to point E ⟨on⟩ the circumference of the circle; let it be line AE. Join BE and make ∠BEZ = ∠BEA.

Then, for any point on line EZ, however far EZ is extended, if it lies | in a visible V 285b object, the form of that point will extend along line EZ, and will be reflected from point E to the eye A on line EA.

Now, the lines drawn from point B to line EZ will be unequal, and the shortest among them will be the perpendicular drawn from B. ⟪Now it [the perpendicular] [can be] equal to BA, or longer or shorter [than it]. If it is longer⟫, then the distance of any point on EZ, however far EZ is extended, from the centre B will be longer than the distance of A from centre B. ⟪If it is equal⟫, then only the point on EZ, forming the foot of the perpendicular, will be the same distance from centre B as point A is from B. All remaining points [on EZ] will be farther from point A than A is ⟨from centre B⟩.

If the perpendicular is shorter than line BA, then only two lines on either side of the perpendicular will be equal to line BA, and only the two points at their extremities will be as distant from centre B as point A is from centre B. The distances of all other points on line EZ from centre B will be unequal to A's distance from B. | And it is possible for V 286a the forms of all points on line EZ to be each reflected to the eye A. And the distances of all these points from the centre of the mirror differ from the distance of the eye's centre to the mirror's centre. And only one point among them, or two, will be as distant [from the centre] as the centre of the eye is; all other points will be unequally [distant].

And, again, a line may be drawn from point A to any point on the circle's circumference other than points G and D, and a line will be reflected from [that point] that corresponds to line EZ, whose situation will be the same as that of line EZ.

⟪**Result**⟫

Thus, it is clear from this proof that, for any point on the circumference of circle DEG other than points G and D, it is possible for the form of some point located on a visible object to be simultaneously reflected from that point to the eye A; and the distance of every one of those points will be unequal to the distance of the eye's centre from the mirror's centre.

⟪**V 2.VII.22**⟫

[262] [Diagram V.36] Let us also put the centre of the eye outside the circle, as point H. And draw lines HT and HK touching the circle. It is clear, as was shown for point A, that
V 286b it is possible for the form of some point on a visible object to be | reflected simultaneously to the eye H, from all points on arc TGK; and the distance of that point from the mirror's centre will be unequal to the distance of the eye's centre to the mirror's centre.

⟪**V 2.VII.23**⟫

[263] [Diagram V.36] Now, let point N, ⟪one of the points on EZ⟫, lie outside diameter DAG, and on the side of ET with respect to line GA. Join BN and continue it in a straight line to meet the circle, say at points L and M. Point E will then be located in sector LBD; and the form of point N will be reflected to the eye A from point E, which is located in sector LBD; and the form of point N will [also] be reflected from arc GM, i.e., the base of sector GMB, as was previously shown; and the same [holds] for every point located on line EZ outside ⟨diameter⟩ GBAD and on the side of ET.

⟪**Result**⟫

[264] It is clear from this demonstration that it is possible for the points located outside the diameter passing through the centre of the eye, and whose distances from the mirror's centre differ from the distance of the eye's centre, to have the forms of each [point] of them reflected from the arc of the sector on whose sides that point and the eye's centre are
V 287a located, | [at the same time] as the reflection of the form of that point from the arc of the sector opposite to this sector.

⟪**V 2.VII.24**⟫

[265] [Diagram V.36] Furthermore, angles BEA and BEZ must be both acute. We then draw from point B a perpendicular to line EZ; let it be BF.

Then, if line BA is perpendicular to line AE, then BA = BF, and angles FEA and FBA are equal to two right angles. We then draw lines BN and BO on either side of perpendicular BF at random. Then neither line BN nor line BO is equal to line BA; and angles OBA and OEA are together less than two right angles; and, therefore, $\angle OEA < \angle OBG$. Similarly, the line drawn from any point on line [segment] FE to point B will not be equal to line BA, and it will contain, with line BG, an angle greater than angle FEA.

And angle NBA together with angle NEA will be greater than two right angles; and,
V 287b therefore, | $\angle NEA > NBG$. And, similarly, the line drawn from any point on line [segment]

FZ to point B will not be equal to line BA. And, therefore, the angle contained by that line and line BG will be smaller than angle FEA.

[266] [Diagram V.36] If BA is not perpendicular to line AE we draw from point B a perpendicular to line AE; let it be perpendicular BQ. This perpendicular will be equal to perpendicular BF, because ∠BEF = ∠BEQ. Make FC = AQ, and join BC. Then BC = BA, and ∠CBF will be equal to ⟪∠QBA, and, therefore, ∠CBA = ∠FBQ⟫.

And ∠FBQ + ∠FEQ will be equal to two right angles, therefore ∠CBA + ∠CEA will be equal to two right angles. Thus, every line drawn from point B to line [segment] CE will be unequal to line BA, except for one line; and every one of those lines will contain with line BG an angle greater than ∠CEA; and every line drawn from point B to line [segment] CZ will be | unequal to line BA, and will contain with line BG an angle smaller V 288a than ∠CEQ.

If perpendicular BQ is located on the side of point E, then we draw line BC on the side of point E.

⟪Result⟫

[267] It is clear from this demonstration that, for the points that lie outside the diameter ⟨passing⟩ through the centre of the eye, and whose distances from the centre of the mirror are unequal to the distance ⟨of the centre of the eye from⟩ the mirror's centre: if the form of ⟨any of⟩ those points is reflected from the arc of the sector on whose sides the eye's centre and that point are located, then [either] the reflection angle may be smaller than the angle adjacent to the angle contained by the diameters passing [respectively] through the eye and through that point, or [the reflection angle] may be larger than this [adjacent] angle.

⟪V 2.VII.25⟫

[268] [Diagrams V.37, 37(a), 37(b)] That having been shown, I say that for any point whose form is reflected from the arc of the sector on whose sides the centre of vision and that point are located, if the distance of that point from the centre of the mirror is unequal to the distance of the centre of the eye from the mirror's centre, then, [either] the reflection angle will be either smaller [Diagram V.37(b)] than the angle adjacent to the angle contained by the diameters, | of which one passes through that point and the other V 288b through the centre of the eye, or [the reflection angle] will be larger [Diagram V.37(a)] than the [adjacent] angle, but they cannot be equal.

Let point A be the centre of the eye, B the visible point, and G the centre of the mirror; and let DEZ be the circle whose plane passes through the three points. Draw diameters AGHE and BGDZ. Let the form of point B be reflected to the eye A from point T, and let line BG be unequal to line AG. Join AB, GT, AT, and BT. I say that ∠ATB [the reflection angle] ≠ ∠AGD [the adjacent angle].

For ⟨if⟩ [∠ATB] were equal to ∠AGD, then ∠ATB + ∠AGB is two right angles.

We then draw the circle that circumscribes triangle ABG: this circle will then pass through point T, because ∠ATB + ⟪∠AGB⟫ is two right angles; let the circle be circle ATBG.

Now, ∠ATG = ∠BTG, therefore arc AG = arc GB, and therefore line AG = line GB. But these two lines are assumed to be unequal, which is impossible. Therefore, ∠ATB ≠

V 289a ∠AGD. | Therefore, for any two points not equally distant from the centre of the mirror: if the form of one of them is reflected to the other from a point on the arc of the sector on whose sides those two points are located, then the reflection angle [∠ATB] cannot be equal to the angle [∠AGD] adjacent to the angle contained between the two diameters passing through those two points, that [latter] angle being the vertex of the sector.

⟨⟨V 2.VII.26⟩⟩

[269] [Diagram V.38][16] I say further that, for any two points [eye, object] not equally distant from the mirror's centre: if the form of either point is reflected to the other from two points on the arc of the sector on whose sides the two points are located, then the two reflection angles will not both be smaller than the angle adjacent to the angle contained between the diameters passing through those two points.

Let us redraw the diagram, and let the form of point B be reflected to the eye A from
V 289b points T and K; and let | ∠ATB < ∠AGD. I say then that ∠AKB is not smaller than ∠AGD.

[270] [Diagram V.38] If possible, let it [∠AKB] be smaller. We join AT, BT, AK, BK, GT, GK, and AB. And let BG > GA. Bisect ∠AGB by line GFN. Then line BF > line FA [since BF:FA = BG:GA], and ∠GFB will be obtuse.

We then bisect AB at point Q. Then the perpendicular drawn from point Q onto line AB meets line GN in the direction of N ⟨relative to AB⟩.

We circumscribe circle ATB about △ATB: this circle will not pass through point G, because ∠ATB and ∠AGB are together less than two right angles, and [the circle] falls short of point G [since ∠ATB < ∠AGD], and therefore cuts line GT, say ⟨at⟩ O.

Now, since ∠ATG = ∠BTG, therefore arc AO = arc OB. We draw line OQ: it will be perpendicular to line AB; this line [OQ] will then meet GN in the direction of N. And this
V 290a circle cuts line GK, say at point M. Then point M is | not identical with point O.

[271] [Diagram V.38] Now, circle ATB does not pass through point K, because if it did, point M would then bisect arc AB, since ∠AKG = ∠BKG, which is impossible, because point O is the point that [bisects] arc AB. Therefore, circle ATB does not pass through point K.

And the circle circumscribing △AKB also falls short of point G [as a result of the assumption ∠ATB < ∠AGD], and cuts line GK at a point below line AB ⟨⟨he means on the side of G⟩⟩, and that point will bisect the arc of that circle subtended by line AB. Line AB will be a chord in that circle; point Q bisects that chord, and QO is perpendicular to this chord; and therefore line QO bisects the arc subtended by line AB of the circle circumscribing △AKB.

Therefore, the point on line GK that bisects this line [i.e., this arc] lies on QO; and, therefore, line QO meets line [segment] GK below line AB, which is impossible [if the second point of reflection K is on arc EN], because line QO does not meet line [segment] FG ⟨⟨below [AB]⟩⟩.

Therefore, the point of the second reflection does not lie between points N and E, if the
V 290b angle of the | second reflection is smaller than ⟨⟨∠AGD⟩⟩.

---

16. Note that the diagram is that drawn by Kamāl al-Dīn, and point K is therefore the point of intersection of two circles, hence the diagram represents the impossible case set out in paragraph [271]. The text also considers positions of K on the circular arc EN, but not on the circle through A, B, and T.

[272] [Diagram V.38] Now, the point of the second reflection must be such that the diameter drawn to it from point G will cut line QO either between points Q and O, or below point O, because this point must be located below line AB. [Let it first intersect QO between points Q and O.] Let the line drawn from point G to the point of the second reflection be line GYS. Point S is then the point of the second reflection.

We draw lines AS, BS, and TS. Then ∠GTS is acute, since lines GT and GS are equal. But line OQ is a diameter of circle BTA, and line GOT cuts this diameter on the circumference of the circle, and, therefore, line OT is a chord in this circle.

And the line drawn from point T and containing a right angle with line TO will meet line OQ on the circumference of the circle, if OQ is produced rectilinearly in the direction of Q.

[273] [Diagram V.38] And [the] line drawn from point T, which meets the extremity | V 291a
of the diameter, is a chord in the circle. And this line drawn from point T containing with line GT a right angle will be tangent to circle ZTK. But ∠GTS is acute, so line TS is below that chord, and, therefore, line TS lies inside ATB.

And line GYS also cuts the diameter at point Y, and, therefore, point S is located between the diameter and the circumference of the circle; point S therefore lies inside circle ATB.

And ⟨the circle⟩ that circumscribes △ASB also falls short of point G [because of the assumption ∠BSA < ∠AGD], and cuts line GS at a point bisecting the arc subtended by line AB. And diameter QO also bisects this arc. Thus the point of intersection between lines GS and QO is the point that bisects the arc subtended by AB. But the point of intersection is point Y. Therefore, point Y lies on the circumference of the circle that passes through points, A, S, and B.

Now the circle passing through points A, Y, and B cuts circle AOBT in points A and B,
and departs from it after passing points A and B. But point S | lies inside circle AOB, while V 291b
it lies on the circumference of the circle passing through points A, Y, and B. Therefore, the circle through points A, Y, B, and S intersects circle AOB after it passes point B. And, therefore, the circle that circumscribes △ASB intersects circle ATB in three points, and this is impossible.

[274] [Diagram V.38] Furthermore, if line GYS lies below point O, then point S will be located between points T and B. And thus, if we draw the circle that passes through points A, Y, and S, then points O and T will be located inside the circle, and the proof will be as was shown ⟪and the impossibility will arise as before⟫, because [the positions of] points O and T, with respect to that circle, will be the same as [the positions of] points Y and S with respect to circle AOT.

Thus, if the forms of points A and B are reflected to each other from two points, then the two reflection angles will not both be smaller than ∠AGD.

Therefore, for every two points [eye, object] not equally distant from the centre of the mirror: if their forms are reflected to each other from two points on the arc of the sector
on whose | sides those two points are located, then the two reflection angles will not V 292a
both be smaller than the angle adjacent to the angle contained by the diameters passing [respectively] through the mirror's centre and those two points.

⟪V 2.VII.27⟫

[275] [Diagram V.38] And it has also been shown by this demonstration that if the reflection angle is smaller than the angle adjacent to the angle contained by the diameters passing through the two points, then the line drawn from the centre of the mirror to the reflection point ⟪will divide the half of the sector's angle that is next to the farther of the two points from the mirror's centre⟫, because it intersects the perpendicular drawn from the point that bisects the line joining the two points, namely those whose forms are reflected to each other; and it intersects [the perpendicular] below that [bisection] point, and that point is [located] inside this half of the angle contained between the two diameters passing through the two points.

⟪V 2.VII.28⟫

[276] [Diagram V.39][17] Further, let us draw on the mirror a circle passing through its centre at random, let the circle ⟨of intersection⟩ be ABG. Draw in it at random two
V 292b diameters intersecting at the centre; let them be diameters ADG and BDK. Bisect | ∠BDG by line DE, produce [DE] in a straight line ⟨in the direction of E⟩.

On diameter DG assume a point behind the foot of the perpendicular drawn from point E to line DG, and short of point G, which is possible, because ∠EGD will always be acute; let it be point N. Cut off DM = DN, and join MN. Circumscribe a circle about △MDN: this circle will cut line DE outside circle BEG, because the lines drawn from points M and N to point E will [respectively] contain two acute angles with lines MD and ND. This circle will then cut arc BEG in two points on either side of point E; let it be circle DMTHLN. Let point H be on line DE, and let points T and L be the points of intersection. Draw lines ⟪TN, TD, TM, LN, LD, and LM⟫. Then ∠NTD = ∠MTD, and ∠NLD = ∠MLD, because arc MD = arc DN. [Thus, M and N are reflected to each other from points T, H, and L.]

V 293a Now, line DT cuts line MN, say at point | F. Let the point of intersection between lines DE and MN be point S. Then ∠DSF is right and ∠DFS is acute. We then draw from point F a line that contains a right angle with line DF: this line will in all cases cut TN, and in all cases cut circle DMHN, because ∠DTN is acute; let it be line WF. Line NL, or a part of it, will then always be [situated] below line FW with respect to point T.

We then assume a point on line NL ⟪below FW⟫, say point Q. And we join TQ and produce it in a straight line: it will cut arc NL, say in point C. Point C will then lie below line FW, because the line joining points N and L will be below line FW, and so point C will be below this line.

Then arc CN is [or is not] shorter than arc THL; if not, we cut off from it an arc shorter
V 293b than arc THL, and join the cut-off point | and point T with a straight line. This line will cut line NL at a point below line FW.

[In this case we take the cut-off point instead of C.] Thus, let arc NC be shorter than arc THL; and point Q is [assumed to lie] below line FW. Then line TC cuts line FW, say in point Z.

⟪We then make ∠MTO = ∠NTC; then ∠CTD = ∠DTO. Now, since arc NC is shorter than arc THL, then ∠NTC, I mean ∠MTO, is smaller than ∠LMT. Therefore, line LM will meet TO, if both are produced; let them meet at O.⟫

17. Note that line DQ drawn in the text is omitted from the diagram here for the sake of clarity.

We then make ∠QNY = ∠NTQ, thus ΔYNQ ~ ΔNTQ. And ∠TNL = ∠LMT, and ∠LMT = ∠MTO + ∠MOT, thus ∠TNL = ∠MTO + ∠MOT. But ∠YNL = ∠MTO, thus by subtraction ∠TNY = ∠MOT. But ∠NTY = ∠MTO, thus ΔNTY ~ ΔMTO, and line TN corresponds to line TO. We join DO and DQ.

Now, since TF is perpendicular to | line FW, and ∠FTN = ∠FTO, therefore line TO, if produced rectilinearly in the direction of O, will meet line WF if produced rectilinearly in the direction of F. And the [segment] of line TO located ⟪between T and line WF will be equal to TZ⟫. V 294a

Thus, line TZ > line TO; and line QT > line TZ, because point Q lies below line FW; therefore line QT is ⟪much⟫ longer than line TO.

Since ΔTNY ~ ΔTMO, therefore NT:TO = NY:OM. And since ΔTNQ ~ ΔYNQ, therefore NT:TQ = YN:NQ. And NT:TO > NT:TQ, thus NY:OM > NY:NQ, therefore line NQ > line MO.

Now, ∠DML + ∠DNL is equal to two right angles, therefore ∠DNL = ∠DMO; and these two [angles] are | both obtuse. Therefore, if a line equal to line MO is cut off from line NQ, and a straight line joins the intersection-point and point D, then ⟨this line⟩ will be equal to line DO, and that is shorter than DQ, because ∠DNQ is obtuse. Therefore, line DQ > line DO. V 294b

Now points O and Q are located [respectively] on diameters DO and DQ; and ∠OTD = ∠QTD, and ∠OLD = ∠QLD. Thus, if either point O or point Q lies on a visible object, and the centre of the eye is at the other point, the form of that point will be reflected to that eye from both points T and L.

Now, ∠NDQ > ∠MDO. We take ∠QDM as a common angle, then ∠NDM > ∠QDO. | V 295a

Therefore, point O does not lie on line QD, and, thus point O is outside diameter DQ; and line DQ, if produced rectilinearly in the direction of Q, will cut arc GL between points L and G, because line DQ lies between lines DL and DG. But DO, if produced in the direction of O, will cut the circle behind point B; and, therefore, points T and L lie on the arc of the sector on whose sides points O and Q are located.

Now ∠MDN + ∠MLN is equal to two right angles, and ∠ODQ < ∠MDL, thus ∠ODQ + ∠OLQ is less than two right angles. Therefore, ∠OLQ is smaller than the angle adjacent to ∠ODQ. Thus, ∠OTQ is greater than the angle adjacent to ∠ODQ, because it is not possible for each of the two reflection angles to be smaller than the angle adjacent to ∠ODQ. Thus, neither [of the two angles] will be equal to it.

⟪**Result**⟫

It is thus clear from this demonstration that the form of either one of two points [eye, object] not equally distant from the centre of the mirror may be reflected to the other | from two points on the arc of the sector on whose sides those two points are located. V 295b

⟪**V 2.VII.29**⟫

[277] [Diagram V.40(a,b)] I say then that the forms of two points [eye, object] not equally distant from the centre of the mirror cannot be reflected to each other from more than two points on the arc of the sector on whose sides those two points are [respectively] located.

Let us redraw the diagram, and let ∠OTQ > ∠ODA [Diagram V.40(a)]. Make ∠OTt = ∠ODA. Bisect ∠QTt by line TE and produce TE in a straight line; and draw line QZ || line Tt. Join OQ, [and] let it cut line DT in point F. And bisect ∠ODQ by line DSN.

Now the circle circumscribing ∆OTQ will cut line TD at a point [H] below point D, because ∠OTQ and ∠ODQ are together greater than two right angles, and therefore the arc subtended by line OQ is bisected by that point [H]. ⟪The [line] joining that point and the point that bisects OQ is then perpendicular to OQ.⟫

It [the midpoint of OQ] lies between points S and Q, because line QD > [line] DO, and
V 296a QD:DO = QS:SO, | therefore, line QS > line SO; thus the midpoint lies between points S and Q.

⟪If [the line] drawn from a point [H] on line DT below point D to a point [i.e., the midpoint] between S and Q⟫ is perpendicular to line OQ, then ∠DFS is acute. But ∠DSO is acute, [since DQ > DO]. Therefore, the perpendicular drawn from point D to line OQ will lie between lines DS and DF; and, therefore, line DF cuts ∠ODS. Therefore, ∠ODF is smaller than half of ∠ODQ, [thus the reflection point T is on arc NB between N and B].

⟪And since ∠QTE = ½∠QTt, and ∠QTD = ½∠QTO, therefore by subtraction, ∠ETD = ½∠tTO, I mean [½]∠ODA. Therefore, ∠TED = [∠TDA – ∠ETD = ∠TDA – ½∠ODA] = ½∠ODA together with the whole of ∠ODT.⟫ But ½∠ODA together with ∠ODN, which is ½∠ODQ, is [a] right [angle], therefore ∠TED is acute, [since T is on arc NB between N and B].

We draw from point Q a perpendicular to line TZ; let it be QK. Then, since QZ || Tt,
V 296b therefore ∠QZT = ∠tTE. And ∠tTE = ∠QTE, | therefore ∠QTE = ∠QZE, and [thus] line QT = line QZ. And line QK is perpendicular [to TZ], thus line KT = KZ.

[278] Now, line OT is either parallel to line DQ, or it meets it [DQ] in the direction of either A or of G.

If OT || DQ [Diagram V.40+], then ∠OTD = ∠TDQ, and therefore ∠TDQ = ∠QTD. Thus, line QT = line QD.

Now ∠DOT = ∠ODA, and therefore, ∠DOT = ∠OTt; thus line DO = line tT, because lines DO and tT are either parallel, thereby forming parallelogram OTtd, and thus DO = tT; or they ⟨meet⟩ in one of the two directions, thereby forming an isosceles triangle having line OT as base. But OT || line Dt, thus [in the isosceles triangle] DO = tT; therefore
V 297a | line QT:Tt = QD:DO.

[279] If line TO meets line QD in the direction of A, let them meet in point C [Diagram V.40(a)]. Then ∆CDO ~ ∆CTt, because ∠CTt = ∠CDO ⟪and ∠C is common⟫. But QT:TC = QD:DC, since ∠QTD = ∠DTC,[18] and CT:Tt = CD:DO [because ∆CDO ~ ∆CTt], therefore QT:Tt = (QD:DC) × (CD:DO), thus QT:Tt = QD:DO.

[280] If line TO meets line QD in the direction of Q, as OTL_1, let them meet in point L_1 [Diagram V.40++].

We draw line DR || TQ [to meet TO extended in point R]:[19] then ∠RDT = ∠RTD ⟪i.e., ∠QTD⟫, thus line RT = line RD. Therefore, $DR:RL_1 = TR:RL_1$.

V 297b But $DR:RL_1 = QT:TL_1$, and $TR:RL_1 = QD:DL_1$, therefore $QT:TL_1 = QD:DL_1$. |

18. The text in MS K resumes here, at fol. 30a

19. Not drawn in the original diagram.

And $L_1D:DO = L_1T:Tt$, because $\angle L_1Tt = \angle L_1DO$ [i.e., triangles DRL_1 and TtL_1 are similar], therefore, *ex aequali*, $QT:Tt = [(QT:TL_1) \times (TL_1:Tt) = (QD:DL_1) \times (DL_1:DO)] = QD:DO$.

Thus, for all positions, QT:Tt = QD:DO.

Now QT:Tt = QE:Et, because ∠QTE = ∠ETt, and QE:Et = ZE:ET, thus ZE:ET = QD:DO. And ∠QKE is right, and ∠DTE is acute, therefore line QK produced in a straight line will meet line TD, say at point H.

[281] We set apart △HTK, in which angle ⟨K⟩ is right, together with line TEZ and with △TQZ, to make the diagram more clear [Diagram V.40(b)].

Then line ZT, now separated from the circle, is cut into two halves at point K, and into two unequal segments at point E.

Now, [1] ZE:ET = QD:DO, these [two latter lines] being the distances of points Q and O from the centre of the circle, | and [2] ∠KTH = ½∠ODA ⟪because it is half of ∠tTO, V 298a
which is equal to ∠ODA⟫, and [3] the ratio of QD in the separated triangle to DT is the ratio of the distance of point Q in the circle from the circle's centre to the circle's radius, namely line DT, which is a known ratio.[20]

We then draw line EC || KH and circumscribe a circle around △ETC; let it be circle ECMTF. Then line CT will be a diameter of the circle, because ∠CET is right. Produce ED to meet the circle, say at point M. And we join EH and TM.

From ∠EMT, which is equal to ∠ECT, we cut off ∠EMF = ∠EHT: line MF will then cut line CT, say in point Y.

Now ∠QHD = ∠ECD and ∠ECD = ∠TMD, therefore △QHD ~ △TMD; therefore HD:DM = QD:DT, which is a known [ratio]. But HD:DM = EH:YM, thus EH:YM is a known ratio.

[If] EH is known,[21] then MY is known. | Line MY is [situated] between diameter CT V 298b
and the circle's circumference; point F [lies] below line CT and on the circumference; and arc EF is known, because ∠EMF = ∠EHT, which is known. Thus point F is known, and line FYM has been drawn from it so that line YM is equal to a known line; this diagram then becomes the same as the third diagram in the lemmas we have proposed [Diagram V.14]. Thus, line MY occurs either once, or twice, or not at all; and it cannot occur more than twice, as this was shown in the third diagram of | the lemmas.[22] V 299a

Therefore, the forms of points O and Q cannot be reflected to each other from more than two points on arc BG.[23]

And it has been shown that the two forms are always reflected to each other from only one point on arc AW.

20. Ibn al-Haytham here constructs Diagram V.40(b), such that these three properties are satisfied, analogous to Lemma III above.

21. This step is unexplained; see below.

22. The argument is valid with the following modification: assume that Diagram V.40(b) is drawn on a different scale than the corresponding part of Diagram V.40(a); choose the scale and draw the new diagram, V.40(b) in such a way that segment EH is known. One can now argue that the diameter CT is a known segment, F is a known point, and segment MY has a known length.

23. As Kamāl al-Dīn points out, Ibn al-Haytham assumed ∠OTQ > ∠ODA and so omits the single point of reflection on arc NG; cf. point T in Diagram V.38 above.

⟨⟨**Result**⟩⟩

Therefore, for the two points not equally distant from the centre of the spherical concave mirror, if one of the two points lies inside the mirror, then their forms cannot be reflected to each other from more than three points on the spherical concave mirror.[24] And that is what we wished to prove.

⟨⟨**V 2.VII.30**⟩⟩

[282] [Diagram V.40(c)] It is clear from this demonstration that, if the two not equally distant points are assumed, how the reflection point is found on the arc of the sector on whose containing diameters those two points are located.

For if the two [assumed] points [in Diagram V.40(a)] are like points Q and O, ⟨and⟩ a line such as ZT is assumed at random [Diagram V.40(c)] then divided in point E so that ZE:ET [in Diagram V.40(c)] = QD:DO [in Diagram V.40(a)] and bisected at point K; and
V 299b [if] a perpendicular such as the perpendicular KH is drawn; | and [if] ∠KTH [Diagram V.40(c)] = ½∠ADO [in Diagram V.40(a)] is constructed; and [if], from point E, a line such as line QED is drawn [in Diagram V.40(c)] such that the ratio QD:DT [Diagram V.40(c)] is equal to the ratio of QD in the circle to the radius of the circle [Diagram V.40(a)]; and [if] then ∠QDT in the circle [Diagram V.40(a)] is made equal to ∠QDT in the separated triangle [Diagram V.40(c)]; and [if] the lines are joined; then, there will be produced in the circle [Diagram V.40(a)] triangles similar to those in the separated triangle [Diagram V.40(c)]. And thus it can be shown, by reversing the demonstration cited in this proposition, that ∠OTD = ∠QTD.

And if two lines are drawn [in Diagram V.40(c)] from point E in the ratio of QD to DT, there will be produced [in Diagram V.40(c)] at line HT two unequal angles. Then if two angles like those two angles are constructed on line QD in the circle [in Diagram V.40(a)], there will be produced two points on the circle's circumference. Then if lines are drawn from points Q and O in the circle [in Diagram V.40(a)] to these two points, there will be produced two equal angles at each one of the[se] two points. In this way, then, it is possible to find the point of reflection in this and similar sectors.

V 300a And in the way | previously [described] it was shown how to find the point of reflection in the sector opposed to this sector. And in the preceding propositions it has been shown how to find the point of reflection when the two [assumed] points are equally distant from the centre of the mirror.

Thus, it has been shown how to find the reflection point on the surfaces of spherical concave mirrors, when the visible point and the eye-point are assumed, for all positions of these two points.[25]

---

24. The maximal number is actually four, and later on, in Diagram V.40, Ibn al-Haytham correctly states that there can be three points of reflection on the arc corresponding to arc GNB.

25. Note that Ibn al-Haytham does not provide a construction for the reflection point on arc NG in Diagram V.40(a).

⟪V 2.VII.31⟫

[283] [Diagram V.41] Furthermore, let points A and B lie outside the surface of the mirror, and let the centre of the mirror be point G. We join AZG and BEG and draw the plane in which lines AG and BG lie; let it produce circle DEZ on the mirror's surface. And let the forms of points A and B be reflected to each other from point T on arc ETZ. I say that the forms of points A and B cannot be reflected to each other from any point on ETZ other than point T.

If possible, let the reflection take place from point H. We draw lines AT, AH, BT, BH, GT, and GH. And let lines AH and AT cut arc ETZ in points M and N; for if they did not cut | the arc, no reflection would take place from points T and H. Let lines BT and BH cut V 300b arc ETZ in points K and L.

Then ∠BTG = ∠GTA, therefore arc KT = arc TN; and ∠BHG = ∠GHA, therefore arc LH = arc HM. But arc LH > arc KT; thus arc HM > arc TN, which is impossible.

Therefore, the forms of points A and B cannot be reflected to each other from more than one point on arc ETZ.

⟪V 2.VII.32⟫

[284] And [the same] can be shown to hold if one of points A and B lies on the circumference of the circle and the other lies outside the circle.

Thus, it has been shown by our explanations and demonstrations that no two points not equally distant from the centre of the mirror can have their forms reflected to each other from more than three points on the arc of the sector on whose sides those two [assumed] points are located; nor can the forms of [the two points] be reflected to each other | from V 301a more than one point on the sector opposite to this sector.

⟪General rule regarding all reflections of points from this [spherical concave] mirror⟫

[285] For any two points [eye, object] not equally distant from the centre of the mirror, the line joining them either cuts the circle whose plane passes through those two points and the mirror's centre, or it touches this circle, or falls outside it.

If the line joining the two points lies outside the circle or is tangent to it, then there can be no reflection of the forms of those two points to each other from the arc of the sector on whose sides the two points lie. The reason is as follows. For any two lines drawn from those two points to a point on the arc of the sector on whose sides the two points lie, [and] provided that the line joining those two points falls wholly outside the circle or is tangent to it: each of the two lines ⟪or one of the lines⟫ will always fall completely outside the circle, and will cut the arc of that sector in a single point. The angle contained by that line and the line drawn | from the mirror's centre to that point will not be located inside V 301b the mirror's surface, and, therefore, the form of neither one of those two points will be reflected to the other from that point [on the arc]. But the forms of those two points will always be reflected to each other from a single point on the arc of the sector opposed to that on whose sides those two points are located. Thus, two points of this description cannot have their forms reflected to each other except from a single point on the surface of the spherical concave mirror.

[286] If the line joining those two points [eye, object] cuts the circle, then the circle circumscribing the triangle whose sides are the line joining the two points and the two lines joining those two points with the mirror's centre either cuts the circumference of the circle [drawn] on the mirror's surface, I mean the circle in whose plane that triangle lies, or it touches [this circumference] internally, or it falls wholly inside the circle on the mirror's surface.

If the circle circumscribing the triangle lies inside the circle on the surface of the
V 302a mirror, then any | two lines drawn from those two points [eye, object] to [any] point on the arc of the sector marked off by the two diameters passing through those two points, [that is, the sector] on whose sides the two points are located, will contain an angle smaller than the angle adjacent to the angle contained by the two diameters.

The reason is that the angle contained by the lines drawn from those two points to [any] point on the arc of the sector on whose sides the two points are located will be smaller than the angle contained by the segment of the circle circumscribing the triangle, [namely the segment] cut off by the line joining the two points. And the angle contained by this segment is equal to the angle adjacent to the angle contained by the diameters passing through the two points, because [that angle] and that of the triangle at the centre are together equal to two right angles.

That being so, such two points cannot have their forms reflected to each other from more than a single point on the arc of the sector on whose sides those two points are
V 302b located. Such two points, however, will always have their forms reflected to each other | from another point on the sector opposed to this sector. Therefore, the forms of two points of this description cannot be reflected to each other from more than two points [on the mirror].

And the same will hold if the circle circumscribing the triangle touches the circle on the surface of the mirror, because there can be no reflection of the forms of those two points to each other from the point of tangency, because the angle contained by the lines drawn from the two points to the point of tangency and the angle of the triangle at the mirror's centre will together be equal to two right angles.

As for all remaining points on the arc of the sector: the two lines drawn to any of them from those two points will contain an angle smaller than that at the point of tangency. Therefore, the case of these two points will be the same as that of the two points for which the circle passing through them and through the mirror's centre will fall inside the circle on the mirror's surface.

[287] If the circle circumscribing the triangle intersects the circle on the surface of the mirror, and the two points [eye, object] lie outside this circle [of the mirror]; or if one of
V 303a them | is outside the circle and the other on the circle's circumference, then neither of the two points of intersection will be on the arc of the sector on whose sides the two points are located, and between the sector's sides; rather, they will both lie outside the sector's arc, or one of them will be [at] the end of the arc and the other will lie beyond the arc. And thus, any two lines drawn from those two points to a point on the arc of that sector ⟨will contain an angle⟩ greater than the angle adjacent to that contained by the two diameters passing through the two points, the reason being that that angle will be greater than the one that is contained by the segment [of the circle] subtended by the line passing through those two points. And it is not possible for two such points to have their forms reflected to each other, except from a single point on the arc of the sector on whose sides those two points are

located, as was shown in the proposition preceding this one [V 2.VII.31 = Diagram V.41]. However, the forms of two such points will be reflected to each other from a second point on the arc of the sector opposed to this sector. Therefore, it is not possible for the forms of two such points to be reflected to one | another from more than two points [on the concave surface of the mirror]. V 303b

[288] If the two points lie inside the circle on the surface of the mirror while the circle circumscribing the triangle cuts the circle on the mirror's surface in two points, then a segment of the sector's arc will fall inside the circle circumscribing the triangle and another segment outside the circle circumscribing the triangle, and [the part] cut off from the sector's arc [will fall] on either side of the circle circumscribing the triangle.

[289] If one of the two points [eye, object] lies inside the circle on the surface of the mirror and the other outside the circle or on its circumference, then some part of the sector's arc may also fall inside the circle circumscribing the triangle and another part outside it, and the outside part will lie on one of its two sides [of the circle].

The forms of two points [eye, object] [located] according to these two descriptions [in 288 and 289] can be reflected to each other from three points on the arc of this sector. The form can be reflected from two points on the part of the sector's arc that lies inside the circle circumscribing the triangle. | V 304a

And it has been shown[26] that the form can be reflected from a single point on the part of the arc of the sector that is outside this circle [circumscribing the triangle]; where it [the part of the arc outside the triangle] is longer than half the sector's arc [the reflection point may be] on the part that is next to the point [eye or object] closer to the ⟨centre of the⟩ mirror;[27] and where two [different] parts are cut off from the sector [by the triangle, as in Diagram V.38], [the reflection point T is] on the part [of the arc] next to the point [B], which is farther away from the centre [of the mirror]. Therefore, it is possible for the forms of two points of these two descriptions to have their forms reflected to each other from three points on the arc of the sector on whose sides these two points lie, and [the forms] may [also] be reflected from two points.

And the forms of two such points will always be reflected to each other from another point located on the arc of the sector opposed to this sector.

Therefore, the forms of two such points may be reflected to each other from three points on the surface of the mirror, and they may be reflected from four points.

⟨⟨**Result**⟩⟩

[290] Thus, the points located outside a diameter of the mirror that passes through the centre of the eye, and whose distances from the mirror's centre are not equal to the distance | of the eye's centre from the mirror's centre, are such that: V 304b

[in some cases] the form of the single point will be reflected to that eye from the surface of that mirror at a single point, but cannot be reflected from any other point;

or [in other cases] it will be reflected from two points, but not from more;

or [in other cases] it will be reflected from three points, but not from more;

or [in other cases] it will be reflected from four points, but not from more.

---

26. Ibn al-Haytham did not provide a geometrical construction of this reflection point.

27. The text is unclear: no such reflection point may exist.

As for the points whose distances from the mirror's centre are equal to the distance of the eye's centre from the mirror's centre:

some will have their form reflected from a single point only;
some will have [their form] reflected from only two points;
some will have [their form] reflected from only four points;
but they cannot be reflected from more than four points.

And, therefore, for any point perceived by the eye in a spherical concave mirror and located outside the diameter passing through the eye's centre, the form of that point cannot be reflected to that eye from more than four points.

[291] The form of any point located on the diameter passing through the centre of the V 305a eye will be reflected to | that eye from [all points on] the circumference of a circle on the surface of the mirror.

[292] If, therefore, the eye faces a spherical concave mirror larger than half a sphere, then the visible objects [placed] opposite the mirror in this condition are as follows: some of them have their form reflected to the eye from a single place on the mirror; [the form of] others will be reflected from two places; [the form of] others will be reflected from three places; [the form of] others will be reflected from four places; and [the form of] others will be reflected from the circumference of a circle.

An[y] object whose form is reflected from a single place will have a single image; an object whose form is reflected from two places will have two images; an object whose form is reflected from three places will have three images; and an object whose form is reflected from four places will have four images. And the image of the point whose form is reflected from the circumference of a circle will be a single point, which is the centre of the eye, and therefore its form will be a circle that is continuous with the form of what surrounds that point on the surface of the object in which it [the point] lies.

Thus, the images perceived by the eye in spherical concave mirrors will be according to the descriptions we have explained and proved. And that is what we wished to prove.

⟨⟨Notice⟩⟩

[293] If, however, the spherical concave mirror is cut on one side, and the eye faces it, then only the surface of the mirror will face the eye, and the visible objects [placed] opposite the mirror will be on the same side as the centre of the eye with respect to the mirror's surface. If the eye's centre [coincides] with the mirror's centre, the eye cannot V 305b perceive any of the objects | [placed] outside the eye; and it cannot perceive the visible point if it [the point] lies on the mirror's centre. Therefore, the eye will not perceive any of the visible objects in a spherical concave mirror unless the centre of the eye lies outside the mirror's centre and unless the objects lie outside the mirror's centre.

Now, between any point on the surface of a visible object and the mirror's centre there is a straight line that is a diameter of the mirror, and if the eye's centre lies outside the mirror's centre, then between it and the mirror's centre there is a straight line that is a diameter of the mirror. Thus, if the eye faces a spherical concave mirror, the diameter passing through any point perceived by the eye in that mirror, provided that [the point] lies outside the diameter passing through the eye's centre, will cut the diameter passing through the eye's centre; and the two diameters will produce a plane that cuts the surface of the mirror, thus producing in it a circle; and the form of that point will be reflected only

from [one or more points on] the circumference of that circle. And those two intersecting diameters | will mark off on that circle two mutually opposed sectors; and that point and V 306a
the eye's centre will lie on the sides of one of those two sectors.

Now, either the centre of the mirror will lie behind the eye's centre or behind that point, and that point and the eye's centre will be closer than the mirror's centre to the surface of the mirror; or that point and the eye's centre will be farther from the mirror's surface than the mirror's centre, and the mirror's centre will lie between [that point] and the mirror's surface. If the sector, on whose two sides the eye's centre and the visible point are located, lies next to the mirror's surface, while the mirror's centre lies behind the eye's centre and behind that point, then the sector opposite this sector will lie wholly behind the eye's centre. This [opposite] sector will then cease to be effective in this case, and nothing will be reflected from it.

If the mirror is cut off on one side, and the eye and the object lie on the cut-off side of the mirror, and [if] the eye's centre and the visible point are farther from the mirror's surface than the mirror's centre, then all or most of the sector on whose | sides the visible V 306b
point and the eye's centre lie will cease to be effective in this situation, and nothing will be reflected from it. Then such a visible point as has been described will not have its form reflected from more than a single point. Most of the visible objects perceived in these mirrors are only perceived in this manner.

If the eye shifts from being [directly] facing the mirror's surface and the mirror is not smaller than half of a sphere, or if it is close to being half a sphere and it is [cut off] from a fairly large sphere, and [if] the object is away from the eye and the eye and the object are closer to the mirror's surface than the mirror's centre, then the form of that object may sometimes be reflected to that eye from three places. Thus, if the spherical concave mirror is a segment of the mirror's sphere, then that object may have its form reflected to that eye from three places. And if the spherical concave mirror is a segment of a sphere that is cut off on | one side, then the eye will rarely perceive from it more than three forms of a V 307a
single object.

⟪Experimental examination⟫

[294] To examine experimentally the images of visible objects in this [spherical concave] mirror, let the experimenter take a spherical concave mirror from a large sphere, and let him cut it off on two opposite sides so that a segment in the shape of a ring remains. To examine the images in this mirror, let the experimenter place his eye close to the mirror's surface, so as to [be able to] view the whole surface of the mirror: the eye will in this situation be on one of the cut-off sides of the mirror; let him put on the other cut-off side a small-sized object of a light colour. He should then look over the whole mirror: he will in this situation see the form of that object in several places, sometimes he sees it in four places, and sometimes he sees it in three places, and sometimes he sees it in two places. If he sees the form of that object in one place [only], then let him move [the object] from its place, or move the mirror, or his eye, or both the mirror and his eye. By doing this | the images of that object will appear to him, and he will perceive V 307b
the [object's] form in several places, but he will not perceive its form in more than four places.

⟪V 2.VII.33⟫

[295] All the images of the spherical concave mirrors are as we have described and proved with respect to a single eye. If, however, the object is perceived with both eyes at once in this mirror, then any point on it whose distances from the centres of both eyes are equal, or nearly so, will have as its image, with respect to both eyes, a single, common point, with the exception of the point that is reflected along a line parallel to the diameter passing through it: the two forms of [that point] will be located at the two points of reflection. The proof that the image-point in this mirror will be common for both eyes is the same as the [corresponding] proof for the case of the spherical convex mirror. And the form of a single object perceived by both eyes in this [spherical concave] mirror will be two overlapping forms, which the viewer will perceive as a single form if the radial lines issuing [i.e., drawn] from both eyes and meeting in this manner are similarly situated, or
V 308a nearly so, if | every point on [the object] has a single image. The reason for this is the one we explained for spherical convex mirrors.

⟪V 2: Eighth Inquiry. On the Images of Cylindrical Concave Mirrors, Consisting of Ten Propositions⟫

⟪V 2.VIII.1⟫

[296] [Diagram V.42] With regard to the right-cylindrical concave mirror: the intersections between its surface and the planes of reflection are all [conic] sections, with the exception of two: one is the intersection produced by the plane passing through the whole axis, and the other is the intersection produced by the plane parallel to the base of the mirror. The intersection made by the plane through the whole axis will be a straight line or two straight lines, and the intersection made by the plane parallel to the base will be a circle. The images of the straight line will be behind the mirror, like the images of plane mirrors. As for the circle produced by the plane parallel to the mirror's base, if [this circle] lies on the mirror's surface itself, its images and reflections will be like the images and reflections of circles: some will be behind the mirror, some in front of the eye, some will be behind the eye, and some at the centre of the eye. All other
V 308b intersections between the surface of the mirror | and the reflection planes will be [conic] sections.

[297] [Diagram V.42] I say then [that] some of the points whose forms are reflected from the peripheries of these [conic] sections will have their images behind the mirror, and some in front of the mirror. The [images] in front of the mirror will be [located as follows]: some will be behind the centre of the eye, others between the [centre of] the eye and the mirror, and some at the centre of the eye itself.

Let the [conic] section ABG be the intersection on the surface of the mirror. It was shown in [our] discussion of cylindrical convex mirrors that one line can be drawn from any one of these [conic] sections that is perpendicular to a plane tangent to the cylinder's surface; that only one perpendicular can be drawn in one and the same [conic] section; and that this perpendicular is one of the diameters of the circles that cut the mirror.

Let AG be the perpendicular drawn in the plane of this [conic] section at right angles to the plane tangent to the cylinder's surface.

Now, from any point on the periphery of this [conic] section a line can be drawn in the plane of this [conic] section that is tangent to the mirror's surface: for a plane tangent to the surface of the mirror will pass through any point on the mirror's surface, | and every line V 309a drawn from that point in that tangent plane will be tangent to the mirror's surface; and the line in plane ABG, and lying in the plane tangent to the mirror's surface, is tangent to the periphery of the [conic] section. And from any point from which a line is drawn tangent to a periphery of the [conic] section, a perpendicular can be drawn to that line; then, from any point on the periphery of [conic] section ABG, a line can be drawn perpendicular to the line tangent to this [conic] section at that point, and the perpendicular will lie in the plane of this [conic] section.

[298] [Diagram V.42] And it was shown in [our] discussion of the [right-]cylindrical convex mirror that any two of these perpendiculars will meet on the side of [i.e., next to] the [conic] section's concavity. Now, the line drawn from point G at right angles to line GA is tangent to the [conic] section, because it lies in the plane tangent to the cylinder's surface. And from any point on the periphery of [conic] section ABG a perpendicular can be drawn to the line tangent to the [conic] section at that point. That perpendicular will meet line GA; and the line drawn from that point to point | G will contain with that V 309b perpendicular an acute angle, and contain with line GA an acute angle, because this line will fall inside the [conic] section, because it lies inside the cylinder. And if it lies inside the [conic] section, then it lies below the two tangent lines.

Now, for any point ⟨on the periphery of the [conic] section⟩ that is closer to point G, the line joining it to point G will contain with line GA a larger angle, and [therefore] the closer the point on the periphery of the [conic] section is to point G, the smaller will be the angle contained between the perpendicular drawn from that point and line GA, and therefore most of these angles will be acute.

Let point B be close to point G, let line BD be the perpendicular drawn from it, let this perpendicular meet perpendicular GA at point D, and let ∠BDG be acute. We draw line GZ || line BD: line GZ then lies inside the [conic] section, because ∠ZGD is acute, because it is equal to the acute ∠GDB. Draw GO between lines GA and GZ and draw GH above line GZ and inside the [conic] section. Produce line BD | in both directions: it will meet GH V 310a towards point B and line GO towards point D; let [BD] meet line GH at point E, and let it meet line GO at point N, ⟨⟨and let O be located between G and N⟩⟩. Construct ∠DGM = ∠DGN, construct ∠DGL = ∠DGZ, and construct ∠DGK = ∠DGH. And produce GN in a straight line, and assume on it a point at random, say point T.

[299] [Diagram V.42] If the centre of the eye is at point H, and point K is in a visible object, then the form of point K will be reflected to the eye H from point G, and the image of [K] will be point E, which lies behind the surface of the mirror.

If the centre of the eye is at point O, and point M in a visible object, then the form of point M will be reflected to the eye O from point G, and the image of [M] will be point N, which lies behind the centre of the eye. | V 310b

If the centre of the eye is at point N, then the form of point M will be reflected to the eye N from point G, and the image of [M] will be point N, which is the centre of the eye. And if the centre of the eye is at point T, then the form of point M is reflected to point T from point G, and its image [N] lies between the eye and the mirror.

If the eye is at point Z, and point L in a visible object, then the form of point L will be reflected to the eye Z from point G, but the image of [L] will not be determinate and its form [i.e., the form of L] will be at point G.

And the same can be shown for point A, ⟪if it is the point of reflection⟫.

[300] Thus, the points whose forms are reflected from the periphery of a [conic] section in this mirror may have their images behind the mirror, or between the [centre of the] eye and the mirror, or at the centre of the eye itself, or behind the [centre of the] eye. And thus some of the images of the [right-]cylindrical concave mirrors will be located behind the mirror, and some in front of the mirror. And of those located in front of the mirror, some will be behind the centre of the eye, some between the [centre of the] eye and the mirror, and some will be at the centre of the eye itself. And all of these images will be images
V 311a of the points that lie outside the perpendicular drawn | from the centre of the eye at right angles to the plane tangent to the mirror's surface.

⟪V 2.VIII.2⟫

[301] As for the points that lie on this perpendicular: [it will be noted that] an unlimited [number] of planes can be drawn from this perpendicular that cut the mirror, each of which will be orthogonal to the plane tangent to the mirror's surface and in which the forms can be reflected; and [only] one of these planes will be parallel to the mirror's base, whose intersection will be a circle. Thus the perpendicular drawn from the centre of the eye will be a diameter to this circle; and, therefore, if the centre of the circle is [located] between the centre of the eye and a point on the diameter, the form of any such point will be reflected from two points only on the circumference of this circle. The form of the point on the surface of the eye through which this perpendicular passes will be reflected from the extremity of this perpendicular, and, in this situation, no point on this perpendicular will have its form reflected in any of the planes cutting the mirror and drawn through this perpendicular, other than the [plane of the] circle parallel to the mirror's base; for the intersection of the plane passing through the whole of the [mirror's] axis will be a straight line, and, as was shown for the plane mirror, no form of
V 311b any of the points on this perpendicular can be reflected | from a straight line other than the point at the extremity of the perpendicular that is common to all the intersections.

[302] As for the planes whose intersections are [conic] sections: no forms can be reflected from any [point] on the peripheries [of these conic sections], since the form[s] cannot be reflected from any [point] on the periphery of [any] one of these [conic] sections, except [the point at] the perpendicular's extremity; and in a single [conic] section there can only be one perpendicular. Thus, if the perpendicular passes through the centre of the eye, then only the form of the point on the eye's surface through which the perpendicular passes can be reflected from [the perpendicular's] extremity. And no forms will be reflected from [the perpendicular's] other extremity, because it lies behind the [centre of the] eye. Therefore, from the extremity of the perpendicular drawn from the eye's centre at right angles to the plane tangent to the mirror's surface, there will be reflected the form of the point on the eye's surface through which this perpendicular passes.

⟪V 2.VIII.3⟫

[303] If, then, any of the visible points are located on [this perpendicular], and if the centre of the circle parallel to the [cylindrical] mirror's base, of which [circle] that perpendicular is a diameter, lies between that point and the centre of | the eye, then the form of any one of those [visible] points will be reflected to that eye from two points on the circumference of that circle. But no form of [any of] those points on this perpendicular will be reflected from any of the intersections other than the circle's circumference alone. And the images of all points on this perpendicular will [coincide with] the centre of eye itself. V 312a

⟪V 2.VIII.4⟫

[304] If, however, the centre of the eye lies at the mirror's centre, then the point on the surface of the eye, through which the perpendicular passes, will be reflected from the whole of the circumference of the circle; and its image will be the circle's circumference.

⟪V 2.VIII.5⟫

[305] [Diagram V.42] Furthermore, we make ∠GAF a random, acute angle: line AF will then fall inside the [conic] section, because the line tangent to the [conic] section at point A will contain a right angle with AG. Then line AF will meet line GH, because ∠AGH and ∠HAG are both acute; let the two lines meet at H. Make ∠GAC = ∠GAH. Then line AC will meet line GK if GK is produced in a straight line, say at | point C. Therefore, if point C lies in a visible object, and the centre of the eye lies at point H, then the form of point C will be reflected to the eye H from two points on the periphery of the [conic] section AB, namely points A and G. V 312b

⟪V 2.VIII.6⟫

[306] And [in this situation], there can be no reflection of the form of any point to any eye from the periphery of any of these [conic] sections in addition to [these] two points [of reflection]. For there cannot be on the periphery of the [conic] section, at the same time and place, a point from which the form of any visible object can be reflected to one [and the same] eye facing that mirror at that time, from more than two points. For only a single perpendicular can be drawn from ⟨a point on⟩ the same [conic] section to a plane tangent to the cylinder's surface. That being so, the form of one [and the same] point cannot be reflected from the periphery of any [conic] section of the [right-]cylindrical concave mirror to one and the same eye from more than two points.

⟪V 2.VIII.7⟫

[307] [Diagram V.43] Furthermore, I say that for [any] two points on the axis of a right-cylindrical concave | mirror, the form of either one of them will be reflected to the other from the surface of that mirror from [all points on] the circumference of a circle, and its image will be the circumference of a circle. V 313a

Let ABGD be [points] on a cylindrical concave mirror. Let circles AG and BD be the bases of the mirror, and EZ its axis, and let points H and T be located on the axis at random. We bisect line HT at point K and draw through K a plane parallel to the cylinder's base: it will produce a circle on the cylinder's surface, say circle LM; and let LKM be a diameter of the circle. We draw from points L and M two lines extending along the length of the cylinder; let them be [lines] ALB and GMD. And we draw lines HL, HM, TL, and TM.

Since the cylinder is right, the axis HKT will be perpendicular to the plane of circle LM. Then, line HL = line TL, and ∠HLK = ∠TLK, and, likewise, ∠HMK = ∠TMK, and
V 313b ∠HLK = ∠HMK, and thus the four angles | are equal. And line HT will be perpendicular to circle LM.

[308] [Diagram V.43] Thus, with line HT being fixed, and while △HLT turns about line HT until it returns to its [original] position, point L will turn on the circumference of circle LM, and there will be produced at an [arbitrary] point on the circumference two equal angles made [respectively] by lines HL and LT and line KL; and line KL will be, at every point on the circumference of the circle, orthogonal to the plane tangent to the cylinder at that point.

[309] [Diagram V.43] We then let pass through point H a plane parallel to circle LM, and imagine it to be expanded in [all] directions; and produce TL to meet this plane, say at point F.

Then, if △HLT moves around axis HT, and TL is thereby made to move, then point F will move on the plane parallel to circle LM, and a circle will be produced by the movement of point F on that plane; let it be circle FN, and let FQHSN be a diameter of this circle. Then, line HQ is perpendicular to line ALB.

If, therefore, △HLT moves about axis HT, then line HQ will move about point H,
V 314a and point Q will | trace out a circle on the surface of the cylinder, and line HQ will be perpendicular ⟪to all [straight] lines on the surface of the cylinder⟫.

Therefore, if the eye is at point T, and point H is in a visible object, and the mirror is [cut as] a complete ring, then the form of point H will be reflected to the eye T from the circumference of circle LM, and its image will be circle FN. And that is what we wished to prove.

⟪V 2.VIII.8⟫

[310] [Diagram V.44] Furthermore, I say that if the points seen in a right-cylindrical concave mirror lie outside the perpendicular drawn from the centre of the eye at right angles to the plane tangent to the mirror's surface, and if these points do not lie with the eye on [a line parallel to] the axis, then some of these [points] will have a single image, some will have two images, some will have three, and some will have four, but no more than that.

Let a [right-]cylindrical concave mirror be cut by a plane parallel to its base, thus producing in it circle GDEZ, with H as centre. Let points A and B lie in the plane of
V 314b this circle and inside the circle, and let | them be two points between which lines can be reflected to each other at equal angles from four points on the circle's circumference GEZ.

We join AH and BH and produce them to cut the circle, thus producing in it two opposed sectors: one next to points A and B, and the other opposed to it. Then, of the [four] points from which lines will be reflected to each other between points A and B, there will be three points on the arc of the sector on [whose sides] points A and B are located, and a single point of reflection on the opposite arc, as was shown earlier [in [290]], in [our discussion of] the images [belonging to] circles.

Let the three points ⟨on the arc of the sector on which points A and B are located⟩ be points G, D, and E, and let the single point be point Z. We draw lines BE, BD, BG, BZ, AG, AD, AE, and AZ, and we draw from the centre lines HG, HD, HE, and HZ. From point A draw line AQ || HG, and line AF || HD, and line AN || HE, and line AC || HZ. Then produce lines BE, BD, BG, and BZ rectilinearly to meet the lines parallel to the diameters, say at points N, F, Q, and C. | V 315a

From point B draw a perpendicular to the plane of the circle, say perpendicular BT: it is clear that this perpendicular will fall within the concave [surface] of the mirror. Assume on perpendicular BT a point at random, say point T. From [point T] draw lines TN, TF, TQ, and TC, thereby producing triangles △TBN, △TBF, △TBQ, and △TBC. From points E, D, G, and Z, and in the planes of the triangles, draw lines parallel to perpendicular BT, say lines EK, DL, GM, and ZO; since these lines are parallel to perpendicular BT, all these lines are therefore perpendicular to the plane of circle GEZ.

Now, points G, D, and E, are on the mirror's surface; and circle GEZ is parallel to the base of the right cylinder. Therefore, lines EK, DL, GM, and ZO are on the surface of the mirror and parallel to its axis, because all perpendiculars drawn from points on the right cylinder's surface must be on the mirror's surface. Therefore, points K, L, M, and O are on the mirror's surface. We draw lines AK, AL, AM, and AO.

Now let HW be the axis of the cylinder. Then | HW is perpendicular to the plane of the V 315b
circle, and, therefore, parallel to lines BT, EK, DL, GM, and ZO.

From point K we draw a line parallel to AN: it will meet the axis of the mirror and will be perpendicular to it. For AN || HE, and thus the line drawn from point K parallel to line AN will be parallel to line HE; and line EK and axis HW are parallel to each other and thus they are in the same plane; and HE joins them and, therefore, is in the same plane with them. Therefore, the line drawn from point K parallel to line HE will meet axis HW and be perpendicular to it; let that line be KW.

[311] [Diagram V.44] We join AT. Then △TAN is in one plane. But line KW || AN, ⟨⟨and point K lies in the plane of the triangle⟩⟩, thus line KW is in the plane of △TAN, and, therefore, it cuts line TA, say in point Y.

I say then that lines AK and KT are reflected at equal angles, and that the form of point T will be reflected to the eye A from point K.

Demonstration. Line AN || line HE, therefore ∠ENA and ∠EAN are [respectively] equal to ∠BEH and ∠HEA; and these two angles are equal; | thus ∠ENA = ∠EAN, and V 316a
therefore, line EA = line EN.

But line EK is perpendicular to these [latter] two lines, because it is perpendicular to the circle's plane, therefore line AK = line NK, and thus ∠KNA = ∠KAN. Line KY || line NA, thus ∠TKY and ∠YKA are [respectively] equal to ∠KNA and ∠KAN, therefore ∠TKY = ∠YKA. And line KW is perpendicular to the axis; therefore it is perpendicular to line KE and it is thus perpendicular to the plane tangent to the mirror's surface through line EK.

Therefore, the form of point T will be reflected to the eye A from point K.

[312] [Diagram V.44] We may similarly show that the form of point T will be reflected to the eye A from every one of the points L, M, and O. For if we draw from point L a line parallel to line AF, it is clear that it will meet the axis and be perpendicular to it, as was shown for line KW. It is then clear that the two angles produced at point L will be equal, as was shown for the angles at point K; thus it is clear that the line drawn from
V 316b point L parallel to line AF | will be perpendicular to the plane tangent to the cylinder, as was shown for line KW; and from that it will be clear that lines AL and LT are reflected at equal angles. In the same way it can be shown that lines AM and MT are reflected at equal angles. Therefore, the form of point T will be reflected to A from the surface of the cylindrical concave mirror from four points.

And thus, if the mirror is a complete ring, or it is greater than a semicircle, then it is possible that the form of point T will be reflected to the eye A from four points on the surface of that mirror.

[313] [Diagram V.44] Moreover, since the axis HW and the perpendicular BT are parallel, they are in the same plane; and line HB joins them, so it is in the same plane with them. And HB is in the plane of the circle, so line HB is the intersection between the plane drawn from the axis, passing through point T, and the plane of the circle.

Now point T lies outside the axis, because it lies on line BT, which is parallel to the axis. And, for any point that lies outside a line, there can be drawn from that line only one plane that passes through that point. Therefore, only the plane containing lines HW and
V 317a HB can be drawn from axis HW through point T. The plane containing | axis HW and line HA therefore does not pass through point T, because point A lies in the plane of the circle and outside line HB.

Now point A is the eye. And there is only one plane that passes through both the axis and the eye A. And the plane containing the axis and line HA is the plane that passes through the axis and the eye A. Therefore, the plane passing through the axis and the eye A does not pass through point T. Therefore, point T lies outside the plane passing through the axis and the eye A. And point T lies outside circle GEZ, which is parallel to the base of the cylinder, because TB is perpendicular to the plane of the circle. The form of point T will not, therefore, be reflected to the eye A, except in the plane of a [conic] section.

And therefore, the intersections produced in the cylinder's surface by the reflection planes in which the form of point T is reflected to the eye A from points K, L, M, and O are [conic] sections.

[314] [Diagram V.44] Moreover, point A lies outside plane TBHW passing through the axis; and point T lies in this plane and not on the axis, because it lies on a line parallel to the axis. Therefore, line AT does not cut plane TH that passes through the axis, except in
V 317b point T, | which is not on the axis. Therefore, line AT does not meet the axis.

We then produce line AT in the direction of T to meet the cylinder's surface, say at point S; and also produce it in the direction of A to meet the cylinder's surface, say at point d. Then Sd is an intersection between all the [conic] sections from which the form of point T is reflected to the eye A.

And line Sd lies within the concave [surface] of the cylinder, because line Sd does not lie in a plane passing through the axis, and, therefore, points T and A lie inside the concave side of the mirror.

I say then that line ATS [Sd] is not perpendicular to a line tangent to all these [conic] sections, nor [is it perpendicular] to two lines tangent to two of these [conic] sections.

For if it [line ATS] were perpendicular to two lines tangent to two of those [conic] sections, then those two lines would intersect at point S and they would therefore both be in ⟨one⟩ plane; and line AS would be perpendicular to that plane; and the two tangent lines would be two intersections between that plane and the two planes of the two [conic] sections. And therefore, that plane would be tangent to the mirror's surface, because if it were not, | it would cut the cylinder's surface; and if it did, then it would cut each V 318a one of those two [conic] sections at a point other than point S. If that point and point S were joined by a straight line, then that line would fall within the cylinder and therefore within the [conic] section. And this line would be the intersection between the plane that cuts the cylinder and the plane of the [conic] section, and the intersection between this ⟨intersecting⟩ plane and the [conic] section's plane was [assumed to be] tangent to the [conic] section, which is impossible. Therefore, the plane that contains the two tangent lines is tangent to the cylinder's surface.

And, if ⟨the plane containing the two lines⟩ is tangent to the cylinder's surface, and line AS is perpendicular to it, then line AS will meet the axis and be perpendicular to it, as was shown in Chapter 5 of this Book ⟨⟨[namely] Chapter 5 of Book IV⟩⟩. But it has now been shown that line AS does not meet the cylinder's axis; therefore line AS is not perpendicular to two lines that are tangents to two of the [conic] sections that intersect in line Sd.

That being so, none of the lines drawn | from point T at right angles to the lines tangent V 318b to the [conic] sections that intersect at line Sd pass through A, except one line, i.e., line ATS, because they all intersect at point T.

[315] [Diagram V.44] And it was shown in the proposition before this one that some of these perpendiculars meet the reflected lines issuing from point A behind the mirror, and some will meet [the reflected line] in front of the mirror ⟨⟨either between the mirror and the centre of the eye, or at the centre of the eye, or behind it⟩⟩, and some will be parallel to it. And, this being so, the meeting point will be dispersed and not a single point, because these perpendiculars meet at point T, and T does not lie on a line drawn from point A other than line TA.

And it was shown previously that the points at which these perpendiculars meet the reflection lines corresponding to the lines drawn from point A are the images of the points whose forms are reflected along those lines; therefore, point T will have on every one of lines AK, AL, AM, and AO a point that is the image of point T, and none of these points will be identical with any of | the others. V 319a

As for the lines that meet the perpendiculars: [their] meeting-points will be the images. And if one of them is parallel to the perpendicular, then the image of point T, which is reflected on that line, will be the reflection point that lies on that line, as was shown in the earlier discussion on image[s].[28]

[316] [Diagram V.44] If, therefore, point T is [located] on a visible object, and the eye is at point A, then the eye will perceive point T at four places, if the mirror is a complete ring or a segment greater than a semicircle.

I say then that the form of point T will not be reflected to the eye A from more than four points and will have no more than four images.

---

28. Above, paragraphs [202–5].

For [assume] that the form of point T is reflected to the eye A from a point other than the four points from which [that point's] form has been reflected to the eye A. Then, if from that [assumed, fifth] point a line is drawn along the length of the cylinder, it will end at the circumference of circle GEZ, and will be perpendicular to the plane of the circle and parallel to perpendicular TB, and will meet the circumference of this circle at a point other than points G, D, E, and Z.

V 319b If a straight line then joins that point to | point B, it will be in the plane containing the two parallel perpendiculars of which one is TB and the other is the line extending from the [assumed, fifth] reflection point to the circle's circumference. Thus, if this line is rectilinearly produced in the direction of that point, and a straight line joins that point and point H, and a line is drawn from point A parallel to this line, then it [the latter line] will meet the first line. The meeting-point will then be in the plane containing the two parallel perpendiculars. The meeting-point will also be in the reflection plane, because the line joining that point and point H is parallel to the perpendicular drawn from the reflection point; and [this line] is parallel to the line drawn from point A to the meeting-point. Therefore, the line drawn from point A to the meeting-point will be parallel to the perpendicular drawn from the reflection point. And line AT cuts these two parallel lines. And line AT and the perpendicular drawn from the reflection point are both in the
V 320a reflection plane. And the line | drawn from point A parallel to the perpendicular drawn from the reflection point is in the reflection plane. Therefore, the meeting-point lies in the reflection plane, as do the reflection point and point T.

Therefore, if the line reflected to point T is produced rectilinearly in the direction of the reflection point, it will end at the meeting-point. And thus, if a straight line is joined between point A and the reflection point, it will be equal to the line that lies between the reflection point and the meeting-point, because two angles will be produced, at point A and [at] the meeting-point, equal to the two similar angles at the reflection point.

[317] [Diagram V.44] Thus, if a straight line is joined between point A and the point on the circle's circumference, it will be equal to the line between that point and the meeting-point, because the line drawn from the reflection point to the circle's circumference is perpendicular to the circle's plane. From this it follows that the two angles at that point
V 320b on the circle's circumference, namely those [angles] divided by the line | drawn from that point to the circle's centre, will be equal; and, therefore, lines will be reflected at equal angles from points A and B at five points on the circumference of circle GEZ, which is impossible, as was shown with reference to the images of the circle in the discussion of the images of the spherical concave mirrors.[29] Therefore, the form of point T will not be reflected to the eye A from the cylindrical concave mirror at more than four points. And that being so, no point perceived by reflection in the cylindrical concave mirror will have more than four images.

Therefore, if A and B are among the points from which lines can be reflected to each other, at equal angles, from three points on the circumference of circle GEZ, or from a part of the circumference if the mirror is not a complete circle, then the form of point T will be reflected to the eye A from three points on the surface of the cylindrical concave mirror,
V 321a and [T] will have three images, and no more than three. And that is shown, | as was shown

29. Above, paragraphs [285–90].

in [the case of] four points, by [the fact] that the form [of that point] can be reflected from three points and not from more than three.

[318] And if from points A and B lines are reflected from [only] two points on the circumference of circle GEZ, or on part of the circumference, then the form of point T will be reflected to the eye A from two points, and [T] will have two images, and no more than two.

And if lines are reflected between points A and B from only one point on the circumference of circle GEZ or on a part of it, then the form of point T will be reflected from only one point, and [T] will have only one image.

And if no lines whatever can be reflected between points A and B at equal angles, then the form of point T cannot be reflected in any way | from this mirror. V 321b

[319] It is thus clear, from what we have shown in this proposition, that the points visible in the cylindrical concave mirror, if they do not lie in the plane of one circle, are as follows: some of them will have a single image, some will have two images, some will have three, and some will have four, but no more.

If the two points [that are reflected to each other] lie in the plane of one of the circles in this mirror, then the form of one of them will not be reflected to the other except from the circumference of that circle. And if their forms are reflected only from the circumference of that circle, then neither point can have more than four images.

Therefore, for the points visible in the [right-]cylindrical concave mirror, some will have a single image, some will have two images, some will have three, and some will have four, but no more than that.

And that is what we wished to prove.[30]

⟨⟨V 2.VIII.9⟩⟩

[320] Now, [regarding] the two points being assumed, I mean the visible point and the eye-point, | [and] concerning how to find the point, or points, of reflection in the cylindrical concave mirror: the way to do that is the same as the one we followed in the [previous] proposition. That is, we draw through one of the two points a plane parallel to the plane of the cylinder's base: it will produce a circle in [the cylinder]. If this plane passes through the other point, then we find the point, or points, of reflection between the two points on the circumference of this circle. Then the forms of the two points will be reflected to each other from that point on the surface of the cylindrical concave mirror, and from no other point. V 322a

If the plane does not pass through the other point, then we drop from the other point a perpendicular to the plane of the circle, and find the point [or points] of reflection between [the point at] the foot of the perpendicular and the first point, then draw the lines we have drawn in the [previous] proposition, then proceed with the proof by following the reverse of the way we followed in the present proposition. We will thereby find the point [or points] of reflection on the surface of the cylindrical concave mirror between the two assumed points. | V 322b

30. A lacuna in MS K begins here.

⟨⟨**V 2.VIII.10**⟩⟩

[321] If the visible object is perceived by both eyes at once, and if every point on the object has a single image, then the form perceived by the eye in this mirror will be two overlapping forms, and the observer will perceive the two overlapping forms as a single form, like the forms perceived by both eyes at once in the spherical mirror; and, of the radial lines [imagined to be] collected on each of these two forms from the two eyes, there will be lines that are similarly positioned [in both eyes] or nearly so, as was shown previously.

⟨⟨V 2: Ninth Inquiry. On the Images of the Conical Concave Mirrors, Consisting of an Introduction and Eight Propositions⟩⟩

⟨⟨Introduction⟩⟩

[322] With regard to the conical concave mirror: the intersections between its surface and the planes of reflection are all [conic] sections, except for the single plane passing through the whole axis, whose section will be a straight line, as was shown in Chapter 5 ⟨of Book IV⟩. And the images [related to] the straight line will be located behind the mirror.

⟨⟨**V 2.IX.1**⟩⟩

[323] For every one of the[se] [conic] sections, the image of what is reflected from it may be located behind the mirror, or between the eye and the mirror, or it may be located
V 323a at the centre of the eye | itself, or behind the eye's centre. This can be ⟨shown in the way already done for⟩ the [conic] sections belonging to the cylindrical concave mirror.

⟨⟨**V 2.IX.2**⟩⟩

[324] With regard to the perpendicular issuing from the eye's centre at right angles to the line tangent to the cone's surface: if the eye's centre lies between the mirror's surface and the point located on the axis, then the eye will only perceive in this mirror the point on the surface ⟨of the eye⟩ through which the perpendicular passes; and the form of [this point] will be reflected ⟨along⟩ the perpendicular.

And if any visible object lies ⟨on⟩ the perpendicular, the eye will not perceive it [by reflection from the mirror], because it will screen the point at the extremity of the perpendicular; and [the object] will not be reflected from any [point] other than that point [at the extremity], because the perpendiculars drawn from the reflection points will in this case be located behind the eye.

⟨⟨**V 2.IX.3**⟩⟩

[325] If the eye is at the point on the perpendicular that lies on the axis, then the form of the point on the eye's surface through which [point] the perpendicular passes will be reflected to the eye from the circumference of a circle, but not from any [place] other than that circle, because perpendiculars can be drawn from any point on the axis to the circumference of a circle on the surface of the cone, and no perpendicular can be drawn from [that point] | to another circle [on the cone]. V 323b

And if any visible object lies on the perpendicular, the eye will not perceive it, because the perpendiculars drawn from all [points on] the surface of the mirror will be located outside the plane in which both the eye and the visible object lie.

⟪V 2.IX.4⟫

[326] If the point of the perpendicular that lies on the axis is located between the eye and the mirror's surface, then the forms of the points on the perpendicular that lie between the axis and the mirror's surface will be reflected to the eye; and their images will be at the centre of the eye.

⟪V 2.IX.5⟫

[327] [Diagram V.45] I say further that if two points are [assumed] ⟨on⟩ the axis of the conical concave mirror, of which one is the centre of the eye and the other lies on a visible object, the form of the latter point will be reflected to the eye from the circumference of a circle, and its image will be the circumference of a circle.

Let A, B, and G be on a conical concave mirror with A as its vertex. Let it be cut by a plane passing through the mirror's axis: it will produce in the mirror two straight lines; let them be lines AB and AG. Let AD be the axis, and let H and T be two points on the axis. We draw from point T a perpendicular to AB, say TK, and we rectilinearly produce TK | V 324a
in the direction of K to L. Make LK = KT. Join HL: it will cut AB, say in point B. Draw from B a perpendicular to line AB, say BD. We draw through point B a plane parallel to the cone's base: it will produce a circle in the cone, say circle BG.

Then line DB is perpendicular to the plane tangent to the cone through point B. For the line touching the circle and drawn from point B in the plane of the circle will be perpendicular to plane ABG, and the plane containing line AB and that tangent line will be tangent to the cone's surface, as was shown in Chapter 5 of Book ⟨IV⟩.

[328] [Diagram V.45] We join TB; then TB = BL, because TK = KL. But BD is parallel to perpendicular TKL, therefore HB:BL = HD:DT; and BL = BT, thus HB:BT = HD:DT, therefore ∠HBD = ∠DBT. Thus, the form of point T will be reflected to the eye H from point B. | V 324b

[329] [Diagram V.45] If we fix line TH and turn △TBH about line TH, then point B will turn on the circumference of circle BG, because axis AD is perpendicular to circle BG and passing through its centre. And if point B turns on the circumference of circle BG, two equal angles will be produced at every point on the circumference of circle BG, so that line DB, as it turns, will be perpendicular to every plane that is tangent to the cone's surface at the extremity of line DB. Therefore, the form of point T will be reflected to the eye H from the circumference of circle BG.

And since line TL is perpendicular to line AB, point L will therefore be the image of point T. And if △TBH turns round line BH, then line TL will move in a circle, thereby producing a circular cone. The motion of point L will produce on the surface of ⟨this⟩ cone a circle forming the cone's base, and this circle will then be the image of point T with respect to the eye H. That is | what we wished to prove. V 325a

⟪V 2.IX.6⟫

[330] [Diagram V.46][31] I say further that, if visible points seen in conical concave mirrors lie outside the perpendicular drawn from the centre of the eye at right angles to the plane tangent to the cone, and do not lie together with the eye on the axis of the cone, then some [of these points] will have a single image; some will have two images; some will have three; and some will have four, but no more than four.

Let A, B, and G be on a conical concave mirror with AD as its axis. Let point E be in a visible object, and point Z the centre of the eye. We draw[32] through point Z a plane parallel to the base of the cone, and let it produce in the cone circle BG: this plane will then either pass through point E, or it will not.

And, first, let [the plane] pass through point E. Then line EZ will lie in the plane of circle BG.

Now, it was shown in [the discussion of] spherical concave mirrors[33] that, for any two points in the plane of a circle, the form of one of them may be reflected to the other from
V 325b the circumference of that circle at a single point, or at | two points, or three, or four points, or not at all. Thus, points E and Z may or may not have their forms reflected to each other from the circumference of circle BG.

Let the form of point E be reflected to the eye Z from the circumference of circle BG at point H. Let T be the centre of the circle. We draw lines ZH, EH, and TH. Then ∠ZHT = ∠EHT, and line TH will cut ∠EHZ, and, therefore, will cut line EZ, say at point K.

We join AH; and draw from point K a perpendicular to line AH, let it be KM, and extend it in the direction of K: it will meet the axis, say at point D. Then line KM is perpendicular to the plane tangent to the surface of the cone and passing through point M, as was shown earlier.[34] We then join EM and ZM.

[331] I say then that lines EM and ZM are reflected to each other at equal angles.

Demonstration. We draw from point Z a line parallel to line KH, and produce EH to
V 326a meet it, say at point L: then line HL = line HZ, | because the angles at points Z and L are [respectively] equal to the angles at point H. We then draw perpendicular HS ⟪at right angles to ZL⟫: it will bisect line LZ. Draw line ZO || line KM and produce EM to meet it, say at point O; and join OL.

Then, since lines ZL and ZO are [respectively] parallel to lines KH and KM, the plane containing lines ZL and ZO is parallel to the plane containing lines KH and KM.

But the plane of △ELO cuts those two planes, therefore line OL || line MH. But since line HS is perpendicular to line ZL, and ZL || KH, therefore ∠SHT is a right angle, and thus line HS is tangent to the circle.

Now line AH extends along the length of the cone; the plane containing lines AH and HS is therefore tangent to the cone; and it passes through point M [and] line MK is therefore perpendicular to this plane.

We then draw from point S line SN in the plane of ⟨△⟩LZO and parallel to LO: it will cut and bisect line ZO, say in point N. And since line SN || line LO, and LO || line HM,
V 326b therefore SN || line HM. |

---

31. Note that HZ is not on the same straight line as ZNO.
32. The lacuna in MS K ends here, and the text resumes at K134a, line 1.
33. Above, paragraphs [285–90].
34. Book IV, Chapter 5, paragraphs [48–53].

[332] We join MN; line KM is then perpendicular to plane MNSH; and line ZO || line KM, therefore line ZO is perpendicular to plane MNSH. And thus ∠ZNM is right, and line ZN = line NO, therefore line MO = line MZ.

Thus, EM:MZ = EM:MO, but EM:MO = EH:HL, and EH:HL = EK:KZ, thus EM:MZ = EK:KZ. Therefore, the angles at point M are equal. Thus, the form of point E will be reflected to the eye Z from point M. And that is what we wished to prove.

And the proof would be similar if point K were the centre of circle BG.

[333] If the form of point E is reflected to point Z from several points on the circumference of circle BG, then the diameters drawn from point T to those points will divide line EZ in different points corresponding to point K. So if we draw from those points perpendiculars to tangent planes | touching the surface of the cone, corresponding V 327a
to perpendicular KM, then the form of point ⟨E⟩ will be reflected to the eye Z from all extremities of those perpendiculars, which can be shown as was done for point M. Then point E will have as many images as the points on the circumference of circle BG from which the form of point E is reflected to the eye Z.

[334] [Diagram V.47] Now, let the plane of circle BG not pass through point E. From point A we draw a line to E and continue it to meet the plane of circle BG, say at point H.[35] Join HZ. Let the form of point H be reflected to the eye Z from point T on the circumference of circle BG; [and] let K be the centre of the circle.

Draw lines HT, ZT, KT, and AT.

Then plane ATK will be orthogonal to the plane of circle BG, and their intersection will be TK; thus only line TK in plane ATK will lie in the plane of the circle. Now, TK cuts line HZ, say in point N; thus only point N on line HZ will lie in plane ATK. Points H and Z therefore lie outside plane ATK, thus line AEH lies outside plane ATK, and, therefore, point E lies | outside plane ATK. Thus, points E and Z lie outside plane ATK, and they lie V 327b
on either side of this plane, because this plane cuts line HZ in point N. But if points E and Z lie outside plane ATK and on either side of it, then line EZ cuts plane ATK. Therefore, only one point on line EZ will lie in plane ATK; let it be point O.

Now, points H and Z lie below the line tangent to the circle and drawn from point T; therefore line HZ lies below the plane tangent to the cone and passing through line AT. Line AH lies then below the tangent plane. Thus, line HZ lies below the tangent plane, and point O lies below the tangent line. Therefore, point O lies inside the cone and below line AT, and point O lies in plane ATK.

We then draw from point O a perpendicular to line AT; let it be ⟨perpendicular⟩ OS. Then line OS lies in plane ATK.

[335] [Diagram V.47] Then I say that the form of point E will be reflected to the eye Z from point S.

Demonstration. We join ES and ZS, and we draw from point Z a line | parallel to line V 328a
TK in the plane of the circle; let it be ZF. Then draw, ⟨also⟩ from [Z], a line parallel to line OS in the plane of △⟨⟨EZS⟩⟩, say ZQ. And draw line HT to meet ZF, say at point F. Then produce ES to meet ZQ, say at point Q, and join FQ.

Then, since point S lies on line AT, line ESQ will be in the plane of △ATH. But line HTF is in the plane of △ATH, therefore lines EQ and HF are in one plane.

---

35. H is assumed to be different from Z.

And since lines ZF and ZQ are parallel to lines KT and SO, and these two lines are in plane ATK, the plane of ΔFZQ is therefore parallel to plane ATK.

But plane EQFH cuts these two [latter] planes, therefore their intersections are parallel; and the intersections are TS and FQ, thus line TS || line FQ.

[336] [Diagram V.47] From point T we draw a perpendicular to line ZF, say TC: it will bisect ZF, because line ZT = line TF, the angles at point T being equal.

V 328b But line ZF || line KT; thus, since TC is perpendicular to ZF, | and ZF || TK, therefore ∠CTK is right. But TK is the diameter of the circle, therefore line TC is a tangent to the circle. And line AT lies on the surface of the cone, therefore plane ATC is tangent to the surface of the cone.

[337] [Diagram V.47] We produce SO in the direction of O: it will meet the axis of the cone, say at point D. Then line DS is perpendicular to the plane tangent to the cone and passing through line AS. But the plane tangent to the cone and passing through line AS is plane ATC; therefore line DS is perpendicular to plane ATC.

Line DS lies in plane ATK, therefore plane ATK is orthogonal to plane ATC; and thus plane ATC is orthogonal to plane ATK, and thus orthogonal to any plane parallel to plane ATK. Therefore, plane ATC is orthogonal to plane FZQ. Thus, plane ATC cuts planes ATK and FZQ, and their intersections are therefore parallel.

Let line CY be the intersection between plane ATC and plane FZQ, and let line TS be
V 329a the intersection between plane ATC and plane ATK; then line CY || line TS. But TS || FQ, |
therefore line CY || line FQ; and it cuts line [side] ZF in ΔZFQ, then it cuts [the other side] ZQ, say in point Y. Then QY:YZ = FC:CZ, but FC = CZ, thus line QY = line YZ.

[338] [Diagram V.47] Now, line CY is in plane ATC. We join SY: it will lie in plane ATC; and line DS is perpendicular to plane ATC, therefore ∠OSY is right.

But line SY is in the plane of ΔEQZ, and line ZY || line OS; therefore ∠ZYS is right.

And line ZY = line YQ, thus line SQ = line SZ. Therefore, ES:SQ = ES:SZ.

But ES:SQ = EO:OZ, thus ES:SZ = EO:OZ, therefore the angles at point S are equal. Therefore, the form of point E will be reflected to the eye Z from point S.

And that is what we wished to prove.

[339] [Diagram V.47] If the forms of points H and Z are reflected to each other from
V 329b several | points on circle BG, then we [should] follow, for each point, the procedure we followed for point T. It will thus be shown that the form of point E will be reflected to the eye Z from several points on the surface of cone ABG; and point E will therefore have as many images as the points from which the form of point H will be reflected to the eye Z from the circumference of circle ABG. The same can be proved if point E lies next to [i.e., outside] the base of the cone, for the procedure of the proof will be the same for both directions.

And thus it has been shown by this proof, when if the two points are assumed, I mean the visible point and the eye-point, how to find the point or points of reflection on the surface of the conical concave mirror.

⟪V 2.IX.7⟫

[340] [Diagram V.47] That having been shown, I say that, for every two points whose forms are reflected to each other from the surface of a conical concave mirror, the reflection of the form cannot take place from more than four points, | and [that point] will not have more than four images: V 330a

If possible, let the form of point E be reflected to the eye Z from five points on the surface of mirror ABG; and let the example be the same preceding diagram [Diagram V.47]. And let the fifth point be point S.

We draw from point S the perpendicular SD ⟪[at right angles] to the tangent plane⟫: it will cut line EZ, say in point O. We draw through point Z a plane parallel to the base of the cone; let it produce in the cone circle BG. Join AS and continue it to meet the circle, say at point T; let K be the centre of the circle.

Join KT: it lies in plane ATK, thus in the same plane as perpendicular SD. Produce AE and continue it to meet the circle, say at point H. Join HT, HZ, and ZT, and join ES and ZS. From point Z draw a line parallel to line OS, and produce ES to meet it, say at point Q. We draw, also from Z, a line parallel to line KT, and produce HT to meet it, say at point F. Join FQ.

Since lines ES and HT are in the plane | of △AHT, lines EQ and HF are therefore in V 330b one plane. And since lines ZQ and ZF are [respectively] parallel to lines OS and KT, and [since] OS and KT are in one plane, i.e., plane ATK, therefore plane ZFQ || plane ATK; and, therefore, line QF || line ST.

From point S we draw the intersection between the reflection plane, namely plane EQF, and the plane tangent to the cone and passing through line AS; [and] let [the intersection] be [line] SY. This line will meet line ZQ, because ZQ || OS and the two of them are in the reflection plane; let the meeting be at point Y. We draw from point T a line in the plane of the circle and tangent to the circle; let it be TC. This line will meet line ZF, because [line ZF] || line KT; let them meet at point C. Therefore, ∠TCF is right, since ∠CTK is right.

And since line TC is tangent to the circle, plane ATC is thus the plane tangent to the cone and passing through line AT. Therefore, lines CT, TS, and SY are in one plane, and this plane cuts the planes | ATK and FZQ, which are parallel to each other. V 331a

[341] [Diagram V.47] We join YC: it will be parallel to ST, because they are the intersections. And QF has been drawn parallel to ST, therefore line YC || line QF.

And since line SY lies in the tangent plane, thus ∠OSY is right, and therefore, ∠QYS is right. And since the two angles at point S are equal, and line ZQ || line OS, therefore line QS = line ZS. But SY is perpendicular [to ZQ], therefore line QY = line YZ. And line QF || line YC, therefore line FC = line CZ. And the angles at points Z and F are equal, therefore the angles at point T are equal.

Therefore, lines HT and TZ are reflected [to each other] at equal angles. And this we can similarly prove for every point on the surface of cone ABG from which the form of point E is reflected to the eye Z.

If the form of point E were reflected to the eye Z from five points on the surface of the conical ⟨concave⟩ mirror, | it would then follow that the form of point H would be reflected V 331b from five points on the circumference of circle BG, but this is impossible, as was shown for the images [produced by] circles.[36] Therefore, the form of point E cannot be reflected to the eye E from more than four points on the surface of the cone; and point E cannot therefore have more than four images. And that is what we wished to prove.

---

36. Book V, Chapter 2, paragraphs [285–90].

[342] [Diagram V.47] Thus, it has been shown from what we have proved, for points H and Z, if their forms cannot be reflected to each other from the circumference of circle BG, that the forms of points E and Z cannot be reflected to each other from the surface of the cone. For if the form of one of them were reflected to the other from the cone's surface, then it would follow, by the method we have shown, that there exists a point on the circumference of the circle from which the forms of points H and Z are reflected to each other. And, therefore, if the forms of points H and Z cannot be reflected to each other from the circumference of circle BG, then it is not possible for the forms of points E and
V 332a Z to be reflected to | each other from the surface of the cone.

[343] [Diagram V.46] If points E and Z are in the plane of circle BG, then it can be shown by the method shown in the proposition preceding this proposition that the form of either one of them cannot be reflected to the other from more than four points on the surface of the cone, and, therefore [neither of them] can have more than four images.

[344] Therefore, with regard to the points seen in the conical concave mirror: some of them have a single image, some have two images, some have three, and some have four, but no more than four.

⟪V 2.IX.8⟫

[345] If the object is perceived by the two eyes at once, and every point on the object has a single image, then the form perceived by both eyes will be two overlapping forms; but the viewer will perceive the two overlapping forms like the form perceived by the two eyes in spherical concave and cylindrical concave mirrors. [For] there would come
V 332b together in each of these forms radial lines | similarly situated in both eyes, or nearly so, as has been shown previously.

[346] We have now shown all that pertains to the images of visible objects perceived by reflection in the mirrors previously specified. That is what we intended to show in this Book. Let us then conclude this Book with this [last] statement.

[Colophon]

[347] Book V of the work on optics has been completed. Praise and thanks are to God, Who is my only and best support. Transcription [of this manuscript volume] was completed during the first ten days of Jumādā al-Ākhira of the lunar [*hijrī*] year 636 [9–18 January 1239 AD].

Diagrams to Book V

It is not our intention to produce diagrams for Book V in accordance with accepted modern norms, especially those of modern descriptive geometry, but rather to produce them in the common medieval style adopted by Ibn al-Haytham and found in the manuscripts of his *Optics* and Kamāl al-Dīn's *Tanqīh*, while maintaining the features found in most of the manuscripts consulted. Our primary aim was to produce diagrams aligned as closely as possible to the arguments developing in the text.

A caption such as 'Diagram V.1 [75–76]' means that Diagram 1 in Book V relates to paragraphs [75–76] of Book V, Chapter 2 (which constitutes almost the entirety of Book V).

Dotted lines are either found in a manuscript as non-dotted lines, or are mentioned or assumed in the text.

Some areas in the diagrams have been shaded.

We have named additional points, placed between single angular brackets, ⟨ ⟩.

Diagrams added by the editor/translator are always distinguished as such.

The manuscripts referred to are F (Fatih Library, no. 3215), K (Köprülü Library, no. 952), and TA (manuscript of the *Tanqīḥ*, in the Ahmed III Library, no. 3340).

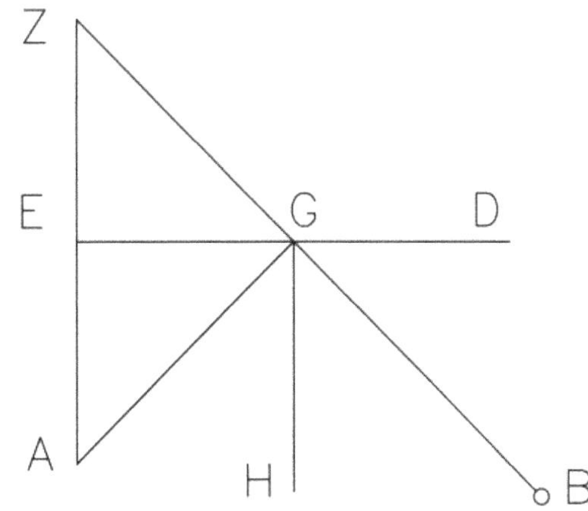

Diagram V.1 [75–76]

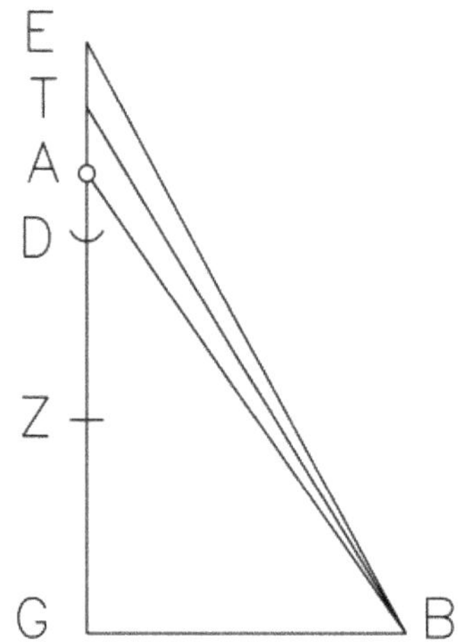

Diagram V.2 [77–81]

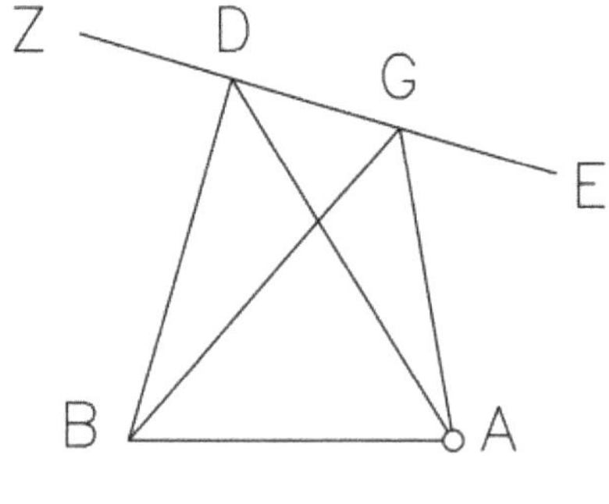

Diagram V.3 [82–83]

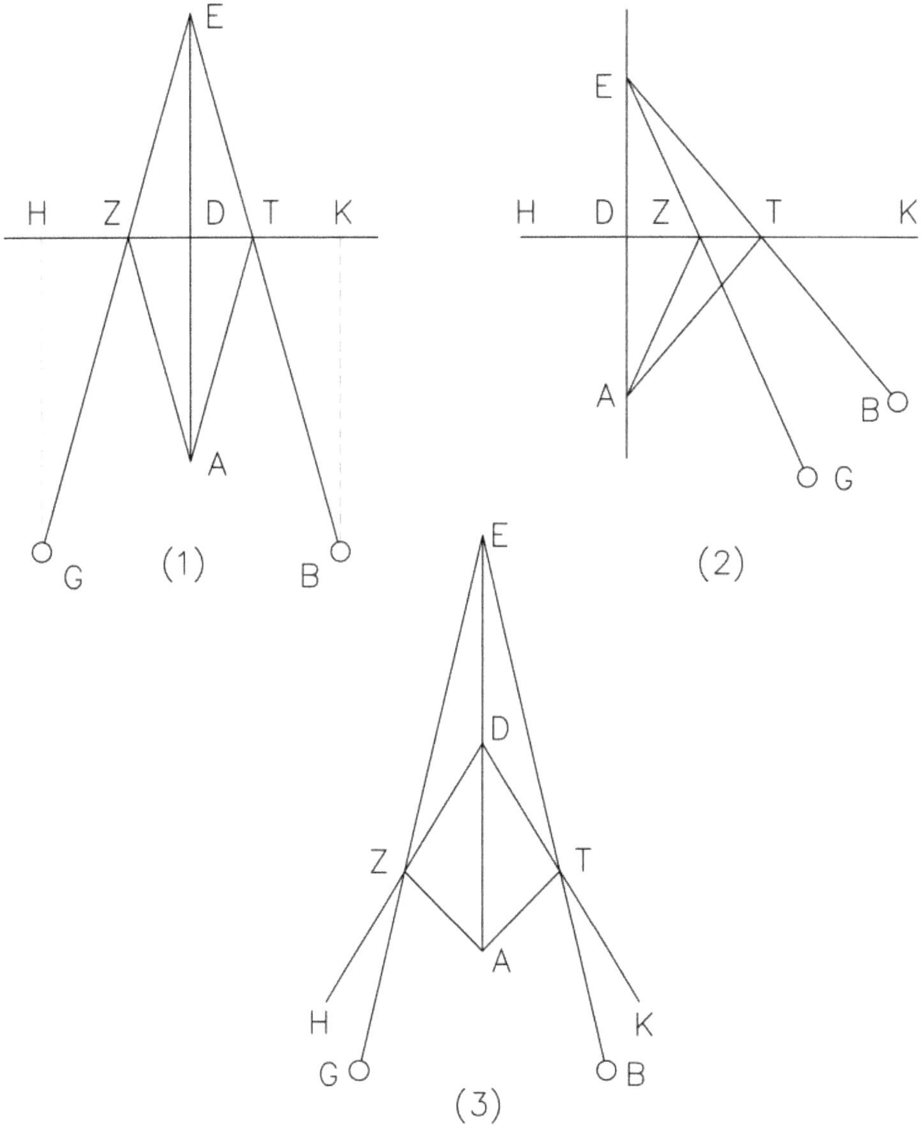

Diagram V.4 (Cases (1) – (3)) [84–89]

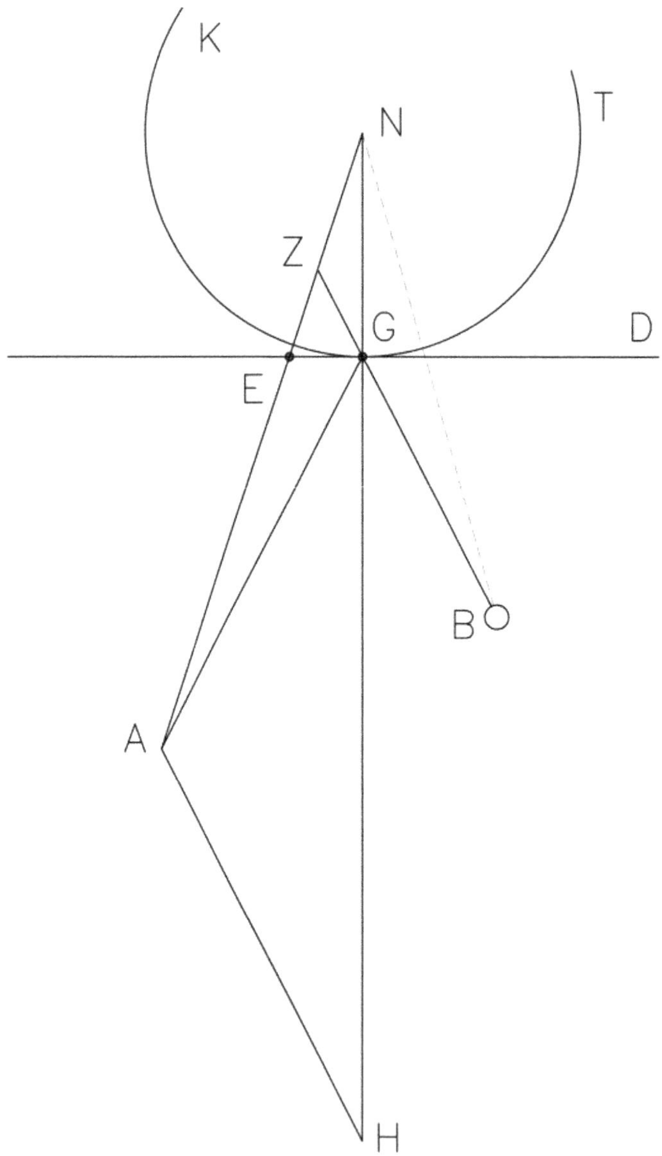

Diagram V.5 [90–95]

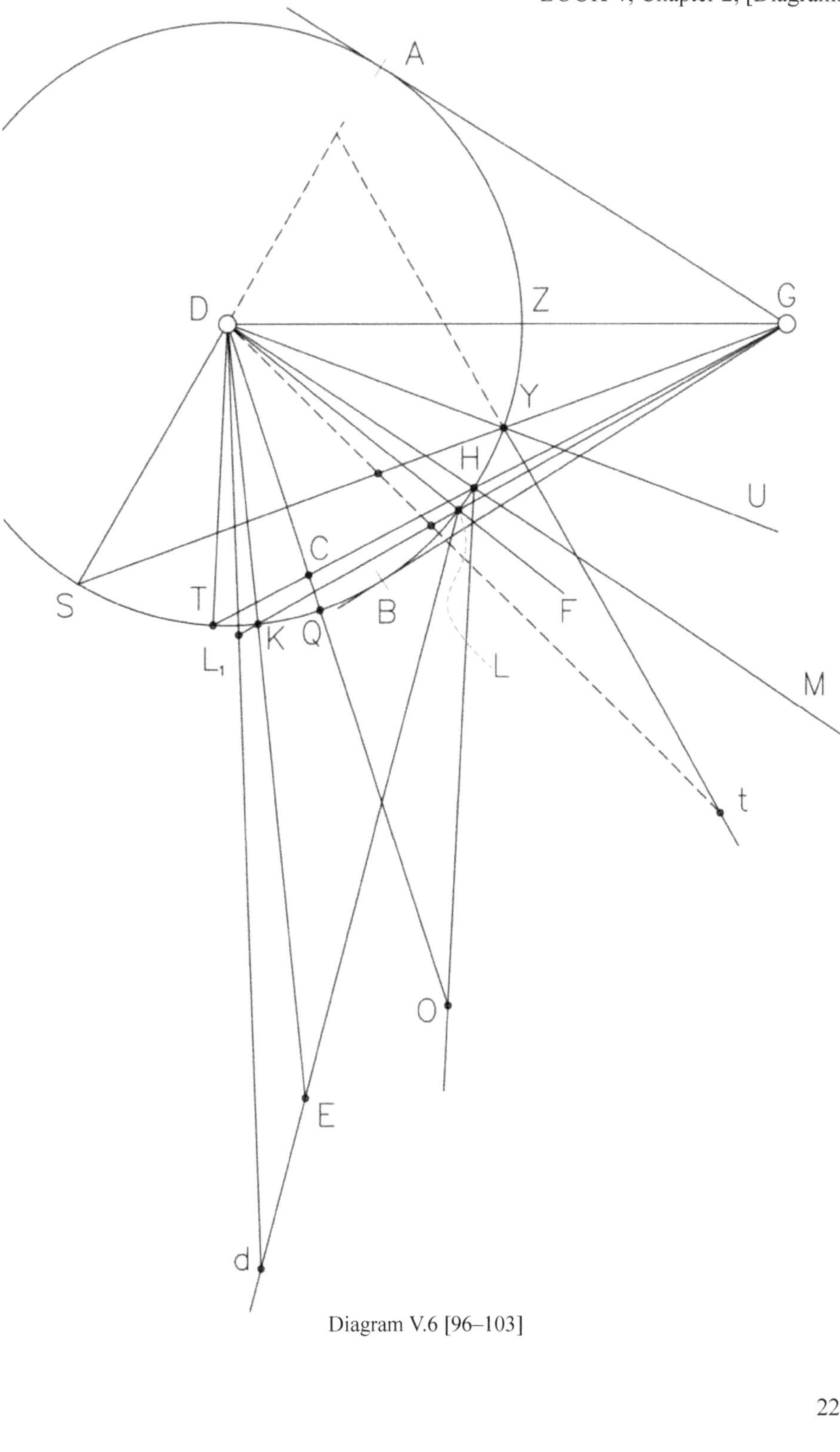

Diagram V.6 [96–103]

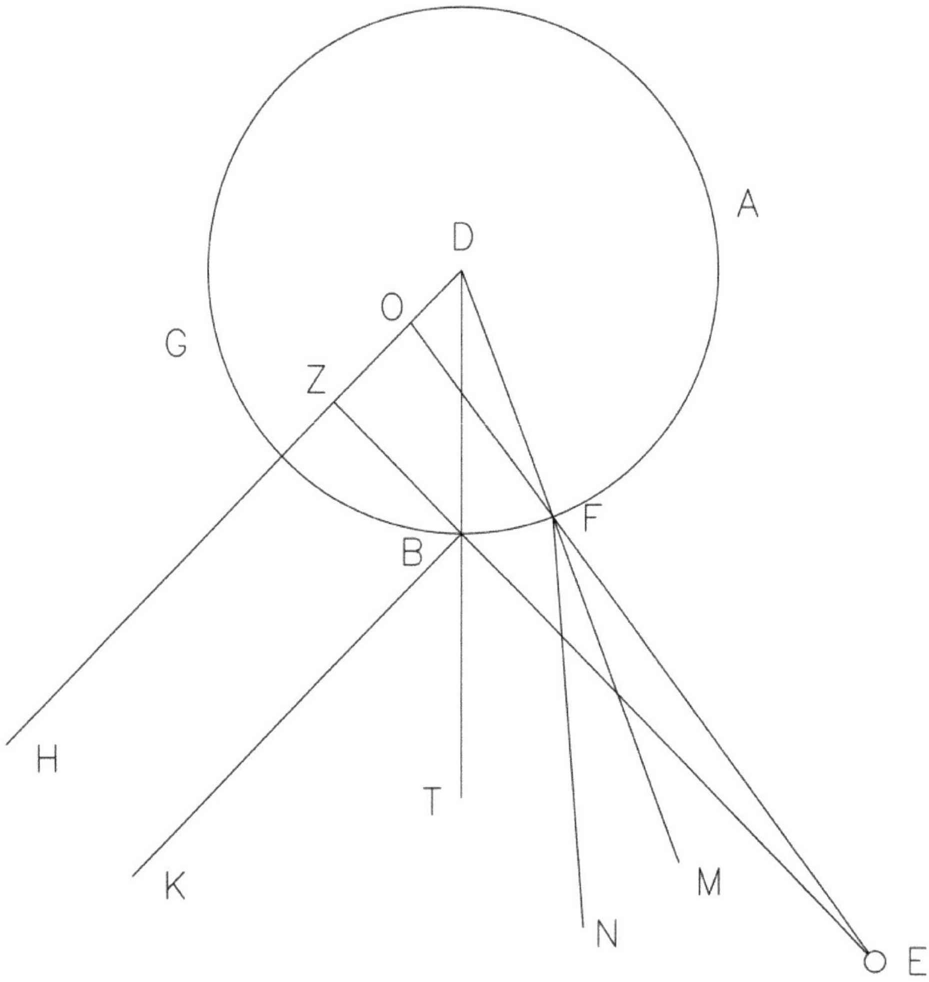

Diagram V.7 [104–5]

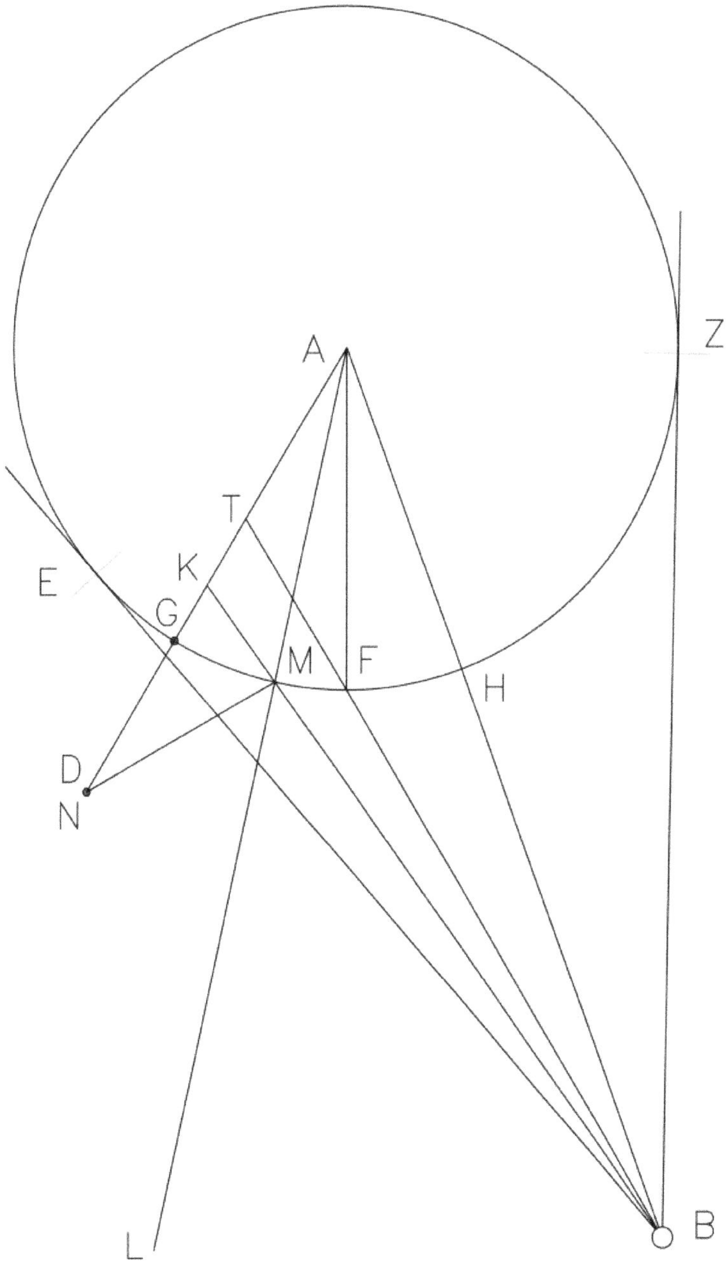

Diagram V.8 [106]

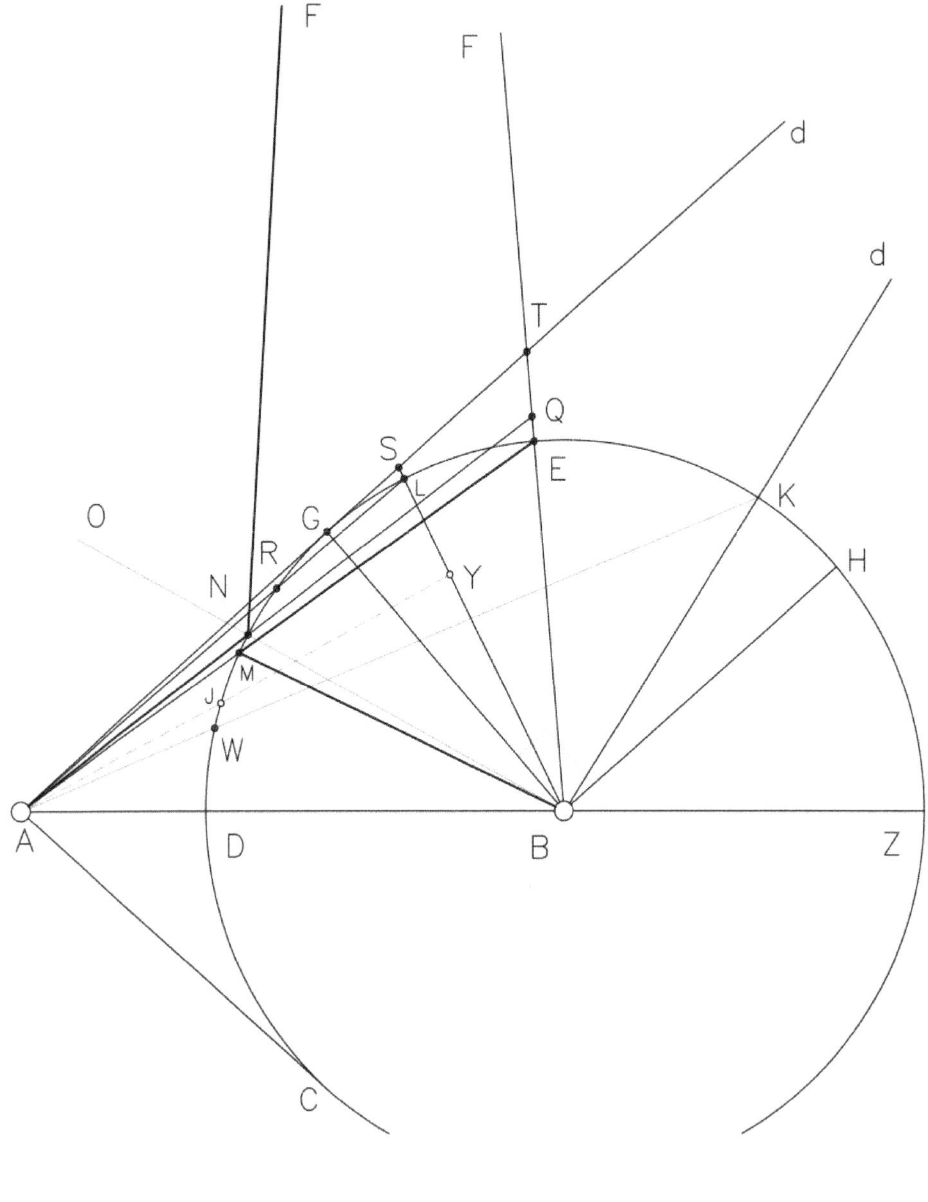

Diagram V.9 [107–16][1]

1. This is the diagram in the manuscript, to which we have added points Y and J, joined by line YJ.

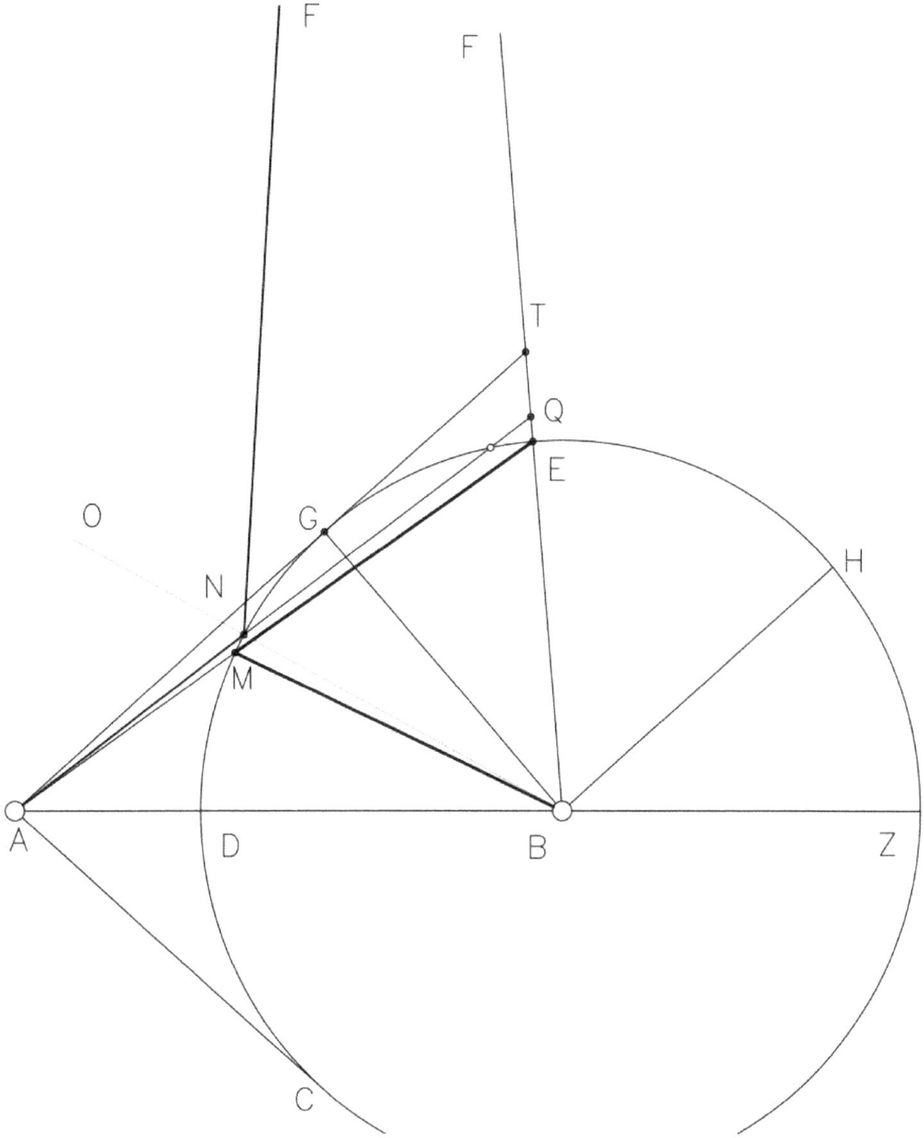

Diagram V.9(2) [107][2]

2. This is a simplified version of Diagram V.9, added for clarification.

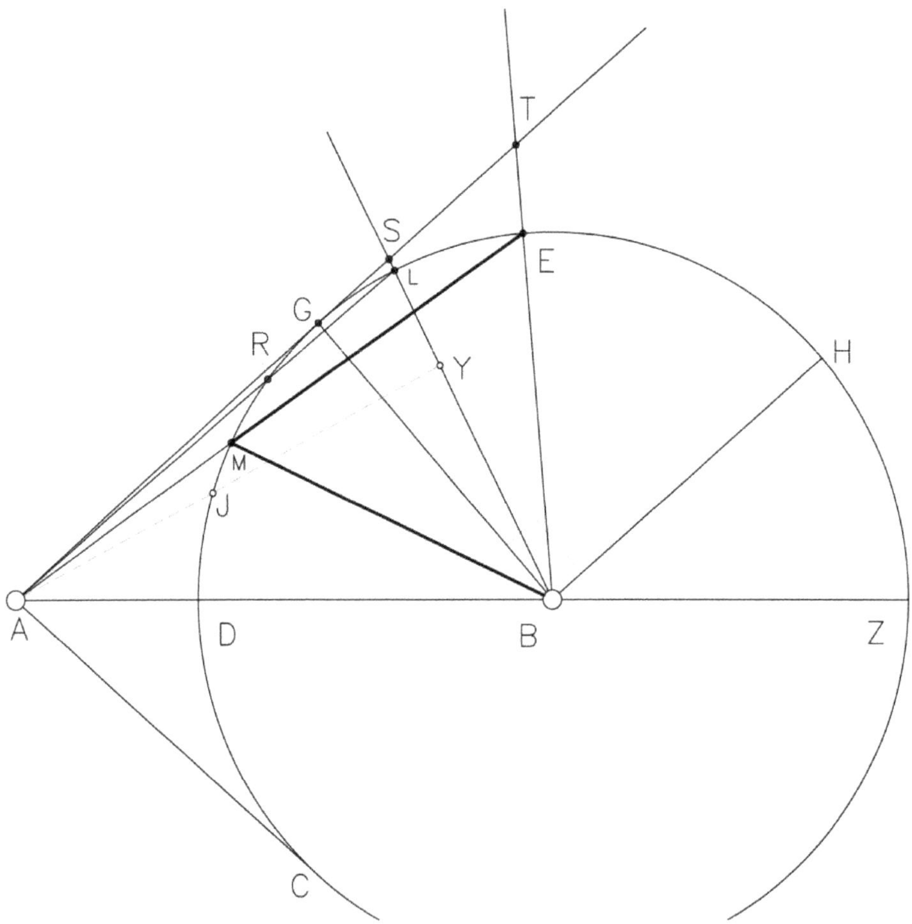

Diagram V.9(3) [108][3]

3. This is a simplified version of Diagram V.9, added for clarification.

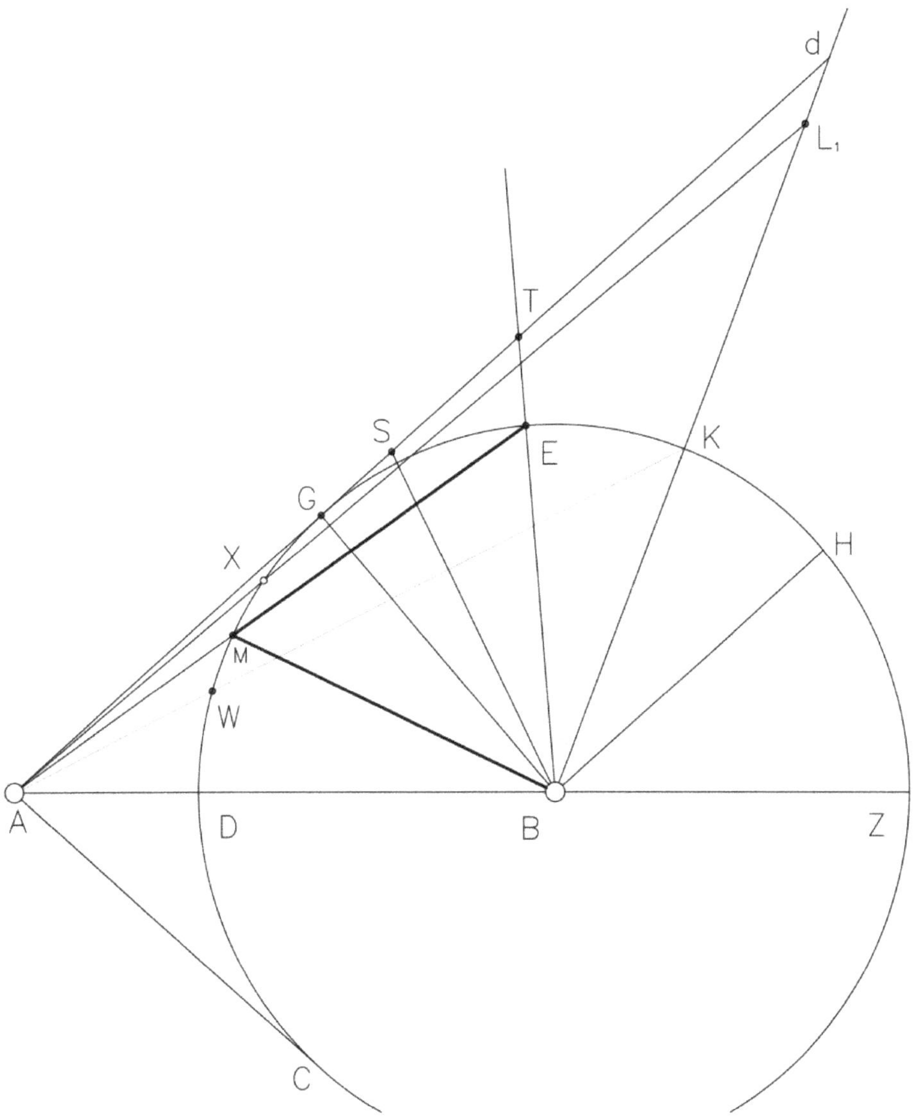

Diagram V.9(4) [109][4]

4. This is a simplified version of Diagram V.9, added for clarification. Point X (Arabic *ḵhā*) is not named in the text.

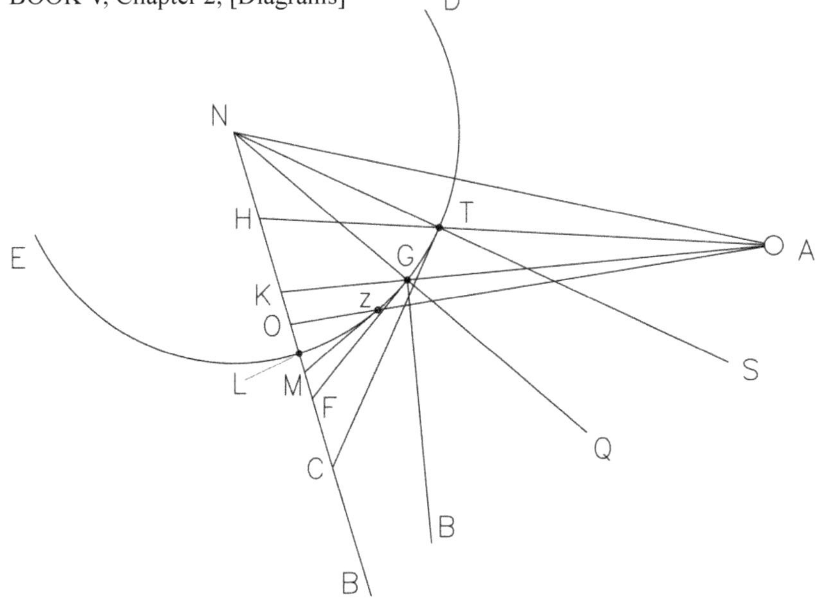

Diagram V.10 [117–23]

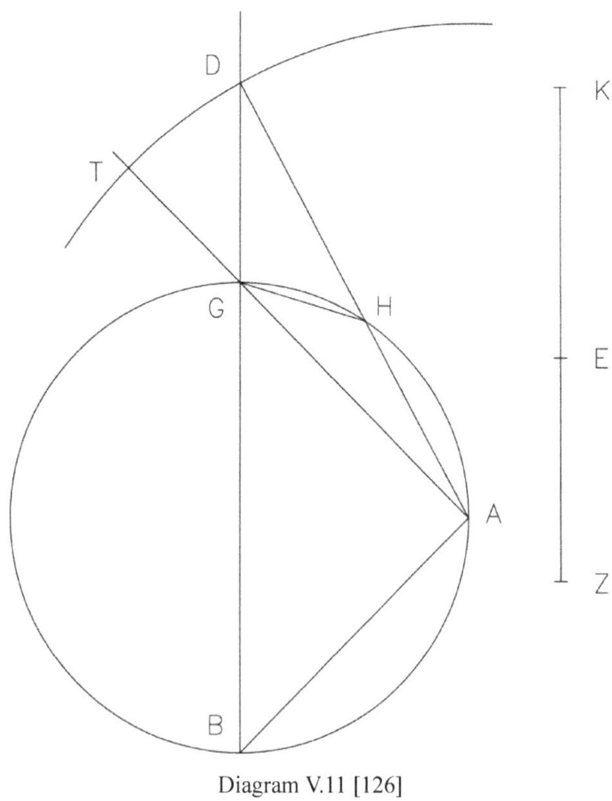

Diagram V.11 [126]

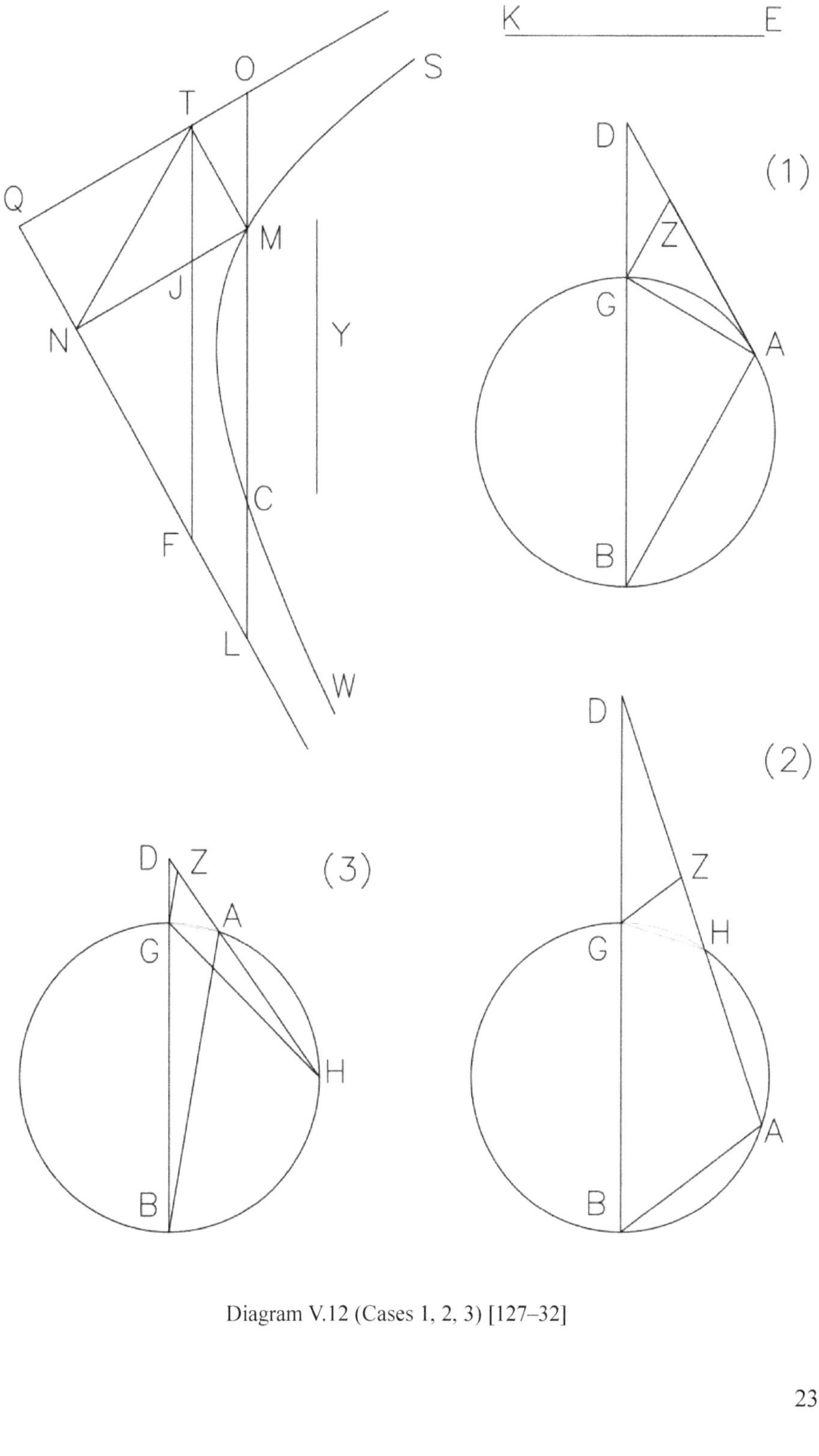

Diagram V.12 (Cases 1, 2, 3) [127–32]

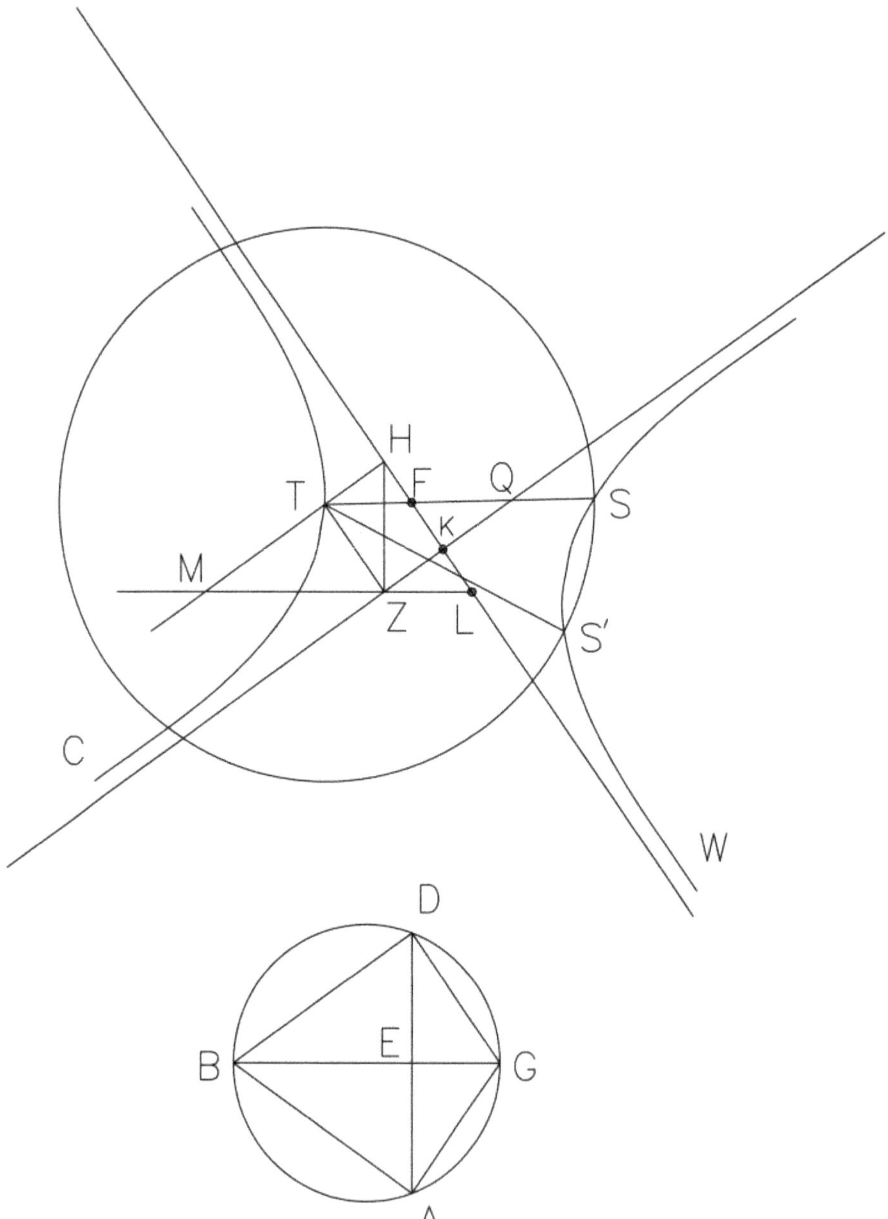

Diagram V.13 [133–37]

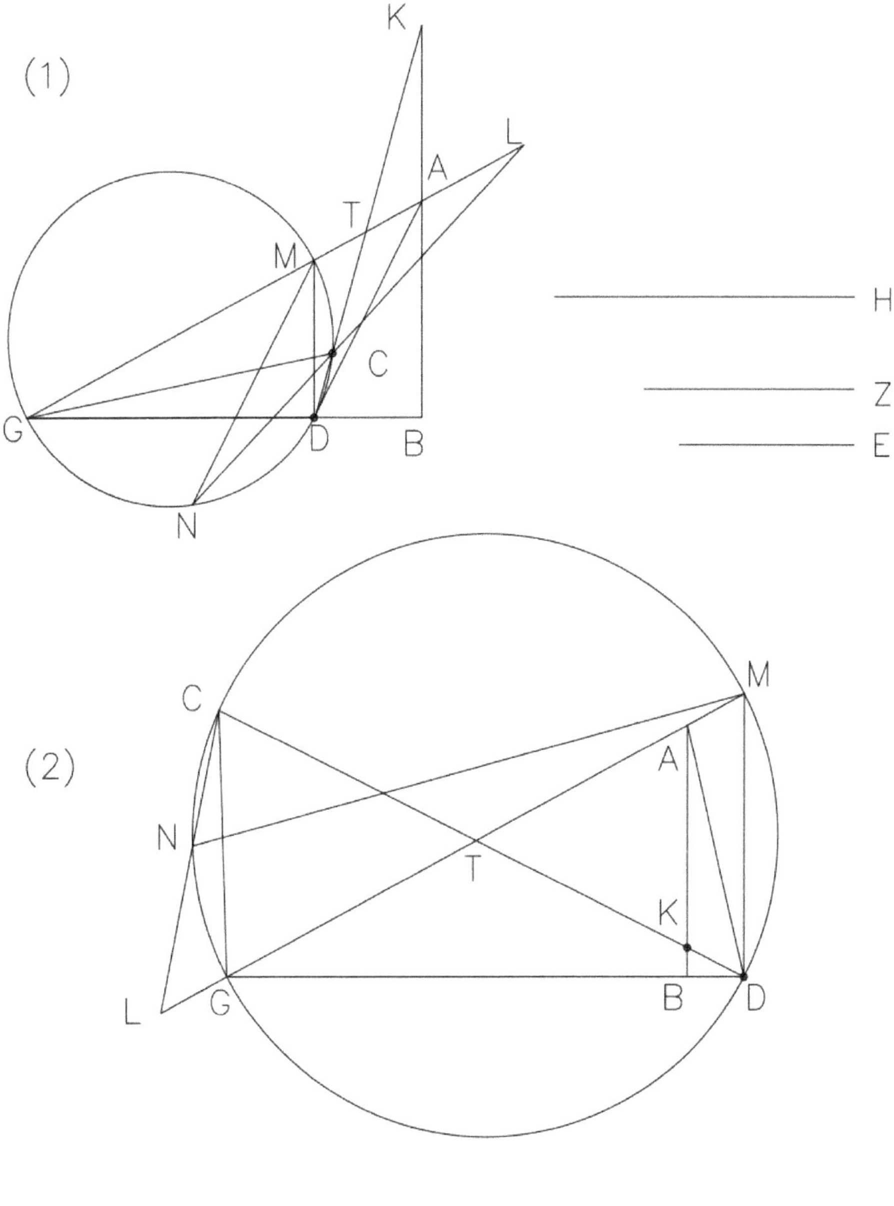

Diagram V.14 (Cases 1, 2) [138]

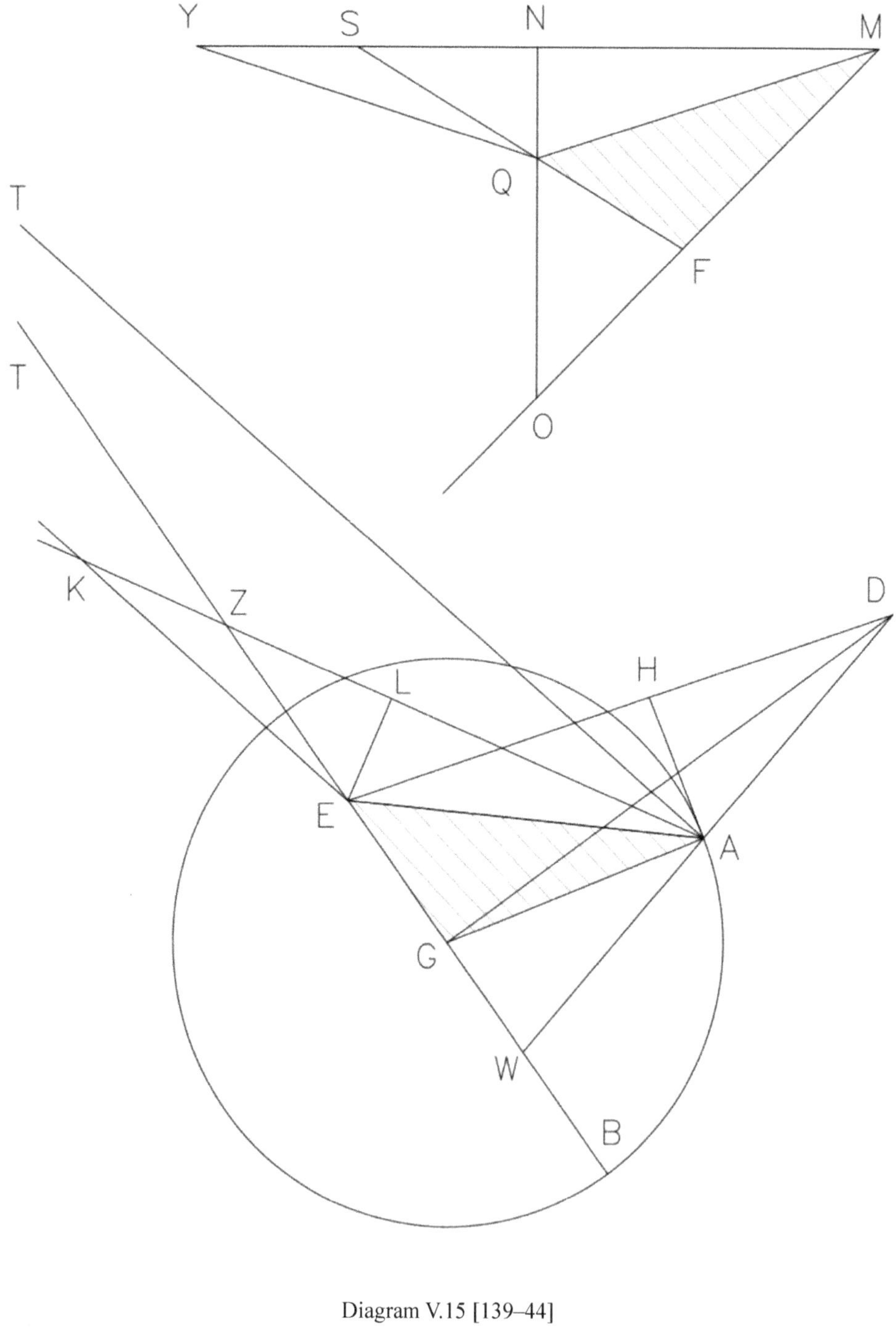

Diagram V.15 [139–44]

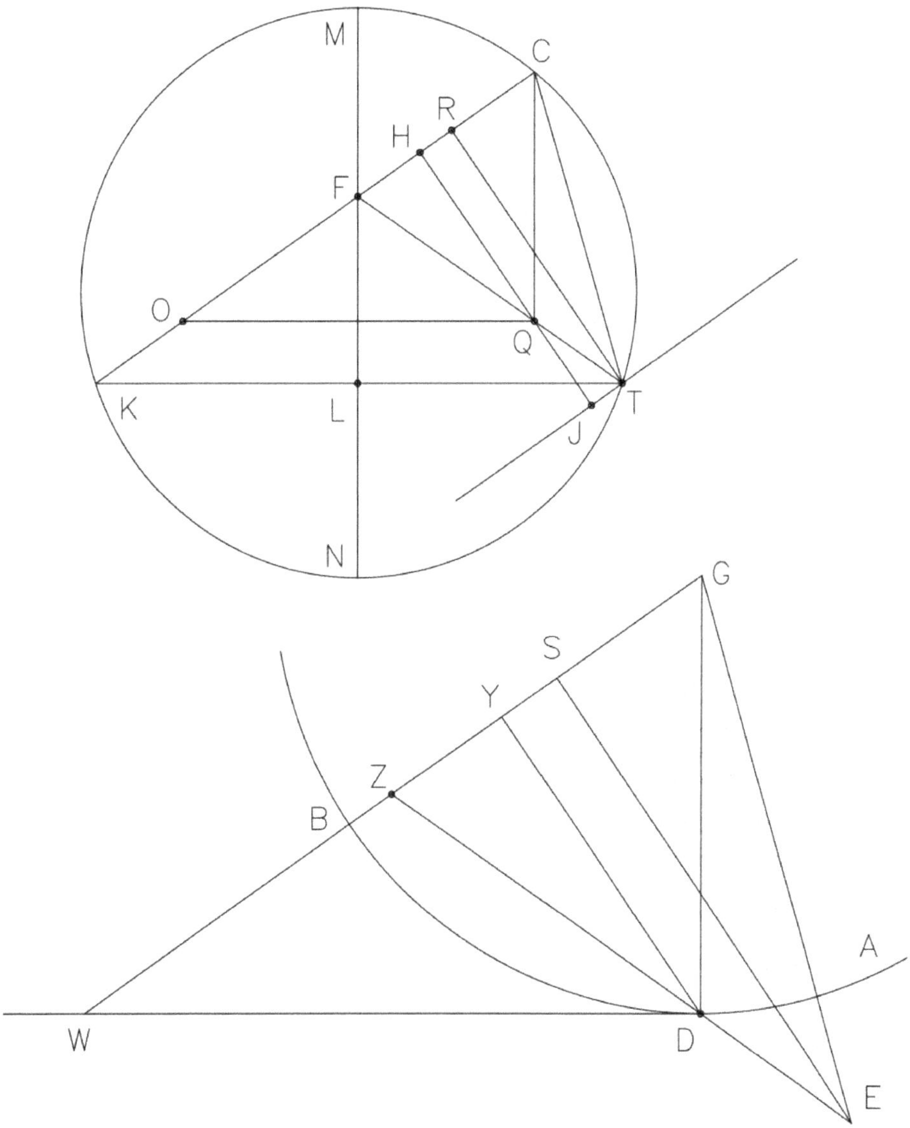

Diagram V.16 [145–47]

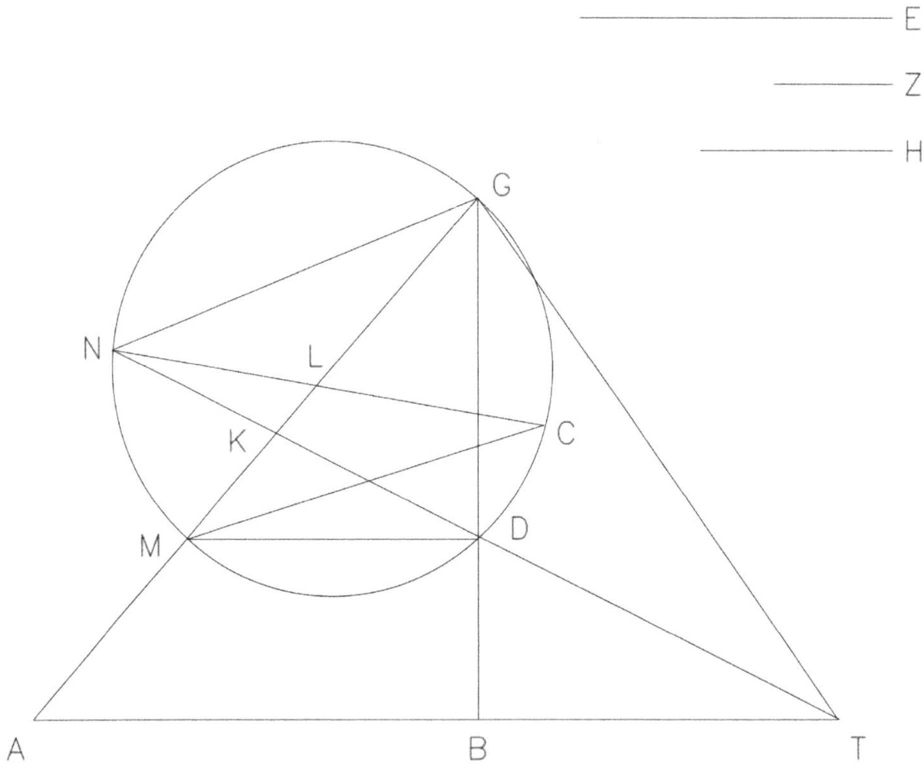

Diagram V.17 [148–49]

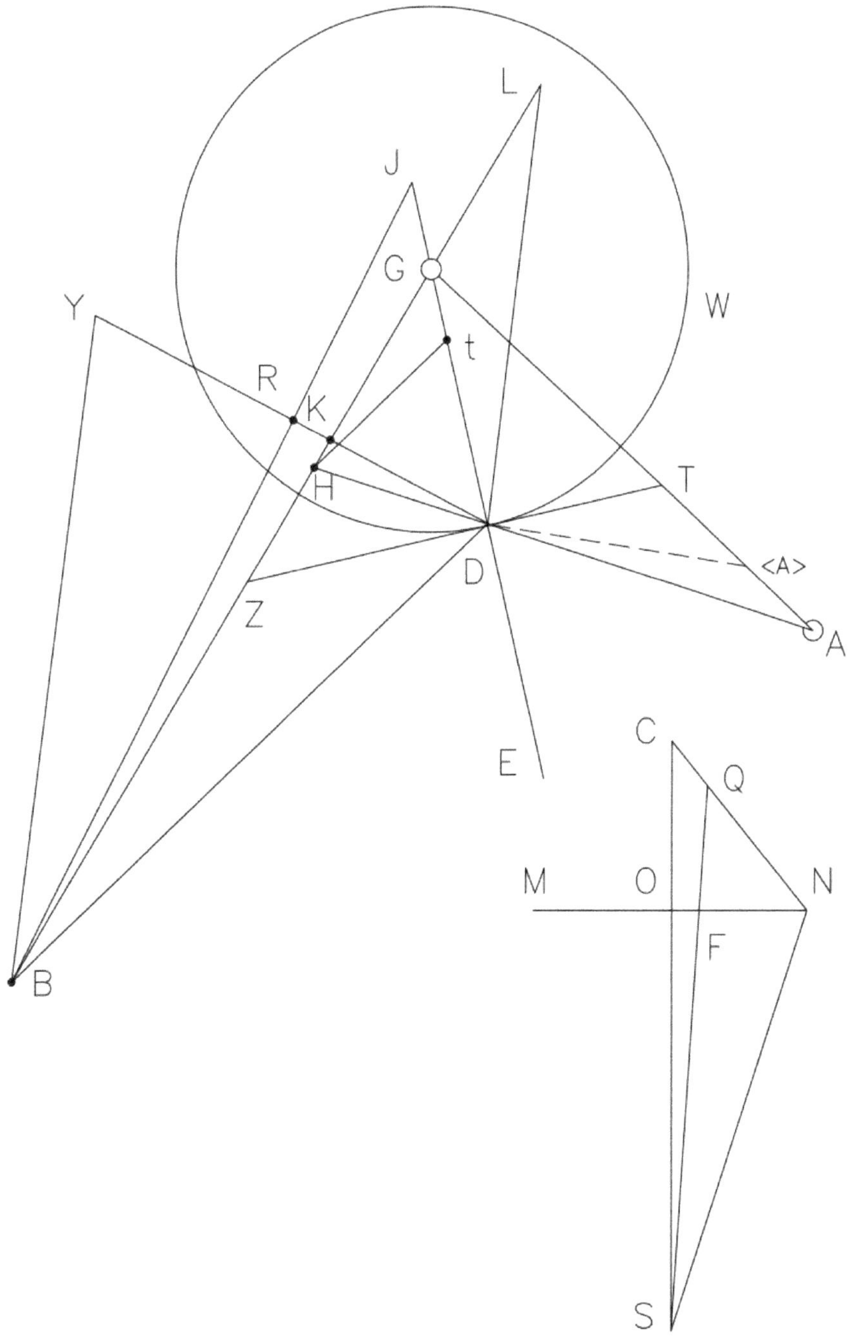

Diagram V.18 [150–54]

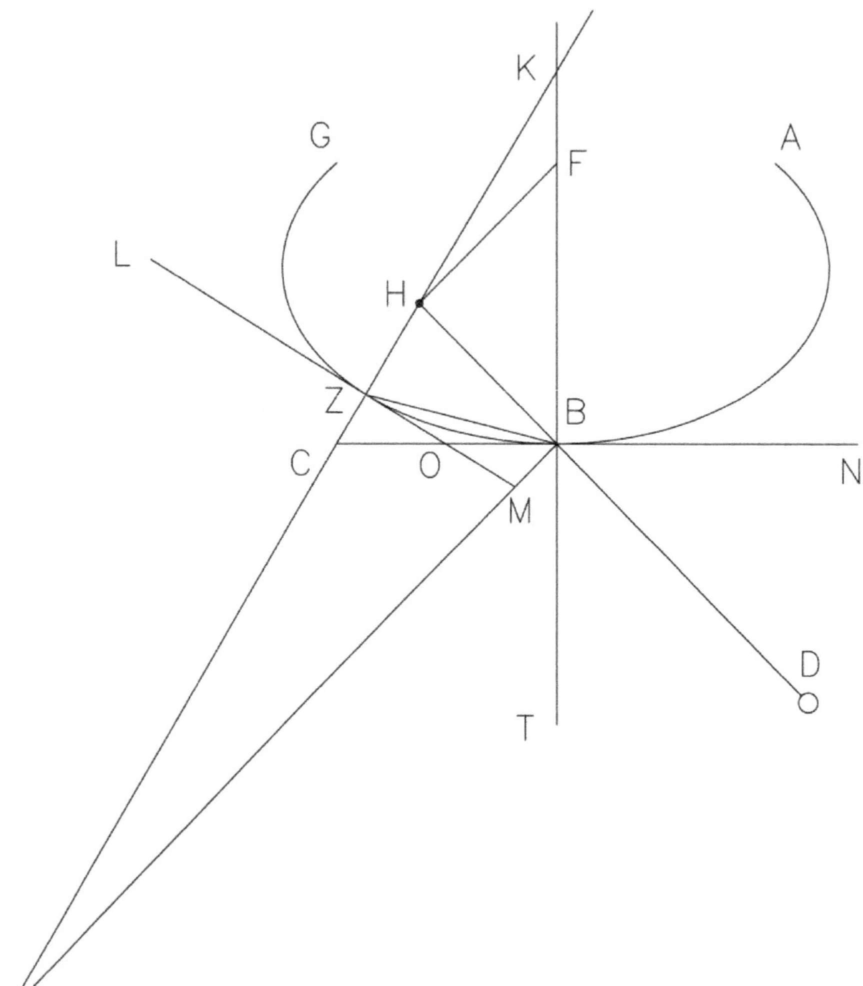

Diagram V.19 [164–66]

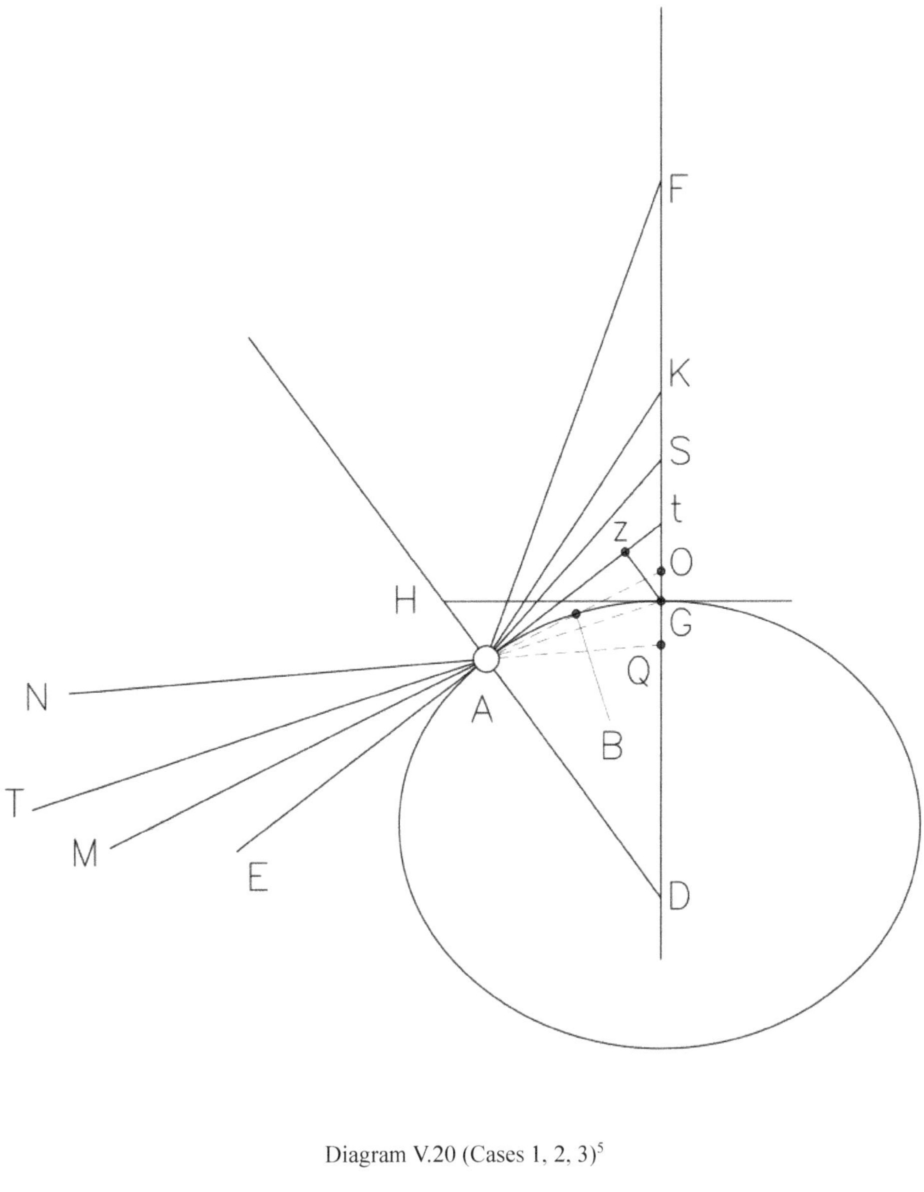

Diagram V.20 (Cases 1, 2, 3)[5]

Diagram V.20(1) [168–69]

5. Diagrams V.20(1–3) are three simplified versions of a single diagram in MS TA [168–72].

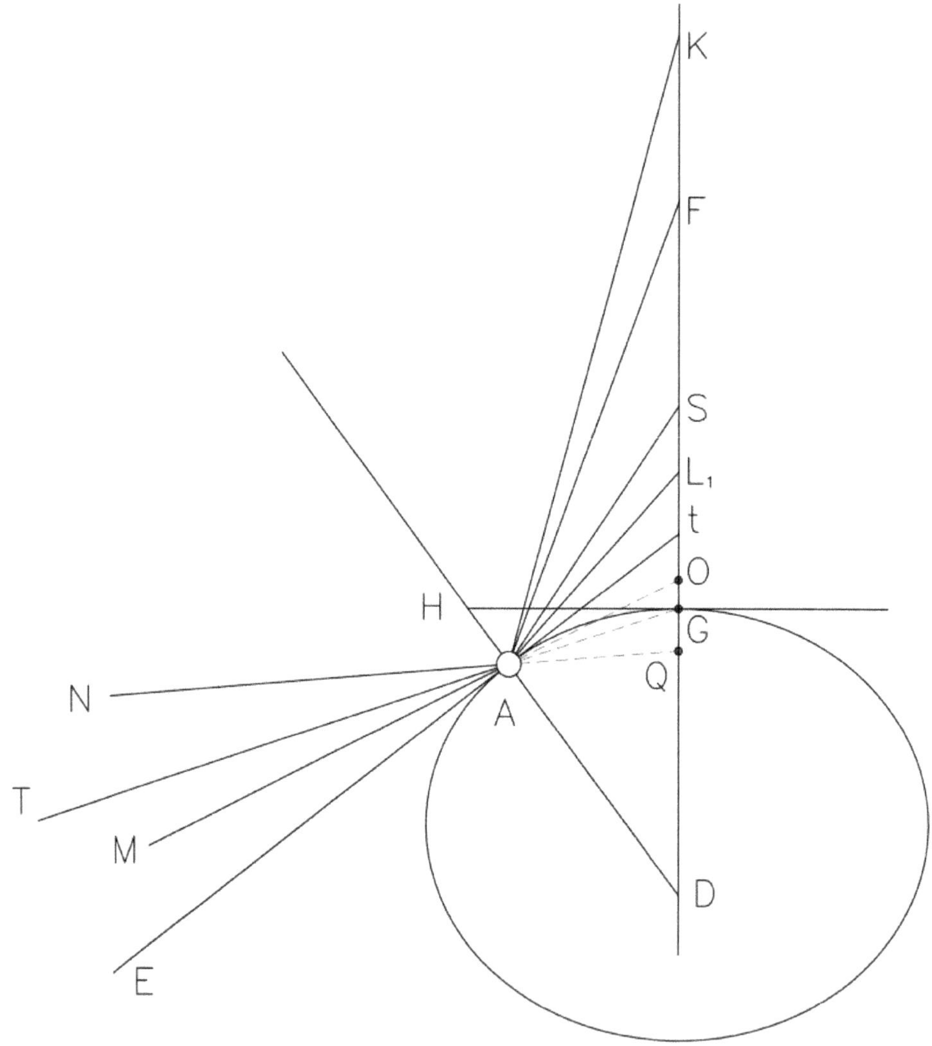

Diagram V.20(2) [170]

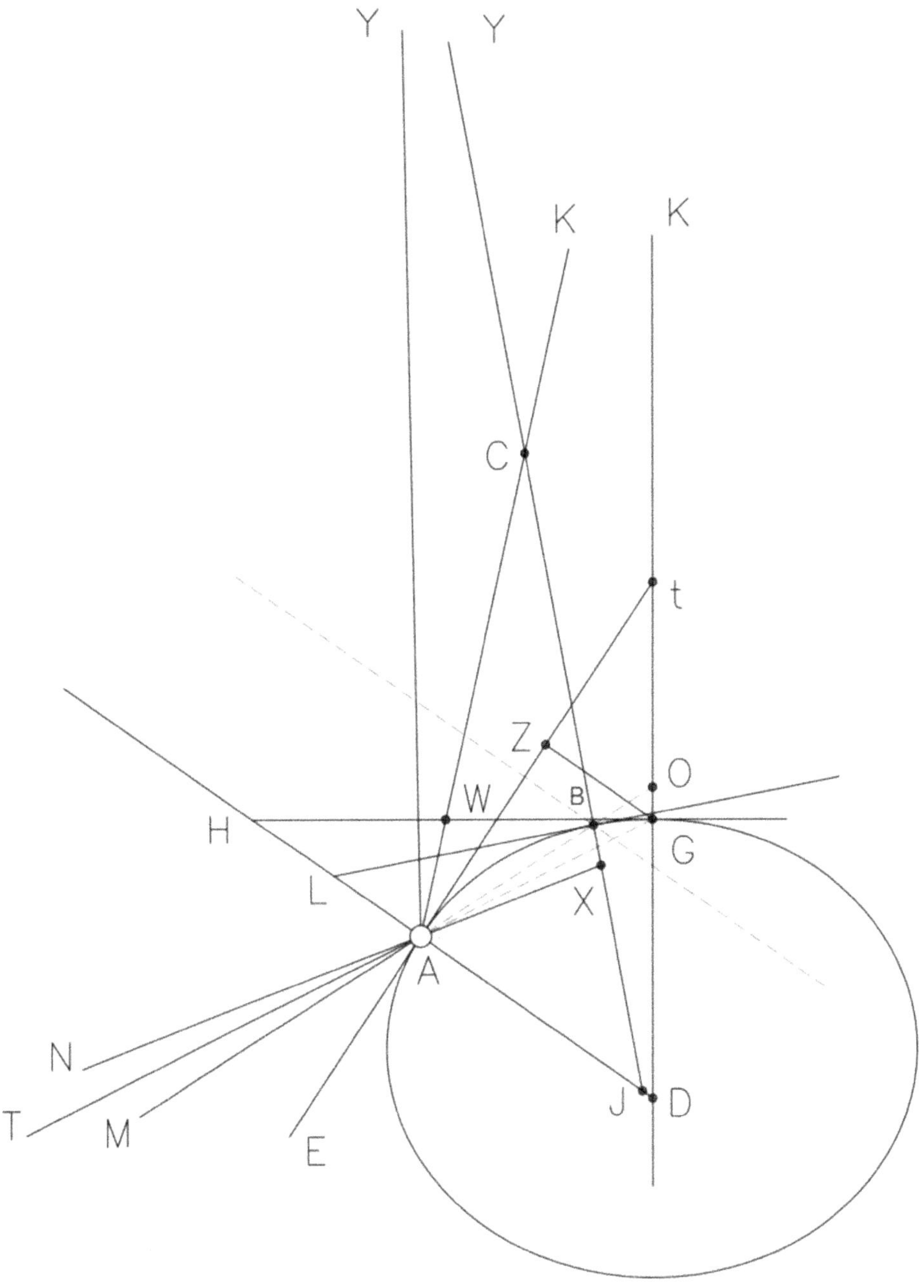

Diagram V.20(3) [171–72]

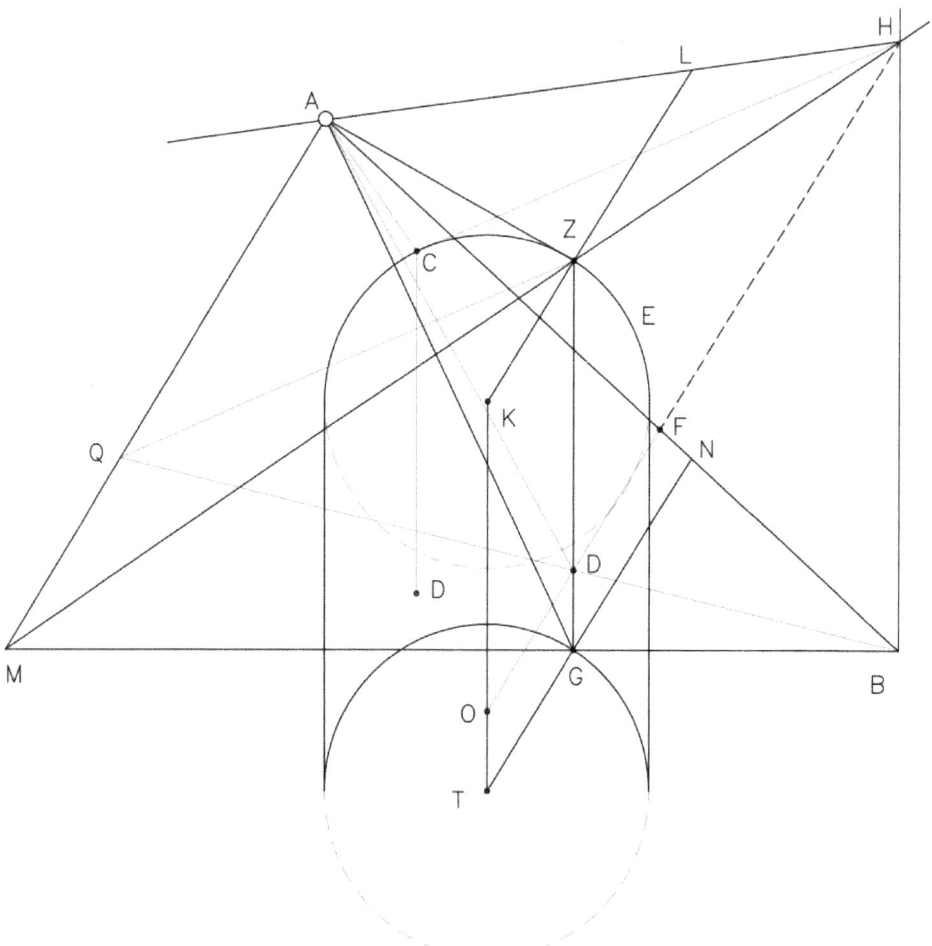

Diagram V.20+ [174–83]

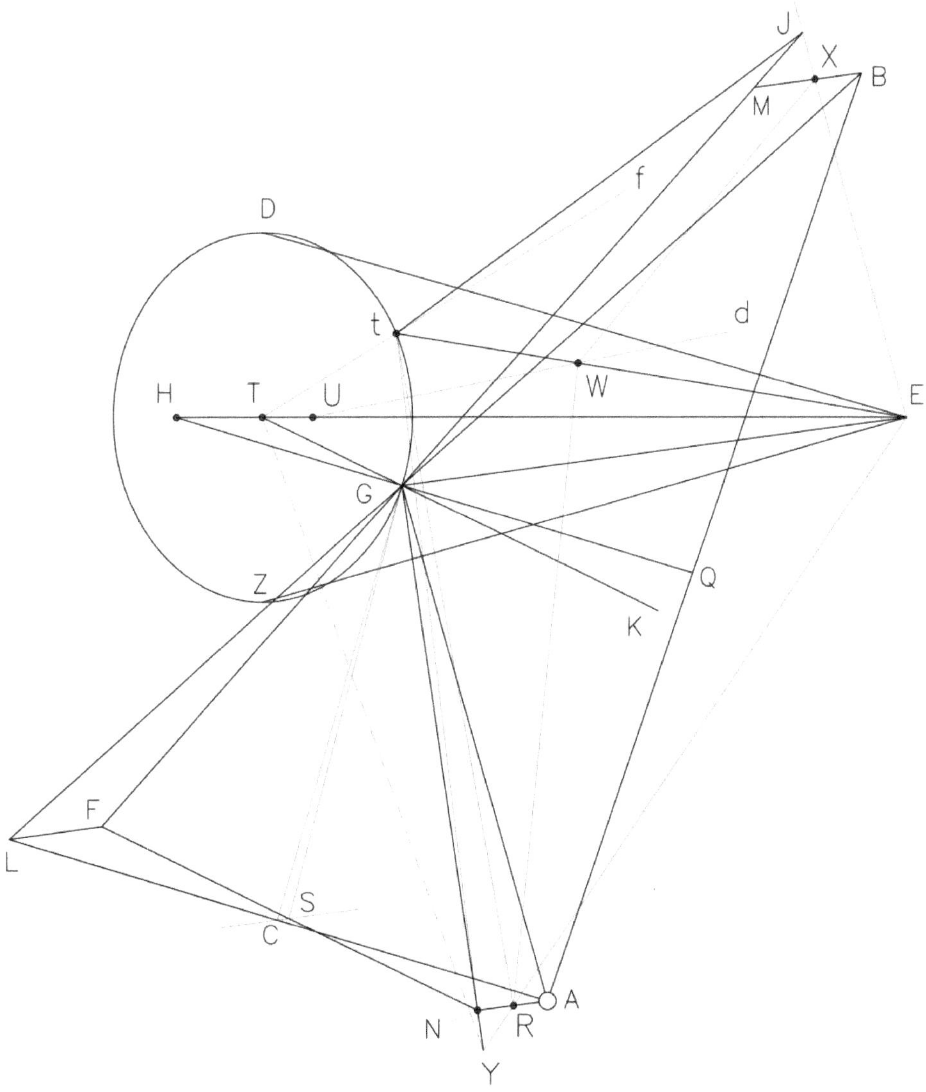

Diagram V.21 [187–90]

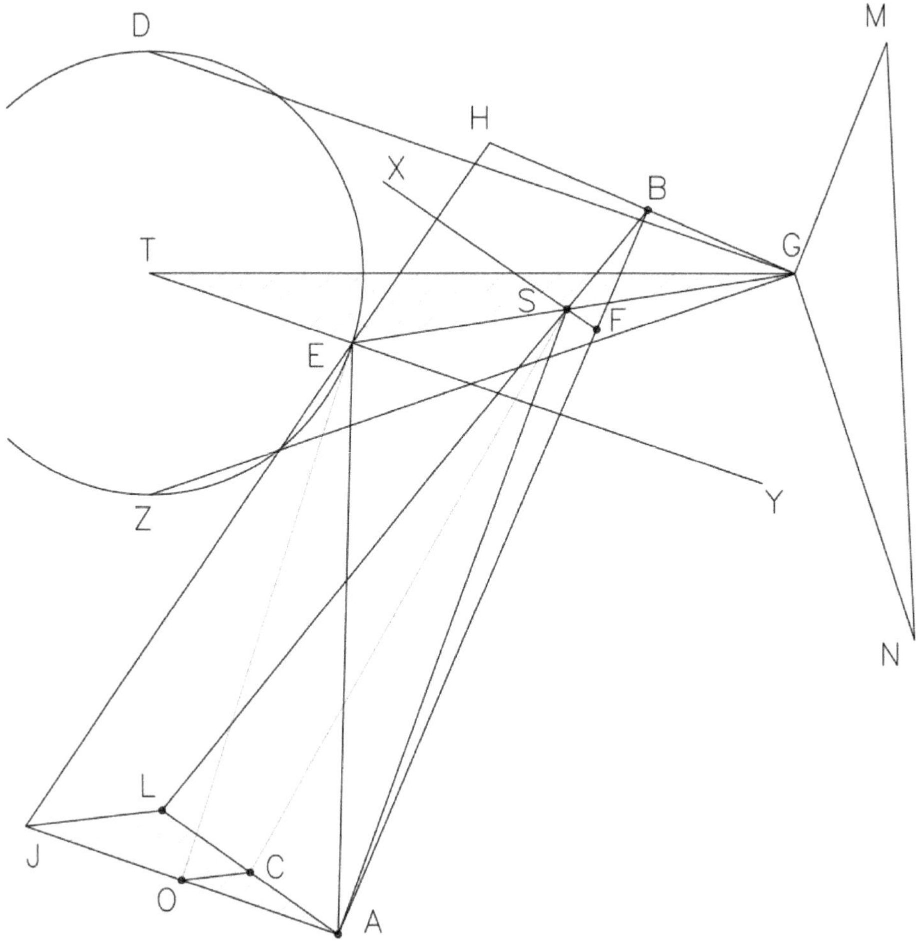

Diagram V.22[6]

Diagram V.22(a) [191]

6. Diagrams V.22(a–d) are four different configurations of two diagrams in MS TA.

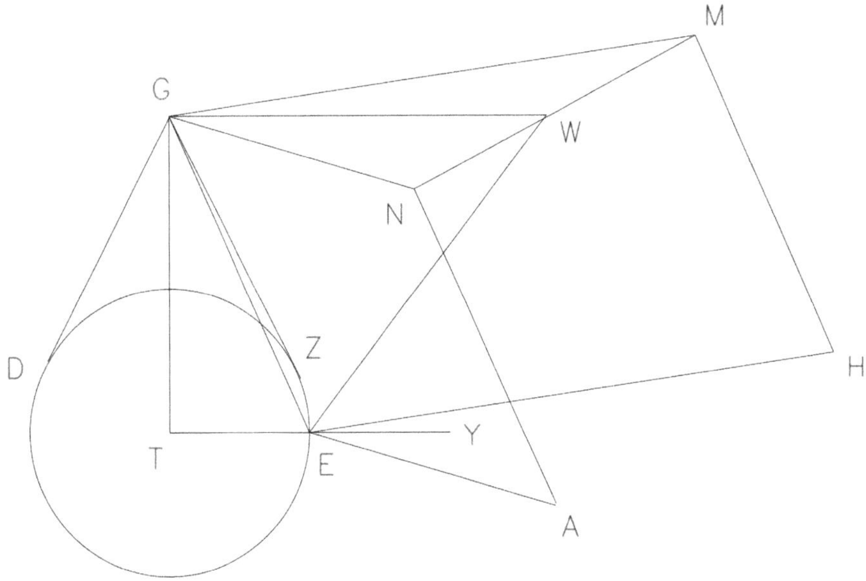

Diagram V.22(b) [192]

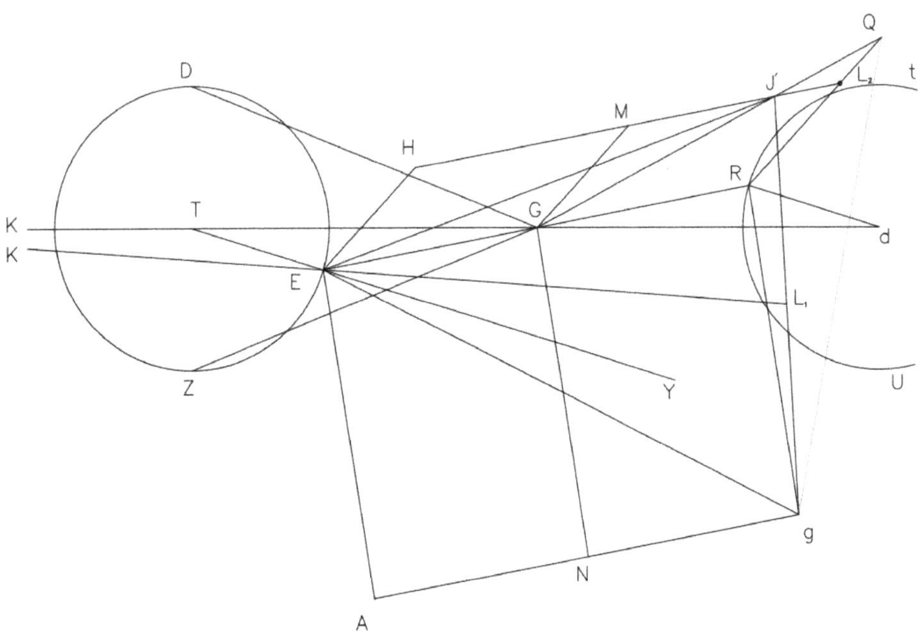

Diagram V.22(c) [193–94]

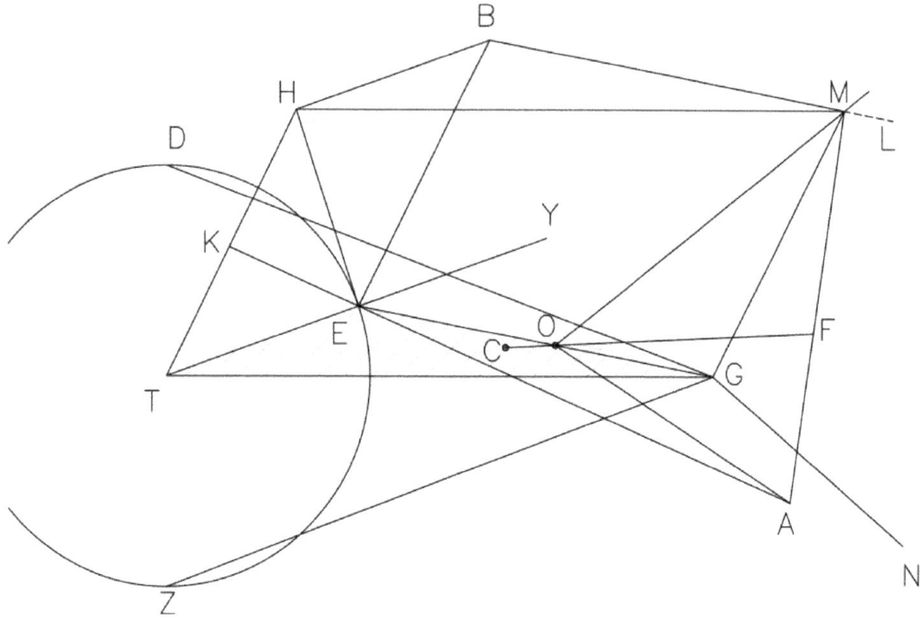

Diagram V.22(d) [195]

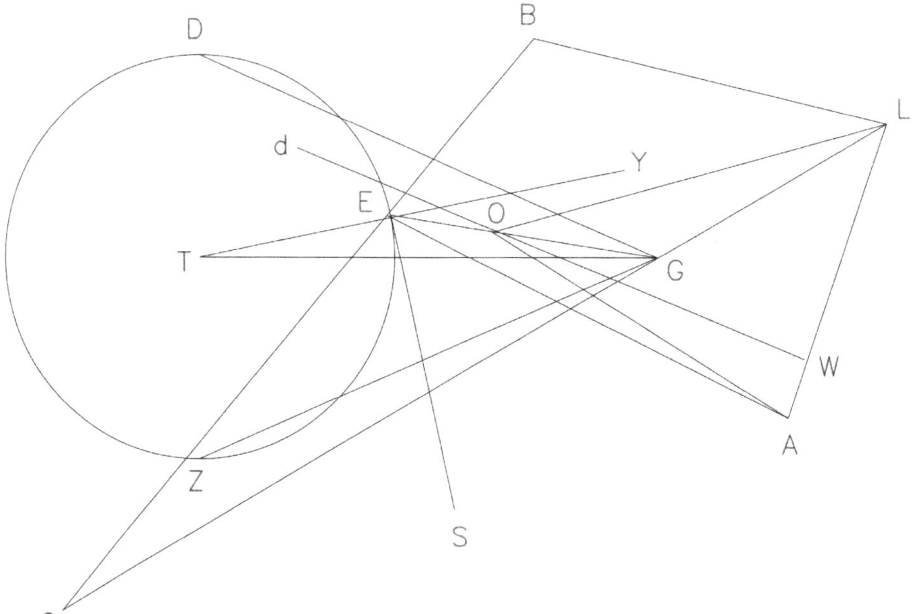

Diagram V.23 [197–98]

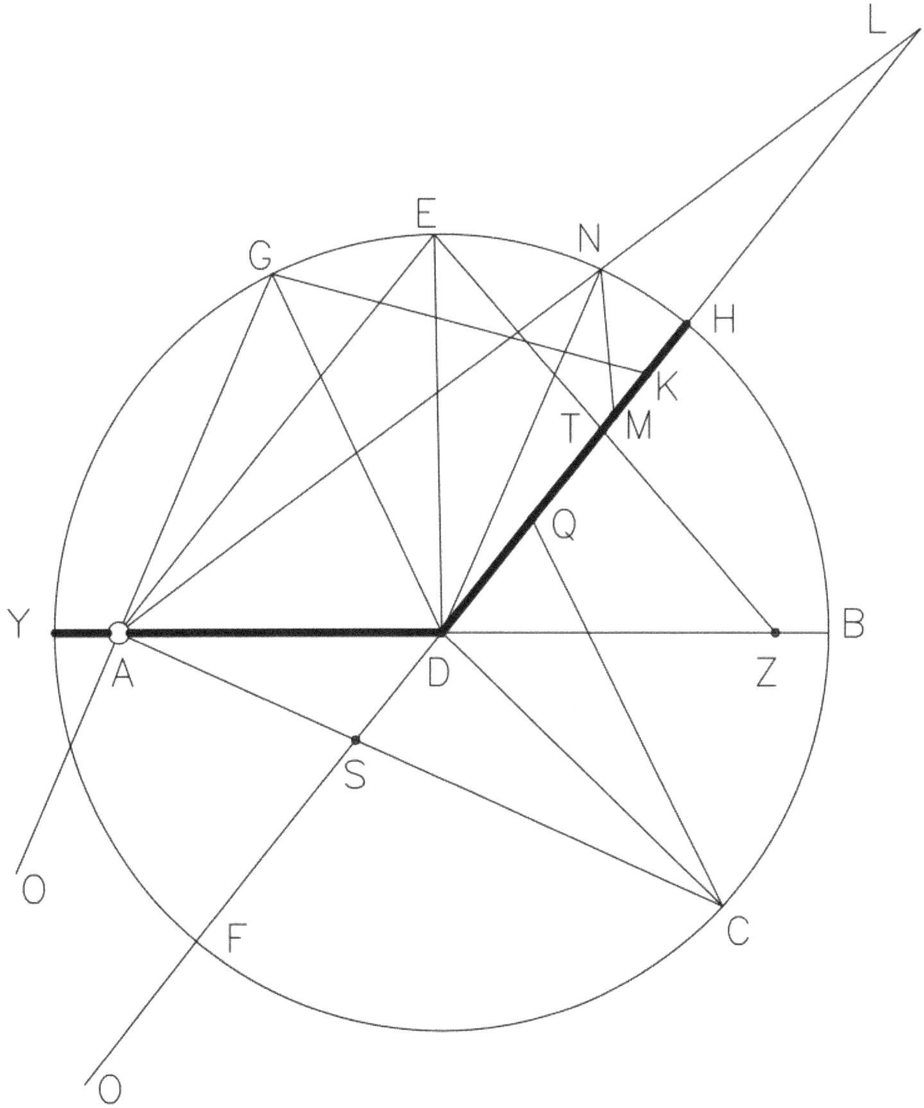

Diagram V.24 [202–5]

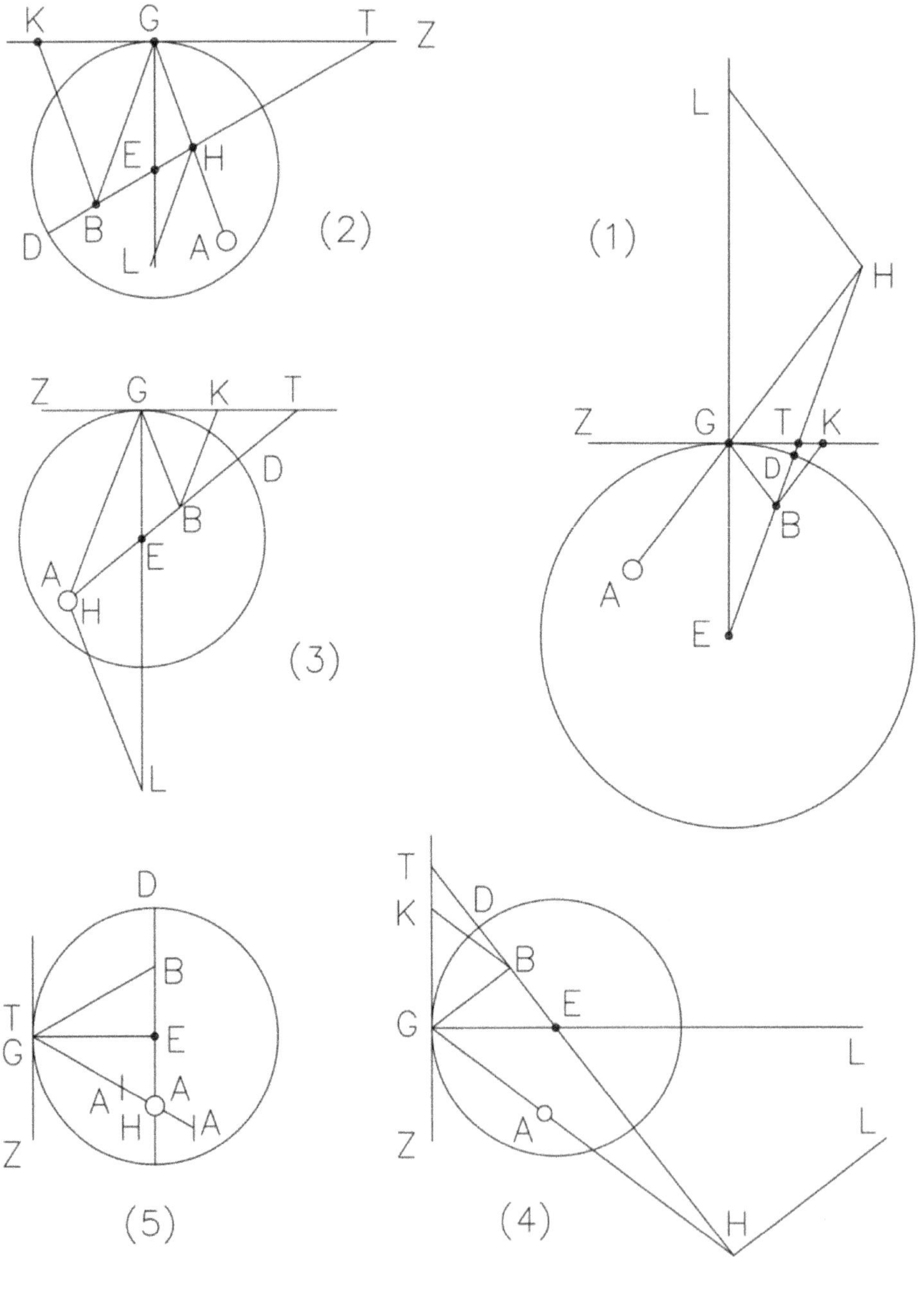

Diagram V.25 (Cases 1, 2, 3, 4, 5) [218–19]

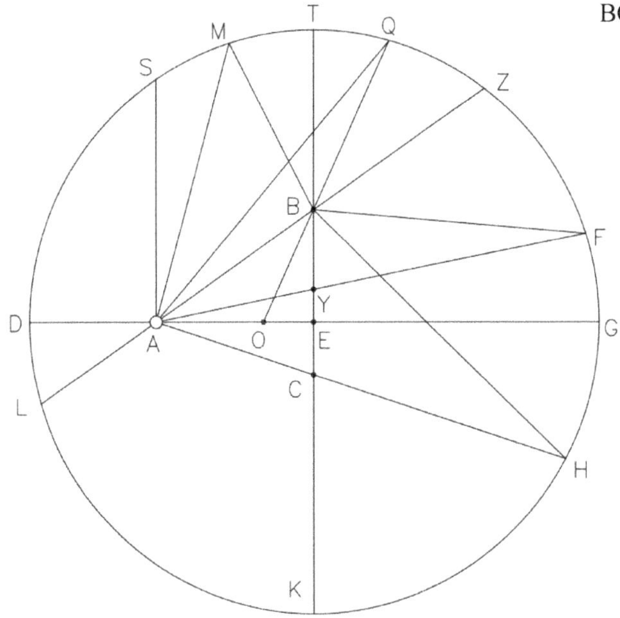

Diagram V.26 [221–23]

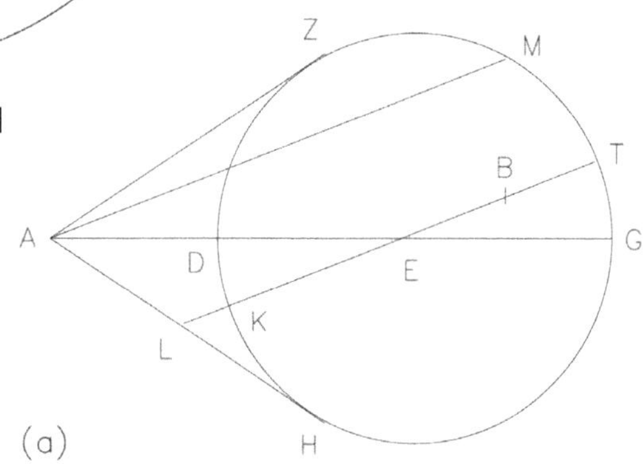

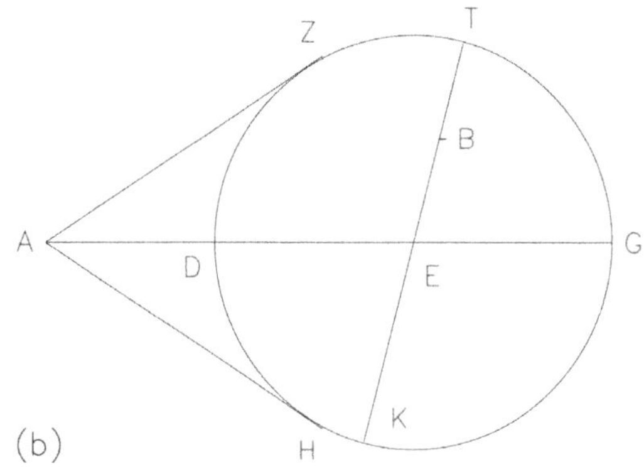

Diagram V.27(a,b) [224–25][7]

7. Diagrams V.27(a) and V.27(b) are two different configurations of a single diagram in MS TA.

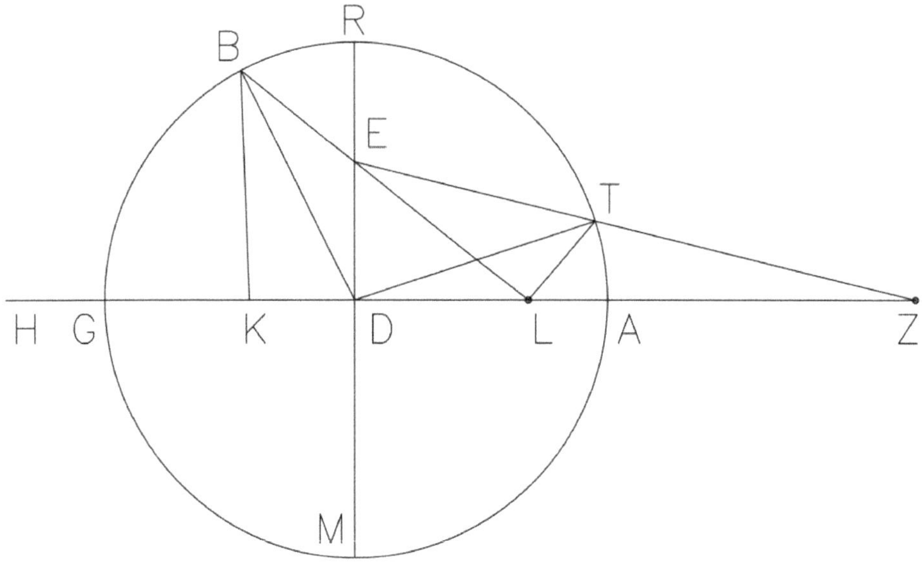

Diagram V.28 [230–31]

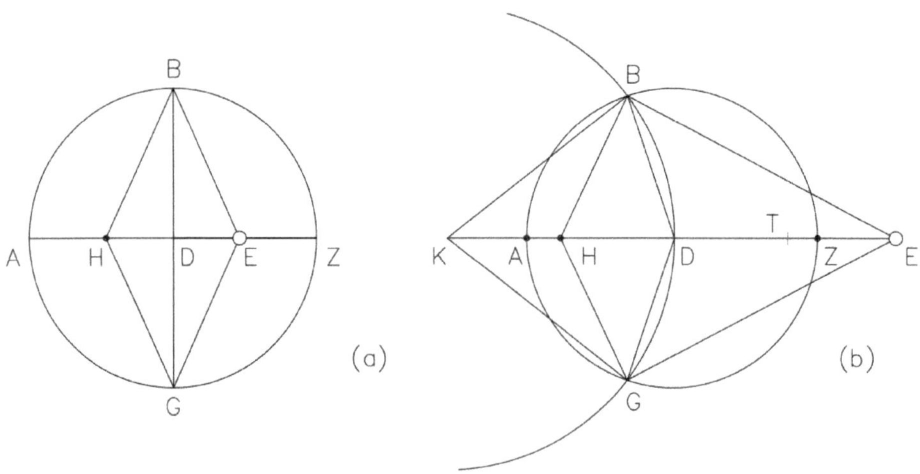

Diagram V.29(a,b) [232–35][8]

8. Diagrams V.29(a) and V.29(b) are two different configurations of a single diagram in MS TA.

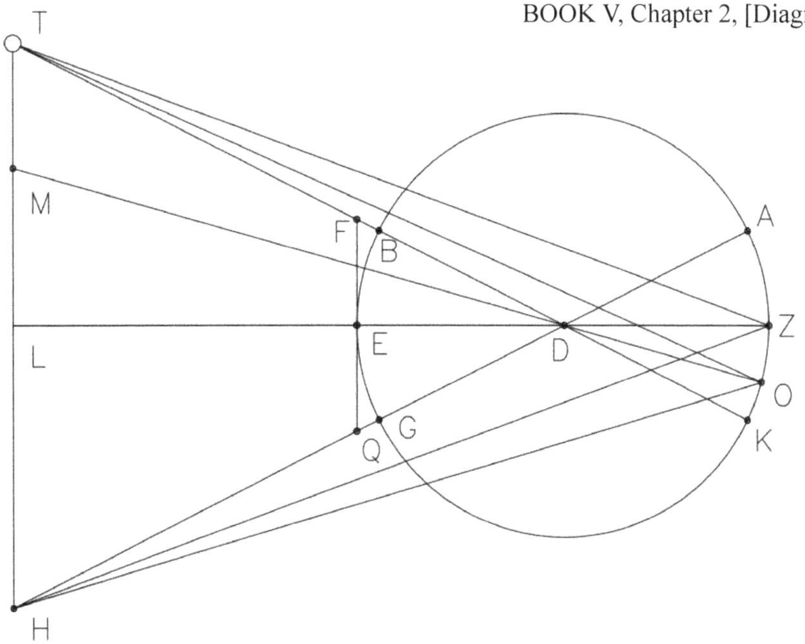

Diagram V.30 [241–44]

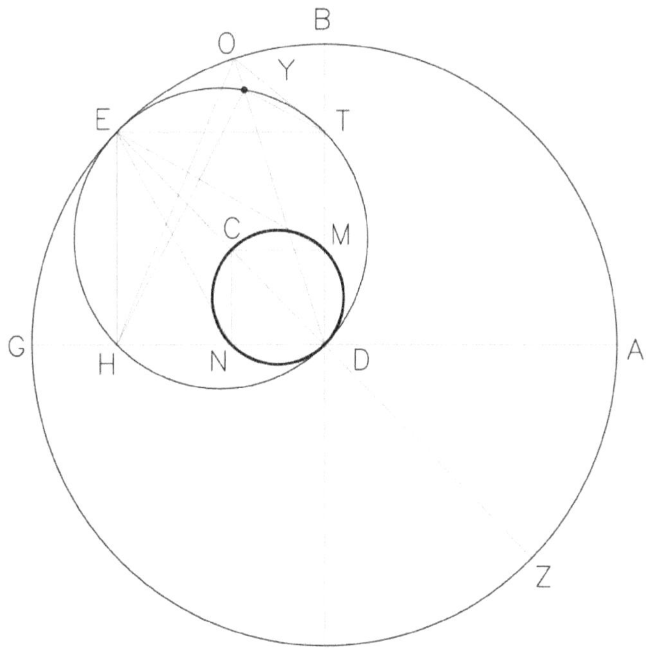

Diagram V.31 [246–47]

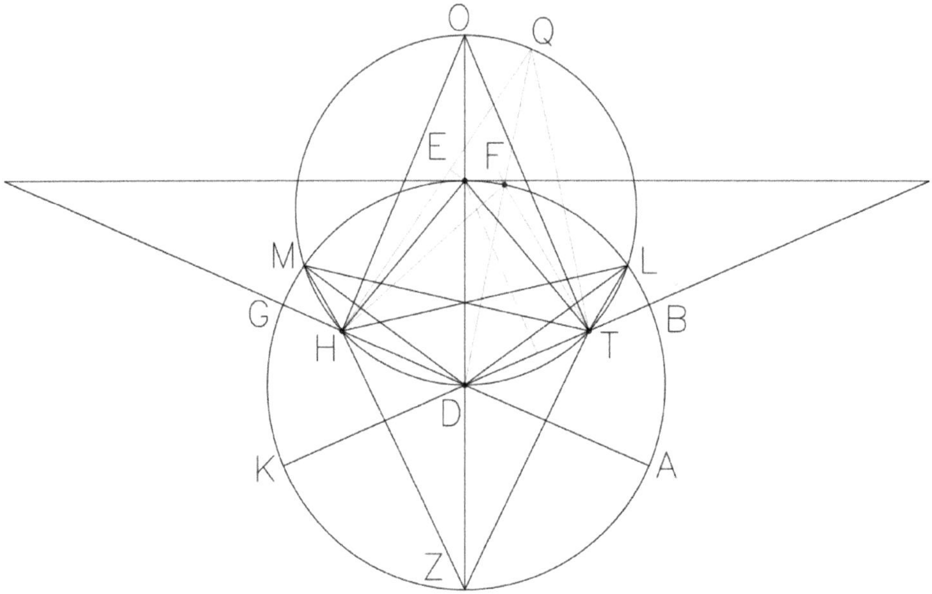

Diagram V.32 [249–51]

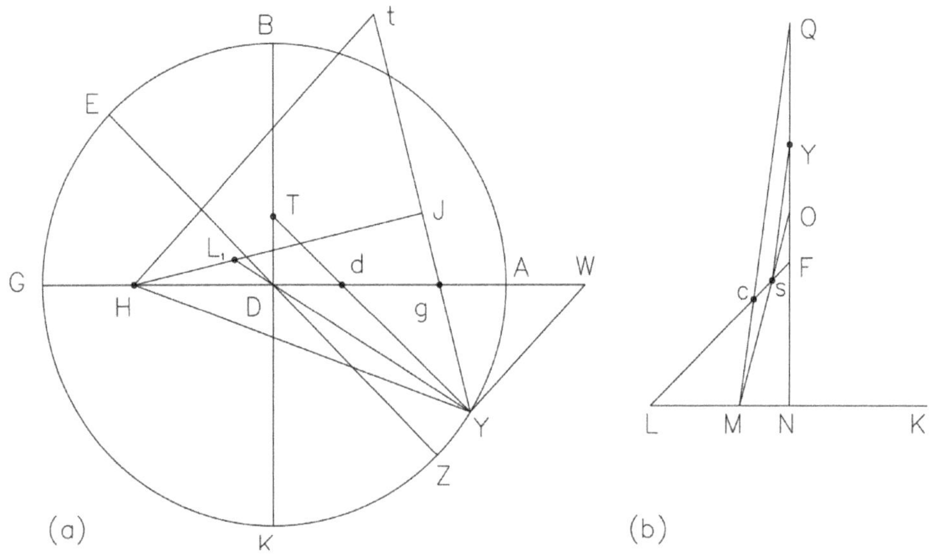

Diagram V.33(a,b) [253–57][9]

9. There are two diagrams under this number in MS TA.

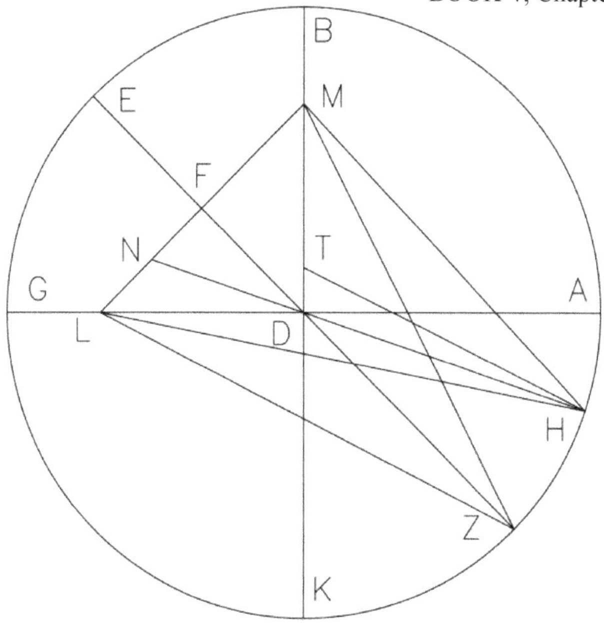

Diagram V.34 [258–59]

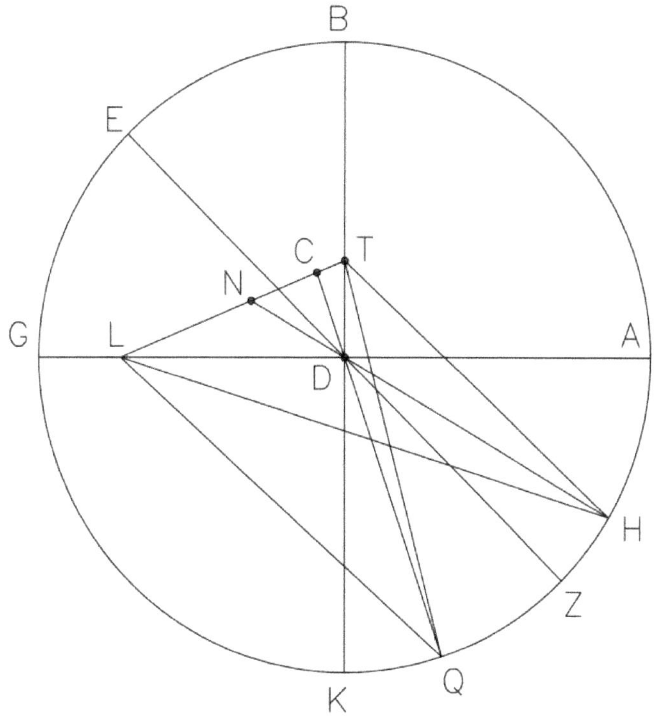

Diagram V.35 [260]

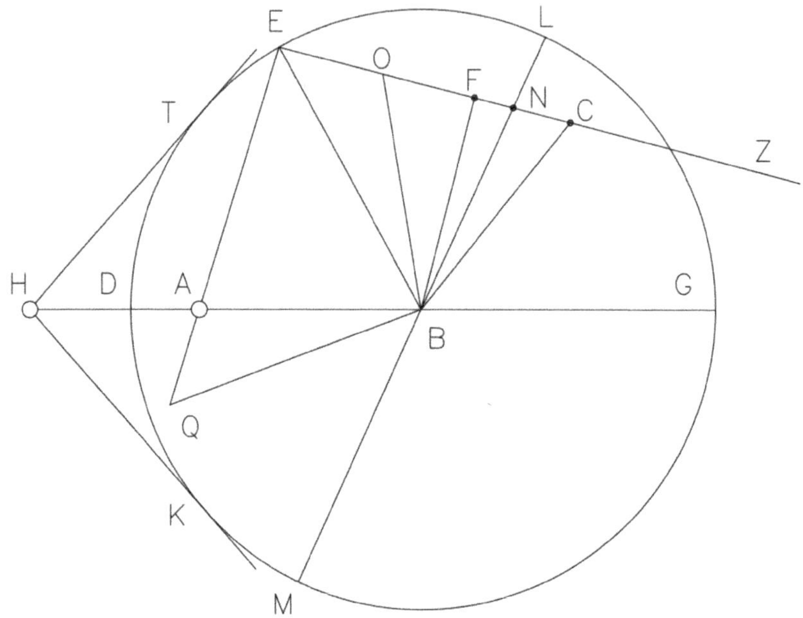

Diagram V.36 [261–67]

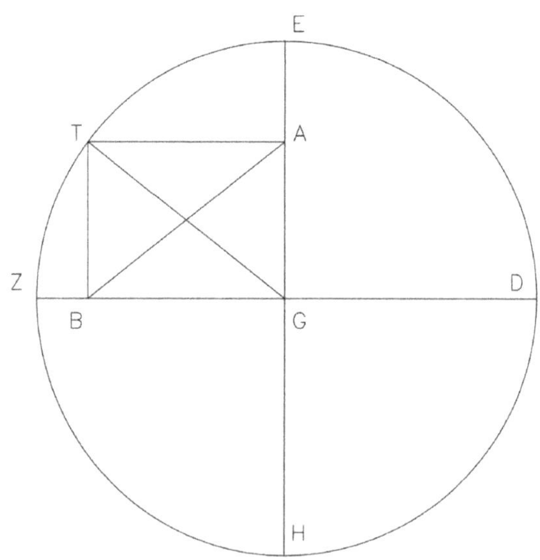

Diagram V.37 [268][10]

---

10. The single diagram that carries this number in MSS TA and TL represents the 'impossible' case in which BG = AG. Kamāl al-Dīn calls ∠ATB 'double the reflection angle', and ∠AGD 'the adjacent [angle] [*tāliyah*]'.

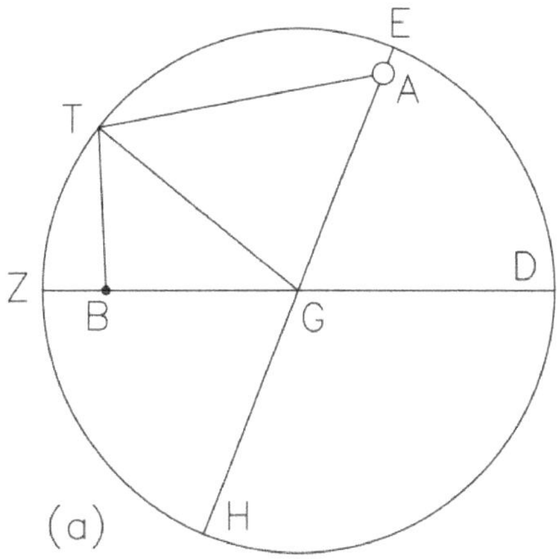

Diagram V.37(a)[11]

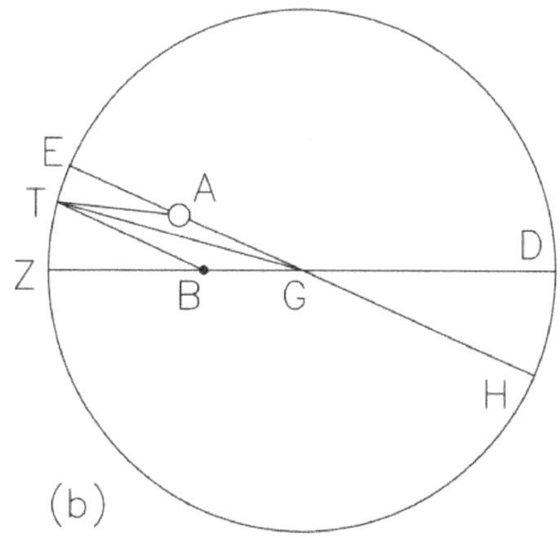

Diagram V.37(b)[12]

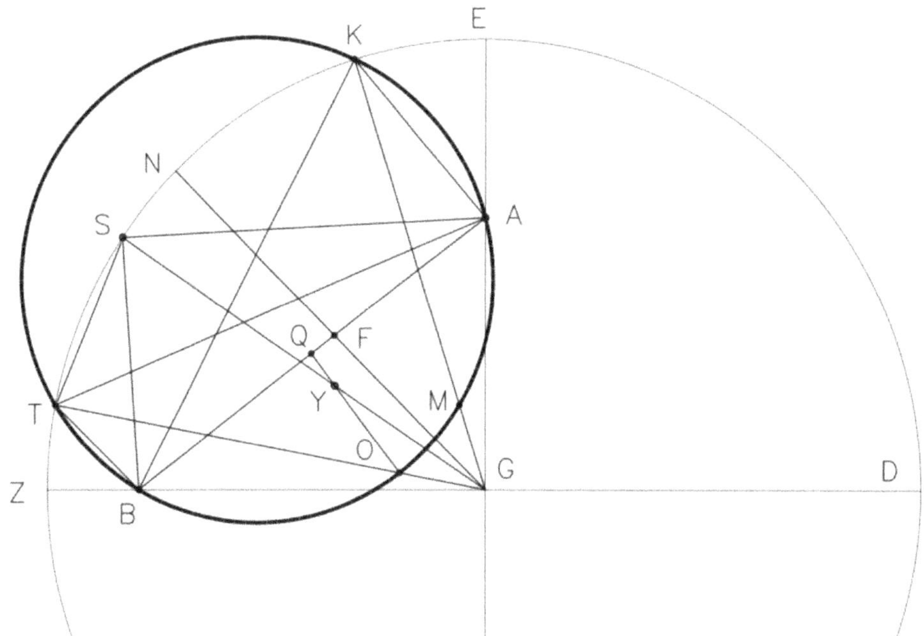

Diagram V.38 [269–75]

11. BG is not equal to AG, and ∠BTA is larger than ∠AGD.
12. BG is not equal to AG, and ∠BTA is smaller than ∠AGD.

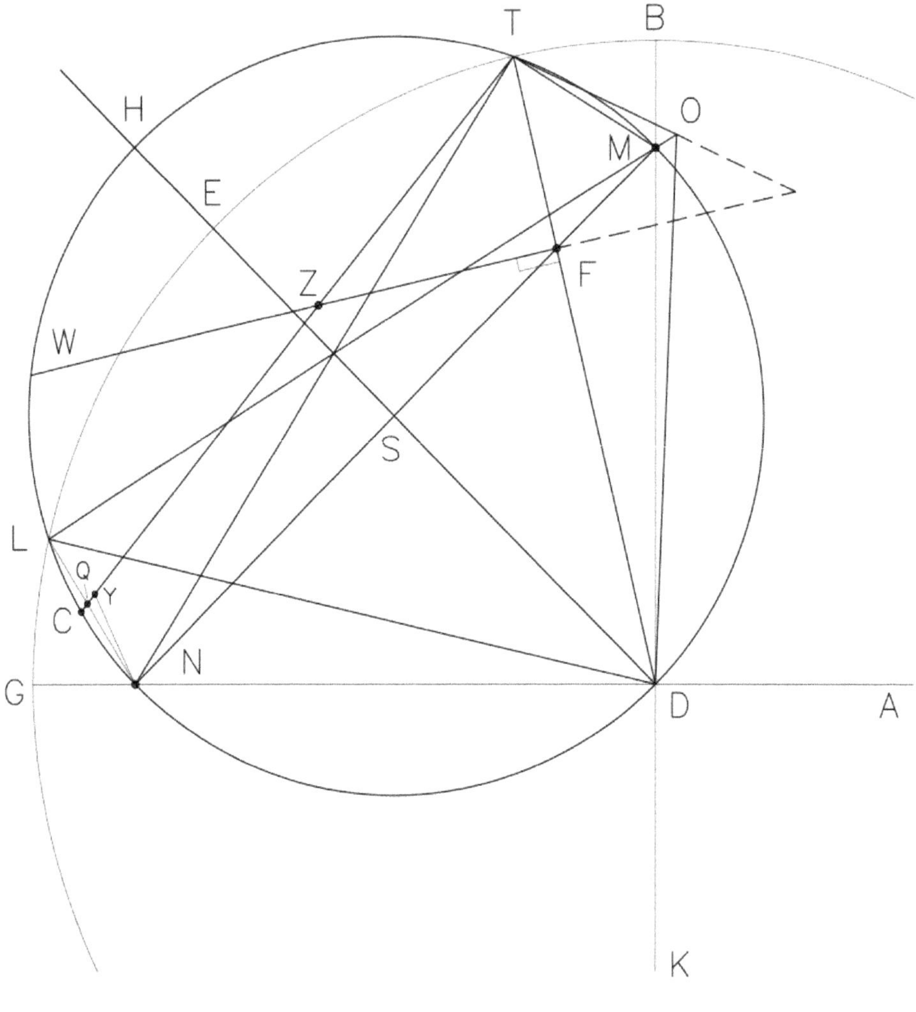

Diagram V.39 [276]

See also additional Diagram V.39+.

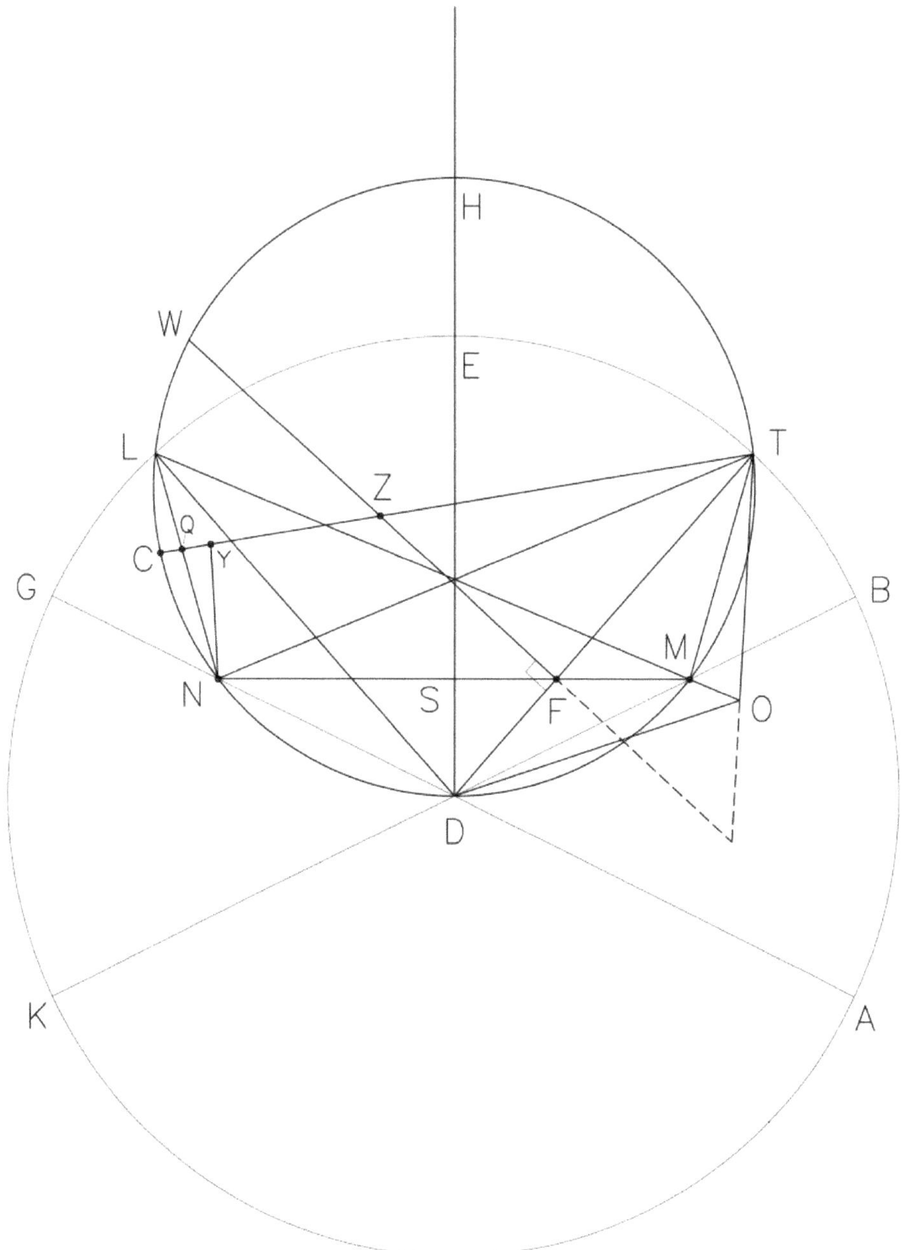

Diagram V.39+ [276]

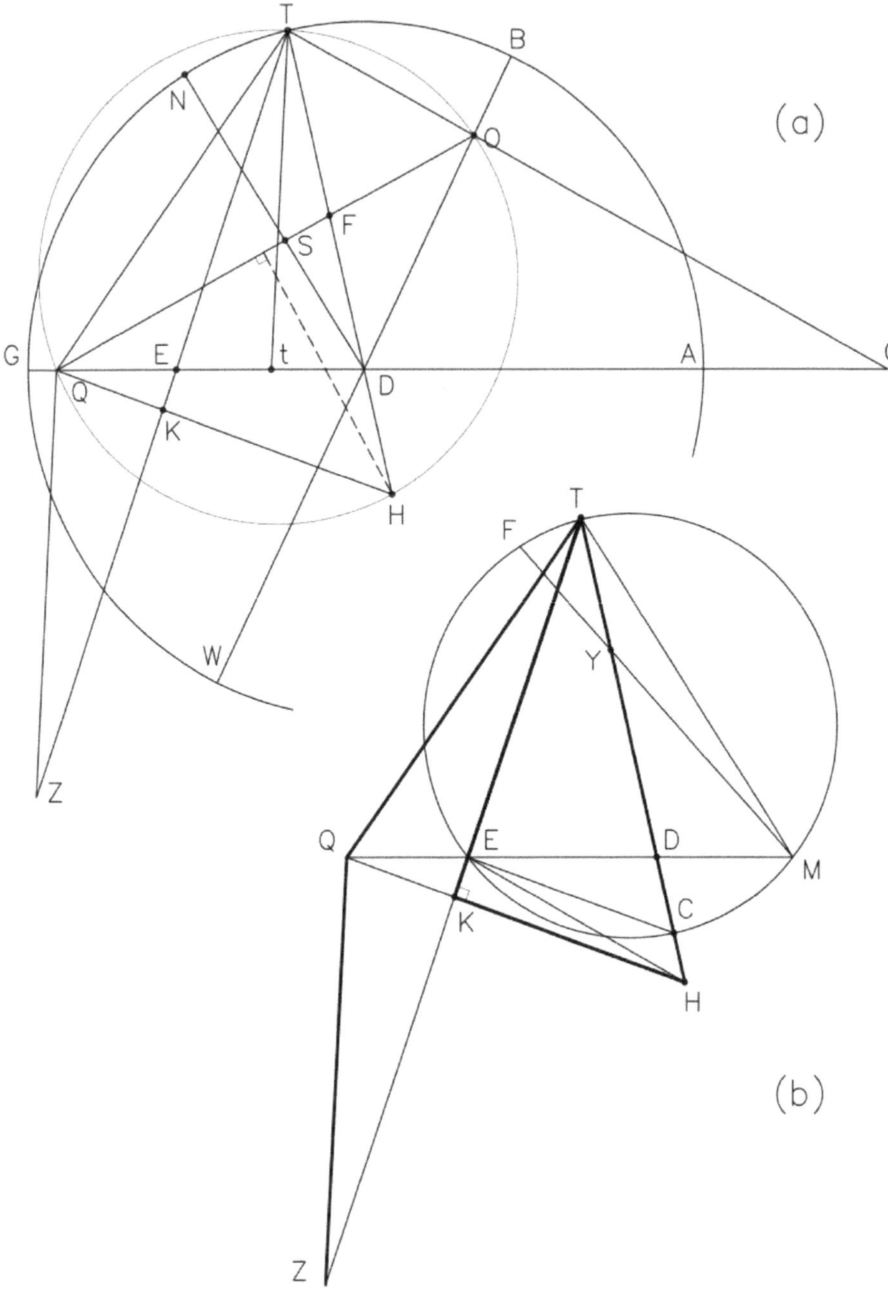

Diagram 40(a,b) [277–81][13]

13. There are two diagrams under this number in MS TA.

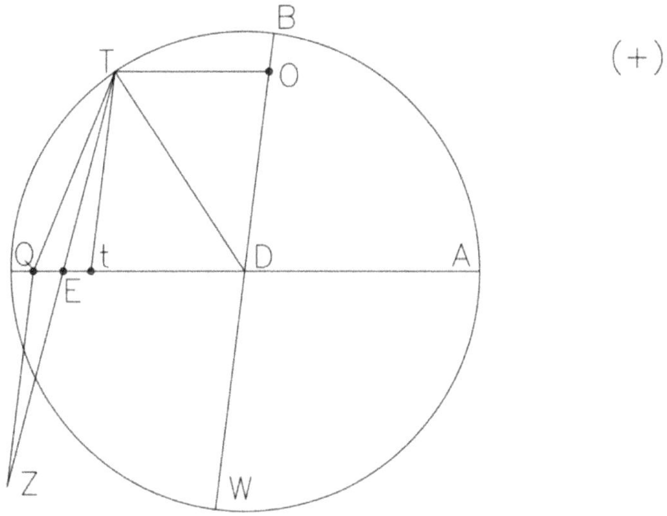

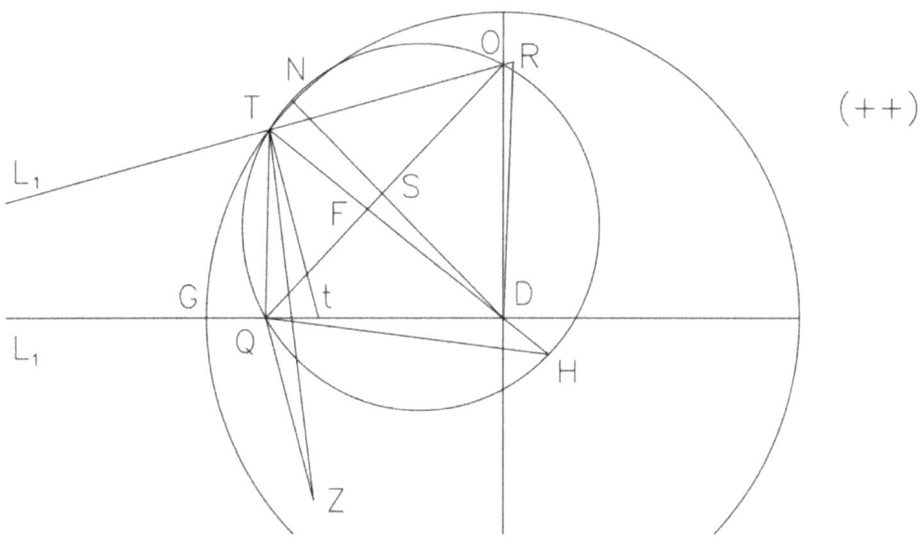

Diagrams V.40(+) and V.40++[14]

14. Diagrams V.40(+) and V.40(++) have been added by the translator. They display two configurations of the problem that are different from the configuration represented by Diagram V.40(a).

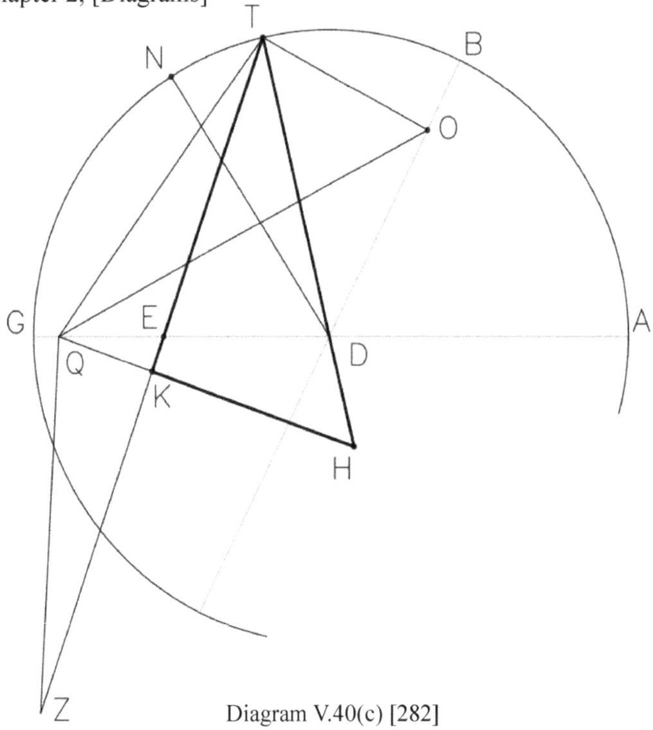

Diagram V.40(c) [282]

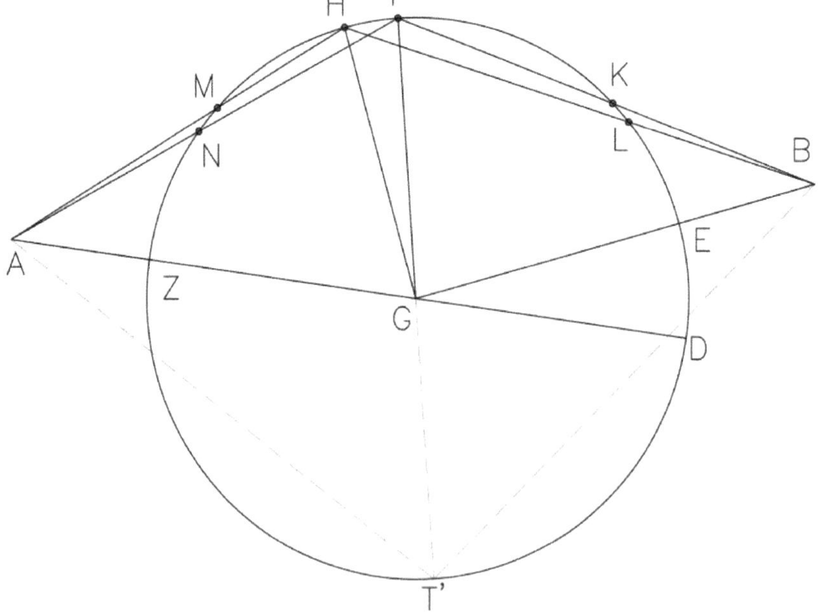

Diagram V.41 [283–84]

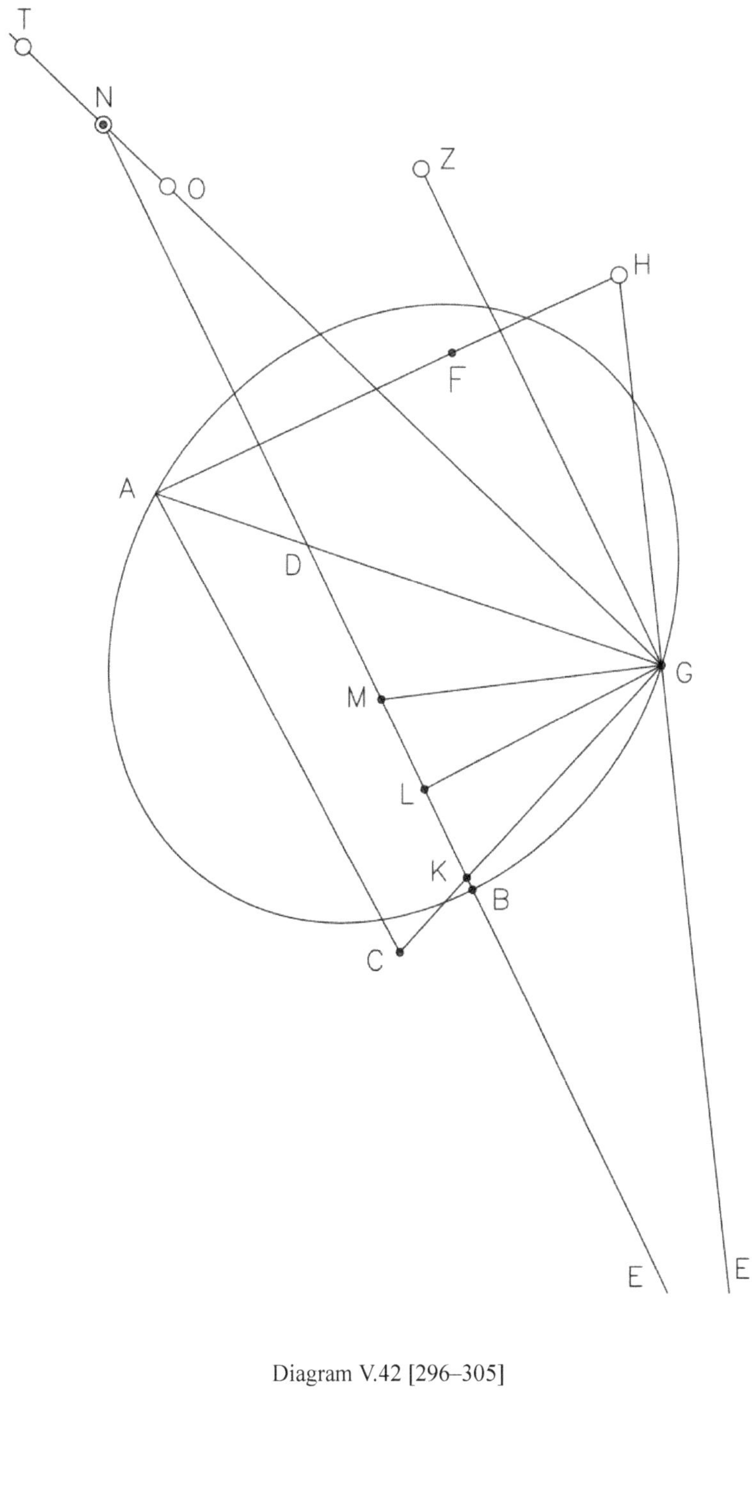

Diagram V.42 [296–305]

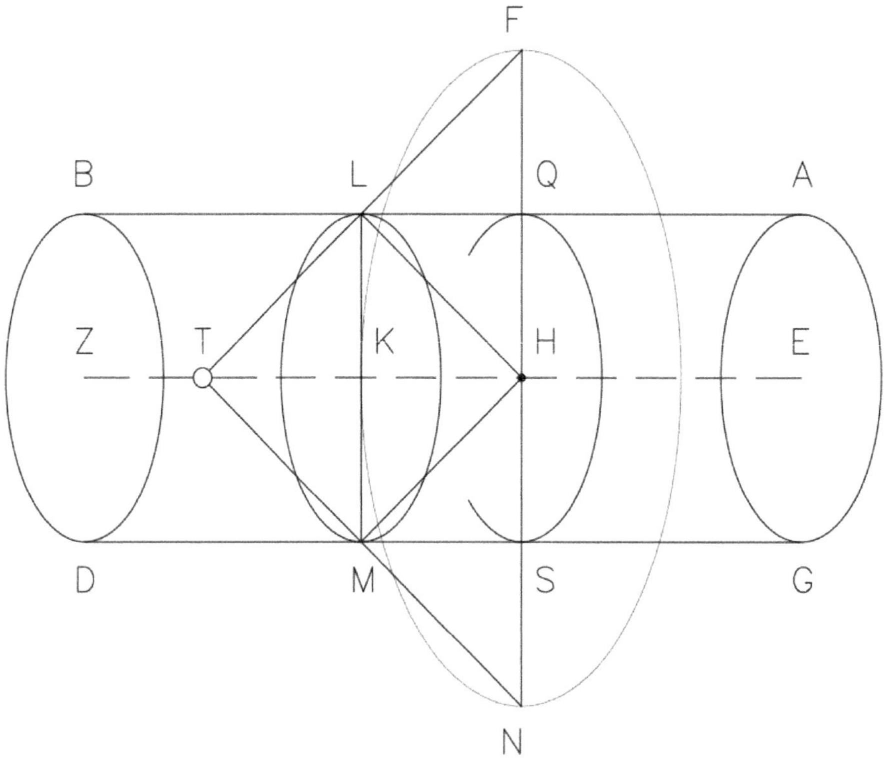

Diagram V.43 [307–9]

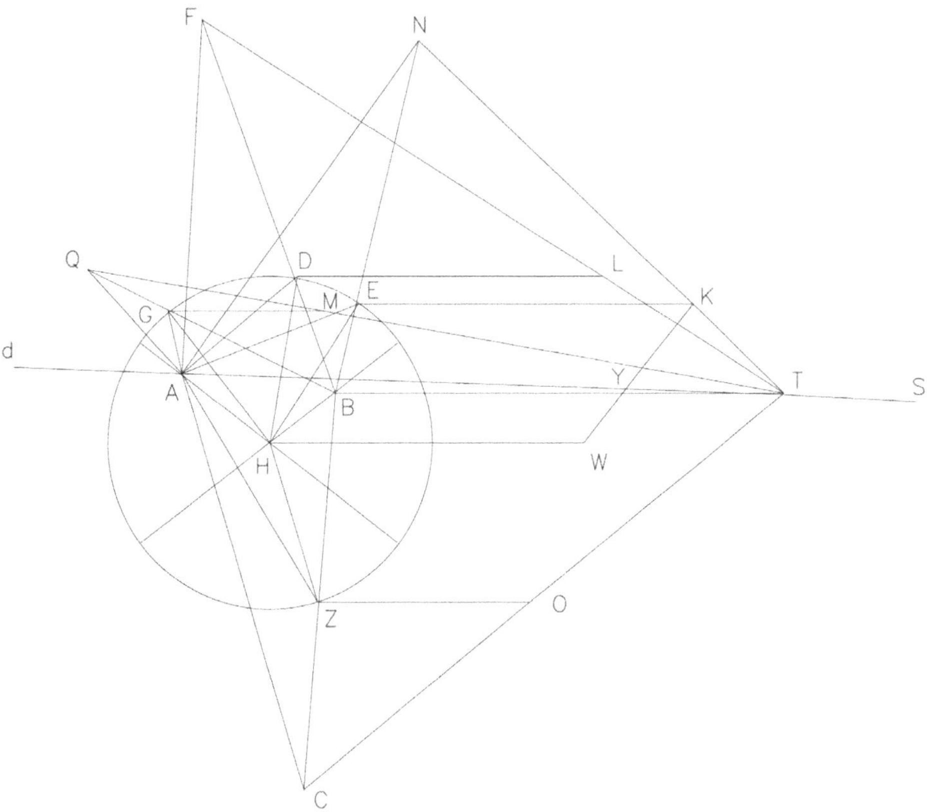

Diagram V.44 [310–20]

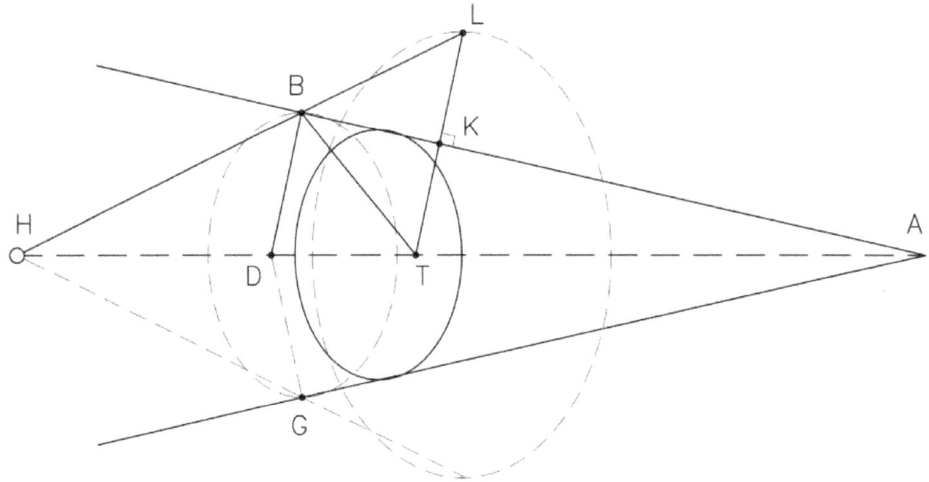

Diagram V.45 [327–29]

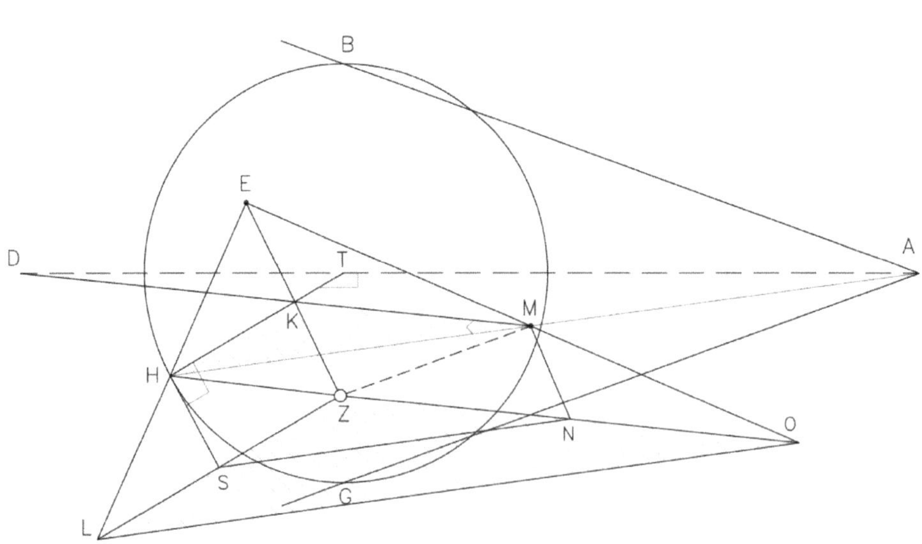

Diagram V.46 [330–33]

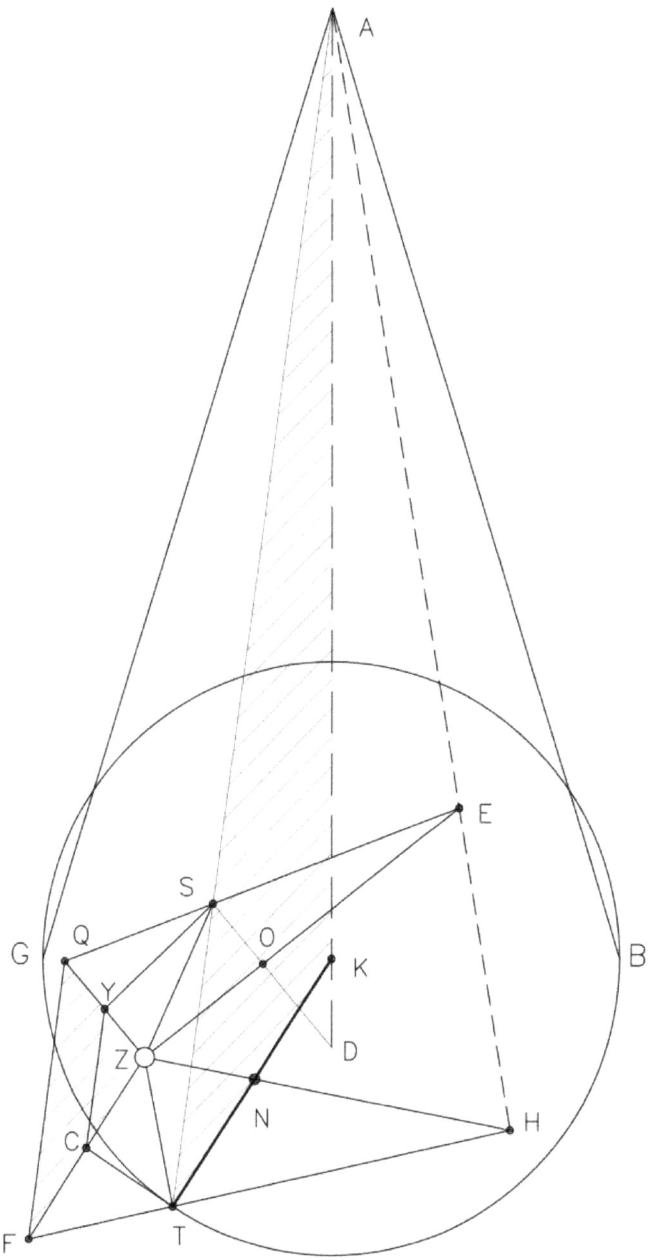

Diagram V.47 [334–43]

List of Works Cited

Hogendijk, Jan P., 'Al-Mu'taman's Simplified Lemmas for Solving "Alhazen's Problem"', in *From Baghdad to Barcelona. Studies in the Islamic Exact Sciences in Honour of Prof. Juan Vernet*, ed. Josep Casulleras and Julio Samsò, 2 vols, Barcelona: Universitat de Barcelona, 1996, I, pp. 59–101 [= *Anuari de Filologia*, XIX (1996), B-2].

Kamāl al-Dīn al-Fārisī, *Al-juzʾ al-ʿawwal min kitāb tanqīḥ al-manāẓir li-l-ʿallāmah Kamāl al-Dīn ibn al-Ḥasan al-Fārisī*, 2 vols, Hyderabad (India): Osmaniya Oriental Publications Bureau, 1347 AH / 1928–9 AD (I), 1348 AH / 1929–30 AD (II).

King, D. A., 'A Handlist of the Arabic and Persian Astronomical Manuscripts in the Maharaja Mansingh II Library in Jaipur', *Journal for the History of Arabic Science*, 4 (1980), pp. 81–86.

Krause, Max, *Stanbuler Handschriften islamischer Mathematiker*, Quellen und Studien zur Geschichte der Mathematik, Astronomie und Physik, Abteilung B: Studien, 3, Berlin, 1936, pp. 437–532.

Naẓīf, Muṣṭafā, *Al-Ḥasan ibn al-Haytham: Buḥūthuhu wa Kushūfuhu al-Baṣariyya*, 2 vols, Cairo, 1942, 1943.

[Risner, F., ed.,] *Opticæ Thesaurus. Alhazeni Arabis libri septem, nunc primùm editi, eiusdem liber De Crepusculis & Nubium ascensionibus. Item Vitellonis Thuringopoloni opticae libri X. Omnes instaurati, figuris illustrati & aucti, adiectis etiam in Alhazenum commentariis, a Federico Risnero*, Basel, 1572.

Rozenfeld, B. A., 'A Medieval Physico-Mathematical Manuscript Newly Discovered in the Kuibyshev Regional Library', *Historia Mathematica*, 2 (1975), pp. 67–73.

Sabra, A. I., 'Ibn al-Haytham's Lemmas for Solving "Alhazen's Problem"', *Archive for History of Exact Sciences*, 26/4 (1982), pp. 299–324, reprinted in A. I. Sabra, *Optics, Astronomy and Logic*, Aldershot, Variorum Reprints, 1994.

Sabra, A. I., *Al-Ḥasan ibn al-Haytham, Kitāb al-Manāẓir. Books I–III: On Direct Vision*, Kuwait, 1983 (in Arabic).

Sabra, A. I., *The Optics of Ibn al-Haytham. Books I–III: On Direct Vision*, 2 vols (I: *Translation*, II: *Introduction, Commentary, Glossaries, Concordance, Indices*), London: Warburg Institute, 1989.

Sabra, A. I., *Kitāb al-Manāẓir (the* Optics*) of al-Ḥasan ibn al-Haytham. Books IV–V: On Reflection and Images Seen by Reflection*, 2 vols (I: *Text, Introduction, Concordance Tables*, II: *Apparatus, Diagrams, Appendices, Analytical Index, Plates*), Kuwait: National Council for Culture, Arts and Letters, 2002 (in Arabic).

Sabra, A. I., 'The Commentary that Saved the Text. The Hazardous Journey of Ibn al-Haytham's Arabic "Optics"', *Early Science and Medicine*, 12 (2007), pp. 117–33.

Smith, A. Mark, *Alhacen on the Principles of Reflection: A Critical Edition, with English Translation and Commentary, of Books 4 and 5 of Alhacen's* De Aspectibus, *the Medieval Latin Version of Ibn al-Haytham's* Kitāb al-Manāẓir, Transactions of the American Philosophical Society 96, nos. 2–3 (Volume One: *Introduction and Latin Text*, pp. i–cxxvii and 1–288, Volume Two: *English Translation*, pp. 289–697), Philadelphia PA, American Philosophical Society, 2006.

SUMMARY OVERVIEW
of Books IV and V

BOOK IV

CHAPTER 1.
PREFACE [1–2]

The subject of Books I–III is the rectilinear perception of visible objects [1]. The subject of Book IV is the perception of visible objects by reflection from polished bodies or mirrors: how it happens, and the location of the lines along which the forms [*ṣuwar*] of points on visible objects are reflected [2].

CHAPTER 2.
THAT THE FORMS OF VISIBLE OBJECTS ARE REFLECTED FROM POLISHED BODIES
⟪In Six Propositions⟫ [1–24]

⟪IV 2.1⟫ A property [*khāṣṣah*] of the polished body is that the light shining upon the body is reflected from it, whether the light is primary, emanating from an essential light, or secondary, emanating from accidental light [1]. Experimental examination of this with reference to strong and weak lights [2–5].

⟪IV 2.2⟫ The shining of secondary light from the polished surface and the reflection of light in straight lines to the place proper to reflection; experimental examination of this [6–9]. Reflection of the forms of colours also takes place along straight lines to the places proper to reflection; experimental examination of this [10–14].

⟪IV 2.3⟫ Reflected forms are weaker than the forms from which they are reflected; examining and refining this experimentally [15]. Reflection itself weakens reflected light, even more than the weakening due to the colour of the mirror [16].

⟪IV 2.4⟫ Experimental examination of the weakening of the forms of colours, and their reduction [*naqṣ*] by reflection [17].

⟪IV 2.5⟫ Answer to the observation that the weakening of reflected forms is due only to their increasing distance from their origin, and explanation that lights and colours are weakened by the effect of reflection [18–20].

⟨⟨IV 2.6⟩⟩ The forms of reflected light and colour are stronger than their secondary forms, if they begin from the same origin, or from two equally remote origins [21–23].

⟨⟨Result of the chapter⟩⟩ [24].

CHAPTER 3.
ON THE MANNER OF REFLECTION FROM POLISHED BODIES
⟨⟨Introduction and Nine Propositions⟩⟩ [1–145]

⟨⟨Introduction⟩⟩ Polish, smoothness, perfect smoothness, perfect polish: their definitions [1].

⟨⟨IV 3.1⟩⟩ All reflections of lights take place in the same manner for all kinds of light without distinction; the general rule for reflection from various shapes of polished surfaces [2]. Description of an instrument for examining the manner of reflection: the copper plate [3–4]; the ring-like wooden piece [5–8]; the wooden panel in the shape of a sector of a circle, the drawing of two arcs on the concave surface of the ring's thickness [9–10]; mounting the plate into the ring [11]; drilling the holes in the ring [12]; mounting the instrument on its base [13–15]; the copper tube [16–17]; construction of the seven mirrors: plane, cylindrical convex, cylindrical concave, conical convex, conical concave, spherical convex, and spherical concave [18–26]; the wooden holding-posts [27–29]; mounting the mirrors on the posts [30–40]. Experimental examination of the manner of reflection of lights from plane mirrors [41–54]. Experimental examination of the reflection of light from the other mirrors [55–61]. Experimental examination of accidental lights [62–65]. Conclusions from the preceding experiments [66–72].

⟨⟨IV 3.2⟩⟩ Conclusion from the preceding experiments that the rules for the reflection of lights constitute a special, natural property [*khāṣṣa ṭabīʿiyya*], inseparable from all lights, whether they are essential or accidental, strong or weak, and therefore inseparable from the least light [*aqall al-qalīl min al-ḍawʾ*], as long as it keeps the form of light, even if the least light cannot be perceived or examined experimentally [73]. Precise determination of the position of the reflection line in each type of mirror: plane [74–80]; cylindrical convex [81–83]; conical convex [84]; cylindrical concave [85]; conical concave [86]; spherical convex [87]; and spherical concave [88–89]. Conclusions [90–91].

⟨⟨IV 3.3⟩⟩ Every point on the surface of any of the mentioned mirrors has the same reflecting property as the remaining points on that surface; experimental examination with reference to cylindrical convex and cylindrical concave mirrors [92–93]. The manner of reflection does not vary with the size of the cylinder [94], and the same is true of conical mirrors [95]. The manner of reflection of all mirrors, regardless of their shape or size, is something proper to polish [*ṣiqāl*] and depends on the perpendicular to the polished surface [96]. Summary of conclusions from the experiments [97]. Light radiates in a conical shape from every point on the shining body upon all points facing it on the polished surface. The lights extending to a point on the polished surface are also reflected in the shape of a cone [98–99].

⟪IV 3.4⟫ Light extends through a fairly wide hole to a polished surface along every line that can be imagined through that hole with one extremity at the luminous body and the other extremity at the polished surface. This light is then reflected from the polished surface along the reflecting lines corresponding to all incident lines through the hole [100–1].

⟪IV 3.5⟫ Shining [i.e., incident] light and reflected light is weakened according to the distance from its origin or from the reflecting surface, and according to its widening and spreading, and according to reflection itself [102].

⟪IV 3.6⟫ The strength of reflected light depends on the light's accumulation [i.e., convergence] after reflection, and yet its strength after reflection will depend both on the effect of accumulation and the weakening effect of its distance from the reflection point [103].

⟪IV 3.7⟫ Concerning accumulated light in the shape of a cone from the whole shining surface, at a point on the polished surface: if the strength due to accumulation exceeds the weakness due to distance, then the light at that point will be stronger than the light arriving at every point along a single line. Detailed examination concerning the three causes of the strength and weakness of the light [104–5]. The concept of least light: the light extending along a straight line is not itself the imagined straight line; rather, it is the smallest [i.e., thinnest] light stretching in a transparent body with [non-zero] width along the line imagined in the middle of it. It is possible to imagine, through this light-carrying body, many other lines parallel to that middle line, and also other lines crossing the middle line. Conclusions regarding reflection when the body carrying the light is divided [106–8].

⟪IV 3.8⟫ Reflection of light when the light passes through a fairly large aperture. Light is continuous in any case [109–10]. The shining [incident] light is not an amalgamation of lines, nor is the reflected light [111–12]. The reason why light is reflected from polished bodies: polished bodies resist the light completely, and resistance is the absence of disposition to be affected, as a result of the body's polish [113]. Resistance is due to polish, not to the hardness of the polished body [114–15]. Reflection and scattering of light from some rough bodies which have large gaps; light is reflected perfectly when the reflecting body has only narrow gaps [116–17]. The reflection of light from some bodies has a parallel in natural bodies; comparison with the reflection of falling heavy bodies and their rebound according to their hardness [118–19]. Why [*lima*, *limmiyya*] lights are reflected in the plane orthogonal to the plane touching the polished surface at an angle equal to the [incident] angle. Comparison with heavy bodies in their natural downwards motion [120–23] and in their accidental motions [124–26]. Argument that rebound is governed by resistance and force [127]. Argument for the equality of the angles of the first (i.e., incident) motion, and the motion of rebound by means of the concepts of motion, its force, resistance, the engendering of motion, and its being composed of two parts, one of which is parallel to the plane tangent to the reflecting surface, and the other orthogonal to it [128–30]. Explanation of the difference between the condition of light's reflection and the reflection of heavy bodies, referring especially to light's straight motion as a permanent property after reflection because of the absence of the power of weight, and light's being characterised by rectilinear motion [131–33].

⟪IV 3.9⟫ Detailed explanation of the manner of reflection of the forms of colours from polished bodies, as examined by narrow and wide apertures [134–40]. Experimental examination of the reflection of the forms of colours, using the instrument previously described in the present chapter [141–45].

CHAPTER 4.
WHAT SIGHT PERCEIVES IN POLISHED BODIES IS PERCEIVED BY REFLECTION
⟪Introduction and One Proposition⟫ [1–11]

⟪Introduction⟫ The opinion of mathematicians and of some natural scientists concerning the manner of perceiving visible objects in mirrors. Referring to Chapter 5, the author explains that this perception takes place by reflection [1].

⟪IV 4.1⟫ There does not exist in the mirror's surface a definite and fixed form of the visible object (as maintained by some natural scientists); rather, the perception of the object in the mirror depends on the object's position with regard to the object and the mirror [2–3]. Explanation by experimental examination of plane and concave mirrors [4–8]. Summary of what has been shown, especially regarding the different perceptions according to the different shapes of the mirrors' surfaces [9], that the objects perceived in the mirror appear as if they were behind the mirror [10], and that there exists no definite and fixed form of the object in the mirror [11].

CHAPTER 5.
THE MANNER IN WHICH SIGHT PERCEIVES VISIBLE OBJECTS BY REFLECTION
⟪Introduction and Eleven Propositions⟫ [1–80]

⟪Introduction⟫ Summary of what has been explained in Chapter 4 [1–4].

⟪IV 5.1⟫ Explanation of how sight perceives visible objects from the reflection of their forms by the mirror's surface in the position proper to reflection [5–8]. Reference to what has been explained in Book I regarding the dismissed doctrine of the ray issuing from the eye [7]. Ibn al-Haytham states that his own theory of visual perception by reflection was neither described by mathematicians, nor mentioned by others, and yet he considers his own doctrine as not being in contradiction with what mathematicians held [9].

⟪IV 5.2⟫ Perception of a single visible object in a single mirror at the same time by a number of observers; and perception of a single visible object by a single eye from different positions [10].

⟪IV 5.3⟫ The occurrence of the form of every point on a visible object in the whole polished surface as a result of the shining of forms of light and colour in the shape of a cone [11].

⟨⟨IV 5.4⟩⟩ General explanation of the shape of forms perceived in mirrors [12–14]; example for plane mirrors [14]. Characteristics of the cone imagined between the eye and the facing part of the mirror's surface, and the positions of the cone's lines [15]; reflection from a single surface in equal angles [16–17].

⟨⟨IV 5.5⟩⟩ Details for plane mirrors [18–19].

⟨⟨IV 5.6⟩⟩ Details for the spherical convex mirror: which points on the mirror can be points of reflection [20–24].

⟨⟨IV 5.7⟩⟩ Details for the right-cylindrical convex mirror [25–35]: which points on the mirror can be points of reflection [25–31]. Different types of intersections of the surface of cylindrical mirrors and the reflection planes for any given position of the eye. These intersections are in one case a straight line and in one case a circle. In all other cases, the intersections are ellipses, because mathematicians have shown that the sections of a cylinder are the same as those of cones [32–34]. Reflection if the intersection between the reflection plane and the cylinder is a straight line [35] or a circle [36]. Reflection if the intersection between the reflection plane and the cylinder is an ellipse; in this case, a visible point will be reflected to the eye in the same reflection plane at only one point on the ellipse [37–38]. Two points on an object will be reflected in different planes, except if the points are in the plane containing the eye and either the axis or a circular section of the cylinder [39]. All reflection planes of this mirror will meet the axis of the mirror; and each reflection plane contains the centre of the eye, the visible point, the reflection point, and the point on the axis of the mirror at which the perpendicular drawn from the reflection point meets the axis [40–41].

⟨⟨IV 5.8⟩⟩ Details for the right-conical convex mirror [42–58]. Geometrical preliminaries, explaining which points on the mirror will face the eye, for any position of the eye outside the cone [42–48]. Any point on the mirror's surface facing the eye can be a point of reflection, and the eye will simultaneously perceive all points reflected from the facing part of the mirror's surface [49–51]. Every perpendicular to a plane tangent to the surface of the right cone will meet the axis of the cone [52]; thus, any reflection plane will contain not only the eye, the object, and the point of reflection, but also the point where the perpendicular to the tangent plane at the point of reflection meets the axis [53]. The intersections between the surface of this mirror and the reflection planes are straight lines if the eye is on the rectilinear extension of the axis [54], or one straight line and conic sections [ellipses, parabolas, hyperbolas] if the centre of eye lies outside the axis [55]. If the reflection plane intersects the cone in a straight line, the object can be reflected to the eye from only one reflection point in this plane [56]. If the reflection plane intersects the cone in a conic section, it is possible for the object to be reflected in this plane to the eye in one or two points of reflection, but not more than two [57–58].

⟨⟨IV 5.9⟩⟩ Details for spherical concave mirror when the eye's centre is at the centre of the sphere or to one side of it [59–61]. All reflection planes will intersect at the diameter passing through the centre of the eye and the centre of the mirror; and all intersections between the reflection planes and the mirror's surface will be intersecting circles through the mirror's centre [62].

⟨⟨IV 5.10⟩⟩ Details for the right-cylindrical concave mirror [63–65]. The whole surface faces the eye if the eye is inside the cylindrical surface or on it; if the eye is outside the surface, then it is possible for the whole surface to face the eye [63]. If the eye faces the mirror, then it is possible that, from every point on the mirror's surface, reflection can take place from the form of the points on the visible objects [64]. Reflection planes from this mirror are similar to the reflection planes from the cylindrical convex mirrors, and all meet the axis; in each of these planes are collected the centre of the eye, the visible point, the reflection point, and the point at which the perpendicular drawn from the reflection point meets the axis of the mirror. The plane through the whole axis and the centre of the eye cuts the mirror's surface in a straight line. All other planes intersect the mirror in circles or conic sections, at some of whose peripheries the forms can be reflected from two points [65].

⟨⟨IV 5.11⟩⟩ Details concerning the reflection from conical concave mirrors [66–76]. If the eye is within the cone, then the whole surface can be facing the eye and reflection will not be limited, whether the mirror is a complete cone or a part of it; also, cases of reflection if the eye is outside the cone [66]. From every point on the facing mirror, it is possible for a point on the objects to be reflected, and the following will be collected in the reflection plane: the centre of the eye, the visible point, the reflection point, and the point on the cone's axis where it meets the perpendicular drawn from the reflection point. The intersections of the reflection planes with this mirror will be the same as with the conical convex mirror [67]. All intersections are straight lines if the centre of vision is on the axis of the mirror, inside the cone. Only one of the intersections will be straight, and the rest will be conics if the centre of vision is inside or outside the surface of the cone or on it; and the reflection takes place from at most two points [68]. If the mirror is a complete cone, then in most cases, the eye will only see the mirror's surface [69]; it will perceive in the mirror many visible objects if it is a long segment of a cone [70], and also if it is a segment of a cone shaped like a ring [71–73].

⟨Recapitulation⟩ [74–80]. From a single point on one of the mirrors, only the form of a single point on a visible object will be reflected to the eye at any time [74–76]. All that sight perceives by reflection from mirrors is perceived [to be] in front of the eye; the reason or cause of this [77–78]. Reflected visible objects are perceived [to be] behind the mirror, or sometimes in front of the mirror, or on the surface of the mirror. The position of the form [i.e., the reflected point] varies with the shape of the mirror and with the position of the eye. The distance of the perceived position of the form varies with the shape of the mirror. This position is called the position of the image, and the form perceived in the mirror is called the image [*khayāl*]. The eye, or sight, perceives the object by reflection at the position of the image [79]. In Book V, the author specifies the position of the image for each of the mirrors and explains [*ta ʿlīl*] sight's perception of the object at that position [81].

End of Book IV [82]

BOOK V

CHAPTER 1.
PREFACE [1]

Book IV is about the manner of perceiving visible objects in polished bodies by reflected forms. The aim of Book V is to show by experimentation [*iʿtibār*] and geometrical demonstration [*burhān*] that sight perceives a visible object by reflection from specific points on the polished body [1].

CHAPTER 2.
ON IMAGES [PERCEIVED BY REFLECTION]
⟪Consisting of an Introduction and Nine Inquiries⟫ [1–345]

[Introduction] An image [*khayāl*] is the form [*ṣūrah*] of a visible object perceived by reflection. The position of the image is where sight perceives the form [to be]. The image of a point on the visible object is (according to Ibn al-Haytham) at the meeting point of (a) the line along which the point's form is reflected, and (b) the perpendicular drawn from that point to the line tangent to the intersection between the polished surface and the reflection plane, as is then explained by experimentation [1]. [By (b), Ibn al-Haytham means a line through the visible point that is perpendicular to the curve in which the polished surface intersects the reflection plane.]

⟪V 2: First Inquiry. Experimental Examination of the Location of Images, Consisting of Seven Experiments⟫

⟪V 2.I.1: Experimental examination of the location of the image in the plane mirror⟫ [2–16]

Experimental examination of the position of the image in a plane mirror, using a fine straight rod, showing the form [i.e., image] of the point assumed on the rod to be on the perpendicular to the mirror's surface [2–3]. The image of every point on the rod, regardless of the different assumed points and the different inclinations of the rod to the mirror's surface, is located behind the mirror on the perpendicular drawn from that point on the mirror's surface [4–6]. This is also found when the rod is attached next to the mirror or when raising the mirror [7–8]. Refining this experiment, using a round cone of wax [9–13]. Result of the experiments [14–16].

⟪V 2.I.2: Experimental examination of the location of the image in the spherical convex mirror⟫ [17–24]

The form of a visible object in the spherical convex mirror is smaller than the object itself [17]. Precise experimental examination of the image's position in this mirror. This position is at the point of intersection between the reflection line and the perpendicular drawn from the visible object to the intersection between the mirror's surface and the reflection plane [18–24].

⟪V 2.I.3–4: Experimental examination of the location of the image in the cylindrical convex and conical convex mirrors⟫ [25–39]

If the visible object is perpendicular to the surface of a cylindrical convex mirror, then (the object's) form will be inclined and bent, even though the object is straight. This is one of the errors of vision (examined in Books VI and VII). The image of a single point is therefore [first] examined (here). Examination using the instrument described in Book IV [27–37]. The images of visible points in this mirror are observed on the perpendiculars drawn from the points to the straight lines tangent to the intersections between the mirror's surface and the reflection plane; thus, the image of the visible point will be at the intersection of the reflection line with the perpendicular drawn from that point to the intersection. This perpendicular is the shortest line from the point to the cylinder's surface and it lies in the reflection plane [38]. The position of the images of the conical convex mirror can be examined experimentally in the same way [39].

⟪V 2.I.5: Experimental examination of the location of the image in the spherical concave mirror⟫ [40–46]

The images of visible points in this mirror are of different types: some are behind the mirror, some are in front of the mirror, and some on the surface of the mirror. Most are distinctly perceived, and a few are indeterminate [i.e., blurred]; and the distinctly perceived images are at the point of intersection between the reflection line and the line between the visible point and the centre of the mirror [40]. Examination using a cone of wax [41–43]. Details of the experiments [44–46].

⟪V 2.I.6–7: Experimental examination of the location of the image in the cylindrical concave and the conical concave mirrors⟫ [47–55]

The images of visible points in cylindrical concave mirrors are also of different types: they may be located behind the mirror, or in front of it, or on its surface. Most of the images will be perceived distinctly, a few are indeterminate. Examination similar to examining the images of the cylindrical convex mirrors [47–53]. Also examination of the images of the conical concave mirror [54]. Conclusion [55].

⟪V 2: Second Inquiry. Explanation of Vision by Reflection, in Two Propositions⟫

⟪V 2.II.1: The reason why the object seen by reflection is perceived to be at the location of the image⟫ [56–57]

It was previously shown that after reflection, sight perceives the form of the reflected object facing the eye. The reason why sight perceives the image at precisely that position is as follows. Sight perceives the form at a glance, then conjectures the form's distance, after which it may contemplate and verify the precise distance, as was made clear in Book II of the *Optics*, where also the role of inference is explained [56], as well as the perception of unfamiliar objects after reflection [57].

⟨⟨V 2.II.2: The reason why the object seen by reflection is perceived to be on the perpendicular drawn from the object to the mirror's surface⟩⟩ [58–69]

The visible point always has a certain magnitude, but does not make an impression by reflection in the same location as the impression by direct vision, unless the point is at the position of the image on the perpendicular. Demonstration in the case of the plane mirror [58–62]. Reason why in the case of the spherical convex mirror the image is on the diameter through the visible point [63–66]. The cases of the spherical concave, cylindrical convex, cylindrical concave, conical convex, and conical concave mirrors can be dealt with in analogous ways [67]. The universal reason [*or* cause] why the images lie on the perpendiculars to the mirrors [68–69].

⟨⟨V 2: Third Inquiry. On the Locations of Images in the Seven Mirrors, in Six Propositions⟩⟩

⟨⟨V 2.III.1: General statement on the locations of images in each of the seven mirrors⟩⟩ [70–74]

General statements, for each of the seven types of mirror, concerning the number of possible images of a visible point, and whether these images can be perceived only behind the mirror, or also on the surface of the mirror or in front of the mirror. Cases of the plane mirror [70], spherical convex mirror [71], cylindrical convex and conical convex mirrors [72], spherical concave mirror [73], and cylindrical concave and conical concave mirrors [74].

⟨⟨V 2.III.2: Demonstrations of the preceding propositions [for the plane mirror]⟩⟩

Construction of the image in the plane mirror for a visible point outside the perpendicular drawn from the centre of the eye to the mirror's surface [75].

⟨⟨V 2.III.3: Determination of the position of the reflection point on a plane mirror⟩⟩ [Diagram V.1]

In the construction it is assumed that the visible point and the centre of vision (eye) are given [76].

⟨⟨V 2.III.4: Only one point on the perpendicular drawn from the centre of the eye to the surface of a plane mirror will be perceived by reflection⟩⟩ [Diagram V.2]

This is the point where the perpendicular intersects the surface of the eye [77–80]. General conclusion [81].

⟨⟨V 2.III.5: A visible point seen by one eye in a plane mirror will have only a single image and a single reflection point⟩⟩ [Diagram V.3]

There cannot be more than one point of reflection [82], and hence there can be only a single image [83].

⟨⟨V 2.III.6: Perception of a visible object with both eyes by reflection from a plane mirror⟩⟩ [Diagram V.4 (three cases)]

Experimental examination [84–86]. Geometrical demonstration [87–88]. Summary [89].

⟨⟨V 2: Fourth Inquiry. On the Images of Spherical Convex Mirrors and Related Matters, in Eighteen Propositions⟩⟩ [90–162]

⟨⟨V 2.IV.1⟩⟩ [Diagram V.5]

If the visible point [A] lies outside the line joining the centre of the eye [B] and the centre of the mirror [N], then the reflection line meets the diameter of the sphere through the visible point at a point [Z], which will be the image of the visible point [90].

⟨⟨V 2.IV.2⟩⟩ [Diagram V.5]

If G is the reflection point, and the tangent to G intersects NA at E, then AN:NZ = AE:EZ, that is, the ratio of the line joining the visible point and the mirror's centre to the line joining the mirror's centre and the image is as the ratio between the segments into which the line between the visible point and the image is divided by the tangent at the reflection point. The dividing point [E] of the line [AZ] between the visible point and image is called the tangent's extremity [91–92].

⟨⟨V 2.IV.3⟩⟩ [Diagram V.5]

The line [NZ] between the image point and the mirror's centre is longer than the line [ZG] between the image point and the reflection point [93].

⟨⟨V 2.IV.4⟩⟩ [Diagram V.5]

The line [ZE] between the image point and the tangent's extremity is shorter than the radius of the sphere [94].

⟨⟨V 2.IV.5⟩⟩

The eye will perceive only one point on the diameter passing through its centre and the centre of the mirror, namely the point where this diameter intersects the surface of the eye. The image of this point will be continuous with the images of the surrounding area on the eye's surface [95].

⟨⟨V 2.IV.6⟩⟩ [Diagram V.6]

Some of the images perceived in the spherical convex mirror by reflection will be behind the mirror, others will be on the mirror's surface, and others in front of the mirror [96–101].

⟨⟨V 2.IV.7 and V 2.IV.8⟩⟩

The image of every visible point will be on the mirror's diameter passing through that point and the mirror's centre [102]. The line from the eye's centre to the reflection point always cuts the mirror's surface at two points, and the reflection point is the closer of these two points to the eye [103].

⟨⟨V 2.IV.9–16: Determination of the locations of images on any diameter of the spherical convex mirror⟩⟩ [Diagram V.7]

⟨⟨V 2.IV.9 and V 2.IV.10⟩⟩

In these inquiries, one considers a spherical convex mirror with the eye in a given position in front of the mirror, and one is interested in the images of all points visible outside the mirror after reflection. To study these images, it is sufficient to consider visible points in an arbitrary plane through the eye and the centre of the mirror. This plane will intersect the mirror in a circle, and the plane will also contain the reflection point and the image. Draw the two tangents from the eye to the intersecting circle, and first consider visible points in front of the mirror on a fixed diameter, which diameter intersects the circle between the two tangents but does not pass through the eye itself. One can then define a limiting point on the fixed diameter inside the circle. In [Diagram V.7] the limiting point Z is defined by the property ZD = ZB, that is, the segment of the fixed diameter between the limiting point [Z] and the centre [D] of the mirror is equal to a certain segment [ZB] on the line [ZE] that joins the limiting point to the eye [E], namely the segment between the limiting point [Z] and the intersection [B] between the circle and the line. The construction of the limiting point [Z] for the fixed diameter [DH] and position of the eye [E] is explained below in Lemma 5 [145–47]. First it is shown that the limiting point itself cannot be an image of a visible point, and no images can be found between the limiting point and the centre of the mirror [104–5].

⟨⟨V 2.IV.11⟩⟩ [Diagram V.8, notation different from Diagram V.7]

Then it is shown that each point on the same fixed diameter between the limiting point [T in Diagram V.8] and the circle can be the image of some visible point. But the intersection of the fixed diameter and the circle [G in Diagram V.8] cannot itself be an image of a visible point [106].

⟨⟨V 2.IV.12⟩⟩ [Diagram V.9]

Even if the fixed diameter does not intersect the circle between the two tangents, the points on the fixed diameter outside the two tangents are visible to the eye. Precise investigation of the images of these points: some are inside the mirror, others on the surface of the mirror, and still others outside the mirror [107–13]. But these images are all between the two tangents and therefore behind the mirror, as seen from the eye [114–16].

⟪V 2.IV.13⟫ [Diagram V.10]

If a visible point is perceived by a single eye in the spherical convex mirror, outside the mirror's diameter passing through the centre of the eye, there is only one image, and the visible point will be reflected from the mirror's surface at only one point of reflection [117–20].

⟪V 2.IV.14⟫

The point on the surface of the eye, through which the diameter is drawn from the eye's centre to the centre of the mirror, will also have a single image. The conclusion is that any visible point will have only a single image [121].

⟪V 2.IV.15⟫

Thus, reflection between two assumed points (eye, visible point) in a spherical convex mirror will be at a single point only [122].

⟪V 2.IV.16⟫ [Diagram V.10]

For every two visible points on the same diameter of the spherical convex mirror and on the same side of the centre of the mirror, the image of the point closer to the surface of the mirror will be farther from the centre of the mirror, and the point on the diameter closer to the surface of the mirror will be reflected at a point farther from the centre of the eye [123].

⟪V 2.IV.17: On determining the point of reflection⟫

Construction of the reflection point on the surface of the spherical convex mirror, if the visible point and the eye are at equal distances from the centre of the mirror. In this case the reflection point is easily found by bisecting the angle formed by the two lines joining the centre of the mirror to the eye and to the visible point [124].

[There follow six lemmas for determining the point of reflection from the spherical convex mirror, if the eye and visible point are at unequal distance, and also for other problems in the rest of the Book.]

⟪Lemma I⟫ [Diagrams V.11–12]

Given: a circle with point A on the circumference and diameter BG; let a line segment be given. To construct through A a straight line that cuts the circle a second time at a point H and the rectilinear extension of BG outside the circle at a point D in such a way that HD will be equal in length to the given line segment. Construction and proof for the case where arcs AB and AG are equal; in this case the line AHD can be constructed by means of ruler and compass [125, 126, Diagram V.11]. If the two arcs AB and AG are not equal, the problem cannot be solved by ruler and compass; Ibn al-Haytham presents a solution by means of a hyperbola intersecting a circle [127–32, Diagram V.12]. References to Propositions 4 and 8 in Book II of the *Conics* by Apollonius [127].

⟨⟨Lemma II⟩⟩ [Diagram V.13]

Given: a circle with a given point A on the circumference and a given diameter BG; let a line segment be given. To construct through A a straight line that cuts the diameter at a point E and the circumference a second time at a point D in such a way that ED will equal in length to the given line segment [133–37]. The problem cannot in general be constructed by ruler and compass; Ibn al-Haytham presents a solution by means of a hyperbola intersecting a circle. References to Propositions 34 and 61 in Book V of the *Conics* by Apollonius to show that the problem can have zero, one, or two solutions [133]. Reference to Proposition 11 in Book II of the *Conics* by Apollonius [134].

⟨⟨Lemma III⟩⟩ [Diagram V.14]

Given: a right-angled triangle ABG with right angle B; a point D, either on side BG, between B and G (Case 1), or on the rectilinear extension of the side next to the right angle B (Case 2); and a ratio E:Z. To construct through D a straight line that cuts the hypotenuse AG at a point T and side AB or its rectilinear extension on the side of A at a point K in such a way that TK:TG = E:Z [138]. Construction and proof based on Lemma I.

⟨⟨Lemma IV⟩⟩ [Diagram V.15]

Given: a circle and two points D and E, of which one is inside the circle and the other outside it. To construct through E and D two straight lines that meet at a point A on the circle in such a way that the tangent to the circle at A bisects angle EAD [139–44]. Construction and proof based on Lemma III.

⟨⟨Lemma V⟩⟩ [Diagram V.16]

Given: a circle with centre G and diameter GB, and a point E outside the circle. To construct through E a straight line that cuts the circumference at a point D and the diameter BG between B and G at a point Z in such a way that DZ = ZG [145–47]. Construction and proof based on Lemma II.

⟨⟨Lemma VI⟩⟩ [Diagram V.17]

Given: a right-angled triangle ABG with right angle B, a point D on side BG, and a ratio E:Z. To construct through D a straight line that meets side AG at a point K and the rectilinear extension of AB at a point T in such a way that TK:KG = E:Z [148–49]. Construction and proof are based on Lemma II. The problem can have 0, 1, or 2 solutions.

[Finding the reflection point on the surface of a spherical convex mirror]
[Diagram V.18]

Given: a spherical convex mirror and the positions of the eye and the visible point outside the spherical mirror; construction of the reflection point on the surface of the mirror. Construction and proof based on Lemma VI [150–52]. Proof that there can be (at most) a single reflection point [153–54].

⟨⟨V 2.IV.18: Binocular perception of a single visible object
in the spherical convex mirror⟩⟩

If a single object is seen by both eyes in this mirror, then the object's image is the same or almost the same for both eyes [155–61]. References to Book III relating to the conditions for the occurrence of a single form in both eyes and in the common nerve [*al-ʿaṣabah al-mushtarakah*] [161–62].

⟨⟨V 2: Fifth Inquiry. On the Images of Cylindrical Convex Mirrors,
Consisting of an Introduction and Seven Propositions⟩⟩ [163–83]

⟨⟨Introduction⟩⟩

The intersections between the surface of a right-cylindrical convex mirror and the reflection planes can be straight lines, circles, or conics [ellipses]: reflection in straight lines is as in plane mirrors, reflection in circles is as before in the circular intersections of the spherical concave mirrors. Reflections in ellipses will now be investigated [163].

⟨⟨V 2.V.1–4⟩⟩ [Diagram V.19]

Preliminaries for showing the locations of images for reflection in ellipses [165–67].

⟨⟨V 2.V.5⟩⟩ [Diagram V.20 (three cases)]

Demonstration that the images for reflection in ellipses may be behind the mirror, or on the surface of the mirror, or in front of the mirror [168–72]. Reference to Apollonius's *Conics* for the construction of a tangent to the ellipse [168]. Sight can only perceive the images as located behind the mirror [173].

⟨⟨V 2.V.6⟩⟩ [Diagram V.20+]

For given positions of the eye and visible point, the reflection in a right-cylindrical convex mirror can only be from a single point and its image can be only a single image [174]. This is shown for: where the eye and visible point are in one plane with the axis [175]; where the eye and visible point are in a plane parallel to the base of the cylinder [176]; and where the reflection plane intersects the cylinder in an ellipse [Diagram V.20+] [177–81]. Reference to Chapter 5 in Book IV [177]. The image will also be a single point [182].

⟪V 2.V.7⟫ [Diagram V.20+]

For a given position of the eye and visible point, to construct the reflection point on the surface of the cylindrical convex mirror. If the eye and visible point are in one plane parallel to the base of the cylinder, one may use the construction for the circular intersection of a spherical convex mirror as above [150–52]. For all other cases, one of the two given points (for example, the visible point) is projected orthogonally onto the plane through the other point (the eye) parallel to the base of the cylinder. If the eye and visible point are not in one plane with the axis, the projected visible point cannot coincide with the eye. Using [150–52], one can find the reflection point for the eye and the orthogonal projection of the visible point in the plane that intersects the cylinder in a circle. The reflection point for the visible point itself can then be constructed, using the method of Inquiry V 2.V.6.

⟪V 2: Sixth Inquiry. On the Images of Right-Conical Convex Mirrors, Consisting of an Introduction and Three Propositions⟫
[184–200]

⟪Introduction⟫

The intersections of reflection planes with this type of mirror are straight lines (in a plane through the axis), or conic sections. Circular intersections are not perpendicular to tangent planes and therefore cannot be reflection planes. Just like the images of the cylindrical convex mirror, the images of the conical convex mirror can be behind the mirror, or in front of it, or on its surface [184–85]. If a perpendicular is drawn from the eye to a straight line on the conic mirror, the foot of this perpendicular can be a reflection point, and the eye will perceive in it the intersection of its own surface with the perpendicular. It cannot perceive other points on this perpendicular at other points of reflection [186].

⟪V 2.VI.1⟫ [Diagram V.21]

For any position of the eye and visible point, the reflection in the conical convex mirror can be only in (at most) one point, and there is only one image [187–90].

⟪V 2.VI.2⟫

For a given position of the eye and the visible point outside a conical convex mirror, construction of the point of reflection in the mirror. Consider the plane through the apex of the cone, parallel to the base of the cone, and call this plane through the apex the parallel plane. Five cases are considered with respect to this parallel plane. Case 1, Diagram V.22(a): the eye and visible point are located on the same side of the parallel plane as the conical mirror; the construction is based on [150–52] for the spherical convex mirror [191]. Case 2, Diagram V.22(b): the eye and visible point are located in the parallel plane; the construction is easy by bisecting the eye / apex / visible-point angle [192]. Case 3, Diagram V.22(c): the eye and the visible point are on the same side of the parallel plane, but not on the same side as the conical mirror; the construction is based on the construction of reflection points in circular intersections of spherical concave mirrors, to be dealt with below in Inquiry V 2.VII [193–94]. Case 4, Diagram V.22(d): one point is in the parallel plane, the other is on the same side of the parallel plane as the conical mirror [195], or not on the same side [196]; the construction is based on Lemma V above [145–47]. Case 5, Diagram V.23: the two points are on different sides of the parallel plane [197–98]; the construction is based on Lemma IV above [139–44].

⟨⟨V 2.VI.3⟩⟩

Binocular perception in a conical convex mirror [199–200].

⟨⟨V 2: Seventh Inquiry. On the Images of Spherical Concave Mirrors, Consisting of Introduction and Thirty-Three Propositions⟩⟩ [201–95]

⟨⟨Introduction⟩⟩

All intersections of reflection planes with spherical concave mirrors are circles through the centre of the mirror, just like the sections of spherical convex mirrors, and the perpendiculars to the tangent planes pass through the centre of the mirror. In most cases, the lines from eye to the point of reflection meet the perpendicular through the visible point (that is, the diameter through the visible point) [201].

⟨⟨V 2.VII.1⟩⟩ [Diagram V.24]

A diagram shows how the image of a visible point may be behind the mirror, or in front of the mirror, and in the latter case between the eye and the mirror, or behind the eye. In some cases, the line from the eye to the reflection point is parallel to the diameter through the visible points, and then the image is at the centre of the eye [202–4]. All of these images are perceived as facing the eye [205].

⟨⟨V 2.VII.2–5⟩⟩

The images seen behind the mirror or between the eye and the mirror's surface are perceived as distinct. Discussion of a point in a visible object with an image precisely at the centre of the eye. Conclusion: images behind the eye or in the centre of the eye will be perceived indistinctly [205–8]. Description of a relevant experiment for investigating the images and colours of visible objects reflected in a spherical concave mirror [209–10].

⟨⟨V 2.VII.6–7⟩⟩

Further investigation of the way in which the eye perceives visual objects whose images are behind the eye [211]. Conclusion [212].

[Note on Inquiries V 2.VII.8–32. Here, Ibn al-Haytham investigates the construction of the reflection point and the image, for all positions of the eye and the visual point (*or* object). He first investigates, in Inquiries V 2.VII.8–10 and V 2.VII.13–14, the case in which the eye and the visible point are on the same diameter of the spherical concave mirror; in V 2.VII.9 he proves a theorem that is also valid for the case where the eye and visual point are not on the same diameter. In Inquiries V 2.VII.11–12 and V 2.VII.15–17, he then considers the eye and visual point on different diameters, but at the same distance from the centre of the mirror. In Inquiries V 2.VII.18–32, he discusses the general case where the eye and visual point are on different diameters of the mirror and at different distances from the centre of the mirror. This is mathematically the most complicated part of Book V.]

⟪V 2.VII.8⟫

If the eye is at the centre of the mirror, it will only perceive in the mirror points on its own surface, that is, part of the eyeball itself [213–14]. If the eye is not at the centre, it cannot perceive in the mirror objects on the diameter on the same side of the centre, but in some cases, it can perceive in the mirror objects on the diameter on the other side of the centre [215–17].[1]

⟪V 2.VII.9⟫ [Diagram V.25 (five cases)]

Demonstration of a general geometrical theorem for a spherical concave mirror: if B is the visible point, H its image, E the centre of the mirror, and T the intersection of the tangent to the reflection point with the diameter (perpendicular) through B, then BE:EH = BT:TH. The theorem is true in any case if the eye A and the visible point B are on the same diameter of the mirror; if A and B are on different diameters, the theorem is true if the ray AG from eye to reflection point G is not parallel to the diameter EB through the visible point B. Then the image H can be defined as the intersection point [218–20].

⟪V 2.VII.10⟫ [Diagram V.26]

If points A and B are on the same diameter of the spherical concave mirror, but at different sides of the centre, it is possible that A is reflected to B. The reflection takes place from points on both sides of the diameter in a circular intersection of the mirror by a plane through the diameter [so the image is a full circle] [221].

⟪V 2.VII.11⟫ [Diagram V.26, Diagram V.27(a,b)]

Consider two points (eye, visible point) A and B inside a spherical concave mirror, not on the same diameter. The diameters through A and B divide the circle into four sectors. It is shown that if point A is reflected to B, the reflection points can be on only two of these sectors: one bounded by the radii through A and B, and the other, diametrically opposite sector, bounded by the rectilinear extensions of the radii through A and B. Reflection is not possible from the two other sectors: the sector bounded by the radius through A and the extension of the radius through B, and the sector bounded by the radius through B and the extension of the radius through A [222–23]. An analogous theorem holds if the eye A is on

1. Details follow in paragraph [221].

the circumference or outside the mirror [224–25]. General statement of the theorem on the possibility of reflection from the four sectors [226]. Further specifications on the positions of the images in these cases [227–29].

⟪V 2.VII.12⟫ [Diagram V.28]

If the diameter of the spherical concave mirror is extended rectilinearly in both directions, any point on the diameter or its extension can be the image of some visible point on the diameter, for a suitable position of the eye. Therefore, the images of this mirror are not limited with respect to their positions [230]. Statement on theorems to the effect that for given positions of the eye and the visible object not on the same diameter, the number of images in a spherical concave mirror can be one, two, three, or four, but not more than four. Reference to the exceptional case where the two points are on the same diameter, and the image can be a full circle [231].

⟪V 2.VII.13⟫ [Diagram V.29(a,b)]

If the eye E and visible object H are on the same diameter of the spherical concave mirror, on different sides of the centre D, construction of the points of reflection. If ED = DH, the points of reflection are all points on a circle on the mirror that is traced out if the extremity of the perpendicular diameter rotates around the diameter EH [232–33]. For ED > DH, Ibn al-Haytham considers a plane of construction through the diameter. In this plane he constructs point K on EH extended such that EK.KH is the square of KD, and then traces, in the plane of construction, the circle with centre K and diameter KD, which intersects the circle on the mirror in the plane of construction in [at most] two symmetrically located points, B and G.[2] All reflection points are contained on a circle on the mirror obtained by rotating B around diameter EDHK [234–35]. Conclusion [236].

⟪V 2.VII.14⟫

Difficulties in the perception of points on a long and thin visible object that is located along the same diameter as the eye. The eye cannot perceive reflection points behind it [if in Diagram V.29 the eye H is looking toward E and the points of intersection B and G lie behind H]. This is one of the reasons why the perception of an object along the diameter will be indistinct, as is confirmed by experimental examination [237–38]. Let E be a point [eye] different from the centre D of a spherical concave mirror, and suppose that the diameter ED extended intersects the mirror at A, so that the centre D is between E and A. As above, consider points H [visible objects] between D and A, such that DH < DE. Reflection between E and H in the mirror as in [234–35] is only possible if such a point H satisfies EA:AH > ED:DH; otherwise no reflection is possible, with the result that H cannot be reflected to the eye E by the mirror. (This may happen if H is too close to the centre D: in this case K is also close to H and the circle in the plane of construction in [234–35] is too small to intersect the mirror [239].) Statement that the eye will distinctly

2. Ibn al-Haytham deals with the conditions for the intersection in paragraph [239].

perceive in the mirror points located outside the diameter through the eye; and that the number of images will be one, two, three, or four, but not more [240].

⟪V 2.VII.15⟫ [Diagram V.30]

In Inquiries V 2.VII.15–17, one considers an eye T and a visible point H on different diameters of a spherical concave mirror, but at the same distance of the centre, thus, TD = HD. Let the radii DT and DH intersect the mirror at points B and G, and let E be the midpoint of arc BG; draw at E the tangent to the circle in which plane TDH intersects the mirror. If T and H are outside the mirror, and either on the tangent at E or further away from the centre D, then H will be reflected to T by the mirror at only one point; this is the point Z at which diameter ED intersects the circle a second time [241–44]. Summary of the proposition [245].

⟪V 2.VII.16⟫ [Diagram V.31]

If T and H are inside the mirror, and T and H are either between the centre D of the mirror and the feet of the perpendiculars from E to DB and DG [246], or T and H are the feet of these perpendiculars themselves [247], then H will be reflected to T from two different points on the mirror. The two reflection points are point E and point Z, which is the second intersection of diameter ED and the mirror. Summary of the proposition [248].

⟪V 2.VII.17⟫ [Diagram V.32]

If T and H are between the feet of the perpendiculars and the circumference of the circle, H will be reflected to T from four points. The four points are E and Z and the intersections of the mirror with the circle through points H, D, and T [249]. The reflection will not be in more than four points [250–51]. If T and H are on the circumference or outside the mirror between the circumference and the tangent at E, they can be reflected from two points [252]. Summary of the three propositions with Diagrams V.30–32, specifying the number of reflection points in all cases where the visible point and eye are at the same distance of the centre [252].

⟪V 2.VII.18⟫ [Diagram V.33(a,b)][3]

Given: the positions of the eye and visible point T and H inside or outside a spherical concave mirror with centre D; suppose HD > TD. Construction of the point of reflection

3. In Inquiries V 2.VII.18–20, Ibn al-Haytham begins with his treatment of the most complicated case, where the eye and visual point [object] are on different diameters, and at different distances from the centre of the spherical concave mirror. As usual, he works in the plane through the centre of the mirror, the eye, and the visible point, because reflection can only occur in this plane. He considers the two radii of the mirror from the centre to the eye and to the point of reflection. According to the previously proved Inquiry V 2.VII.11, reflection points can only occur on the arc between these two radii, and on the diametrically opposite arc; in these inquiries, Ibn al-Haytham deals with the diametrically opposite arc. On this arc there is always a point of reflection, which is constructed in V 2.VII.18. Every point on the opposite arc can occur as a solution, except the midpoint (VII.19), and for a given position of the eye and visible point, the point of reflection on the opposite arc is unique (VII.20).

on the sector opposite to the sector bounded by HD and TD. That is to say, let GHD and BTD be the radii through H and T and extend them to diameters GHDA and BTDK; the desired point of reflection should then be located on arc AK [253–57]. The construction is based on Case 1 of Lemma III [138].

⟪V 2.VII.19⟫ [Diagram V.34]

Consider a spherical concave mirror with centre D, and a plane through the centre intersecting the mirror in a circle, and choose two diameters ADG and BDK of the circle. Let Z be the midpoint of arc AK. Any point H on arc AK can be a point of reflection for an eye L and visible point T on radii DG and DB; because H is in the sector opposite to the sector bounded by DL and DT, the type of reflection is the same as in Inquiry V 2.VII.18. If H is not the same as Z, the distances DL and DT must be unequal [258]. For any given point H other than Z on arc AK, it is possible to choose an unlimited number of pairs of points [eye, visual object] on the radii DB and DG in such a way that H is the point of reflection; the distances of the two members of a pair to the centre of the mirror cannot be the same [259].

⟪V 2.VII.20⟫ [Diagram V.35]

Suppose, as in Inquiry V 2.VII.18, that we are given the positions of the eye and visible point T and H inside or outside a spherical concave mirror with centre D, and suppose HD > TD. The point of reflection on the opposite sector AK, which point was constructed in Inquiry VII.18, is then unique [260].

⟪V 2.VII.21⟫ [Diagram V.36][4]

For any position of the eye inside a spherical concave mirror, and every point on the surface of the mirror (not on the same diameter as the eye), an infinite number of visible objects can be reflected from that point to the eye, and [in all but at most two cases] the distance from the visible object to the centre of the mirror is different from the distance of the eye to the centre [261].

⟪V 2.VII.22⟫ [Diagram V.36]

The same as in Inquiry V 2.VII.21, if the eye is outside the circle and the reflection is from the concave arc facing the eye [262].

⟪V 2.VII.23⟫ [Diagram V.36]

If the position of the eye A is given inside the mirror and also the point of reflection E, then one can find a visible point N that is reflected to the eye from two different points;

4. Inquiries V 2.VII.21–32 deal with reflection points on what is here denoted the arc between the two radii, meaning the radii from the centre to the eye and the visible point. In this Summary Overview, reference is made to the 'auxiliary circle' through the centre, the eye, and the visible point; this auxiliary circle is mentioned by Ibn al-Haytham himself in V 2.VII.32 below.

one reflection point is E, in the sector bounded by the radii through A and N; the other reflection point is the one constructed in Inquiry V 2.VII.18 [253–57], which reflection point is in the opposite sector [263]. Statement of this theorem in general terms [264].

⟪V 2.VII.24⟫ [Diagram V.36]

In a spherical concave mirror with centre B, let the eye A be on diameter DBG, which intersects the mirror at D and G, and suppose that A is between D and B. Let the visible point N not be on the same diameter, and suppose N is reflected to A from a reflection point E, which is located on the sector bounded by the radii BA and BN. Ibn al-Haytham calls ∠AEN the reflection angle, and ∠NBG the adjacent angle (that is, adjacent to ∠ABN between the two radii through the eye and the visible point). If the distances from the visible point and the eye to the centre are not equal (i.e., NB is not equal to AB), the reflection angle cannot be equal to the adjacent angle; it may be greater or less [265–67]. [Thus, if the distances of N and A to the centre are not equal, the point of reflection cannot be on the auxiliary circle through A, B, and N.]

⟪V 2.VII.25⟫ [Diagram V.37(a,b)]

General and simple proof that the reflection angle cannot be equal to the adjacent angle if the eye and visible point are at unequal distances from the centre and the reflection is from the arc bounded by the radii through the eye and the visible point [268].

⟪V 2.VII.26⟫ [Diagram V.38]

Assume the position of the eye A and the visible point B at unequal distances from the centre G in a spherical concave mirror. Let GAE and GBZ be the radii through A and B. Suppose that B is reflected to A from two points on arc ZE (in the sector bounded by the radii). If the reflection angle of one of the points of reflection is less than the adjacent angle (as above, that is, the supplement of the angle between the radii), then the reflection angle of the second reflection point must be greater than the adjacent angle [269–74]. Hence there can be only one reflection point whose reflection angle is less than the adjacent angle. The proof is complicated. [Thus, only one reflection point can be outside the auxiliary circle through A, B, and the centre G; the others must be inside this auxiliary circle.]

⟪V 2.VII.27⟫ [Diagram V.38]

If a reflection point on arc ZE as in Inquiry V 2.VII.26 is such that the angle of reflection is smaller than the adjacent angle (i.e., the supplement of the angle between the two radii through the eye and visible point), then the reflection point (i.e., point T in Diagram V.38) will be on the half of arc ZE closer to whichever point (eye or visible point) is farther from the centre (point B in Diagram V.38). This is concluded from the previous proposition and the diagram [275]. [Thus, the reflection point outside the auxiliary circle cannot be anywhere on the arc between the radii, it must be on the half arc closer to the farther of the two given points (i.e., eye, visible point).]

《V 2.VII.28》 [Diagram V.39, Diagram V.39+]

In a spherical convex mirror, two arbitrary diameters ADG and BDK through the centre D are given. Ibn al-Haytham constructs a specific example of two points O and Q that are reflected to one another from two reflection points T and L, which he also constructs and which are both on the arc BG between the radii. In the example, it is shown that DQ > DO, and for point T the reflection angle will be greater than the adjacent angle (BDA); for point L the reflection angle will be less. The geometry is complicated, but the specific example does not relate to the six lemmas, nor to construction by means of conic sections [276]. [Thus, points O and Q will be reflected to one another from two reflection points L and T; L will be outside, and T inside the auxiliary circle through O, Q and the centre D.]

《V 2.VII.29》 [Diagram V.40++, Diagram V.40(c)]

In a spherical concave mirror, suppose that the positions of the eye and visible objects are given at different distances from the centre. Ibn al-Haytham argues that the object can be reflected to the eye from at most two points on the arc between the radii through the object and the eye [277–81]. In his rather unclearly worded argument, he reduces the problem to Lemma II [133–37], which has two solutions. Part of the reasoning is analogous to Lemma VI above.[5]

《V 2.VII.30》 [Diagram V.40(c)]

For given positions of the eye and the object not at an equal distance from the centre of a spherical concave mirror, construction of reflection points on the arc contained by the two radii through the eye and the centre [282]. The construction is based on Lemma VI above, which in turn is based on Lemma II, with a maximum of two solutions.[6]

《V 2.VII.31》 [Diagram V.41]

If the eye and the visible point are outside a spherical concave mirror, the reflection cannot be from more than a single point on the arc bounded by the two radii (it is assumed that the mirror is cut at suitable places so that the rays can pass through the surface) [283]. Another reflection point may exist on the diametrically opposite arc.

---

5. Ibn al-Haytham assumes that the points are such that the reflection angle is greater than the adjacent angle, but he does not mention this condition explicitly. The condition is pointed out by Kamāl al-Dīn, and also by M. Naẓīf, in *Al-Ḥasan Ibn al-Haytham*, I, pp. 463–67.

6. In the terminology of Greek geometry, V 2.VII.30 is the synthesis belonging to the analysis of V 2.VII.29. Ibn al-Haytham concludes that there are at most two reflection points on the arc between the two radii. It was shown in Inquiry V 2.VII.18 that there is also a reflection point on the opposite arc. Ibn al-Haytham concludes that the total number of reflection points inside the mirror is one, two, or three points, but not more. The conclusion is incorrect, because Ibn al-Haytham omitted the single reflection point outside the auxiliary circle, even though this point was mentioned in Inquiry V 2.VII.27. This extra point can be found by an analogous construction, which always has one solution, but such a construction would require a new diagram.

⟪V 2.VII.32⟫[7]

The same statement as in [283] for the case where the eye is outside the mirror and the visible object is at the circumference of the mirror, or vice versa. Ibn al-Haytham does not discuss the situation where one of these two points is outside the sphere of the mirror and the other point is inside it. He states, correctly this time, that the number of points of reflection may be one, two, three, or four if the eye and visible point are not at equal distances from the centre of the mirror [284]. This is followed by a lengthy summary in the form of detailed general rules for the reflection of points in spherical concave mirrors not at equal distances from the centre of the mirror: two points outside it, such that the straight line through them does not intersect the circle [285], or such that the straight line through them intersects the circle [286]. In the case where one or both of them is inside the circle, Ibn al-Haytham considers the intersections between the mirror and the auxiliary circle through the eye, the visible object, and the centre of the mirror. Only in some cases, where the latter circle intersects the mirror in two points, can the maximum number of four reflection points be reached [287–89]. Summaries of these rules [290–92]. Notice on the spherical concave mirror when it is cut at one side [293]. Experimental examination of the images in this mirror [294].

⟪V 2.VII.33⟫

Binocular perception in the spherical convex mirror [295].

⟪V 2: Eighth Inquiry. On the Images of Cylindrical Concave Mirrors, Consisting of Ten Propositions⟫ [296–321]

⟪V 2.VIII.1⟫ [Diagram V.42]

All intersections between reflection planes and the surface of the right-cylindrical concave mirror are conic sections [ellipses], except if the reflection plane passes through the axis of the cylinder (in which case the intersection is a single straight line or two such lines), or if the reflection plane is parallel to the base plane of the cylinder (in which case the intersection is a circle). If the eye and visible point are in the plane with a straight intersection, but not on the same perpendicular to the axis, reflection will produce images behind the mirror. Reflection in a plane parallel to the base of the cylinder will produce images, as was explained previously for the circle [296]. Demonstration that if a visible point and the eye are not on the same perpendicular to the axis, and the visible point is reflected in a plane with elliptical intersection, the image can be either behind the mirror or in front of it, and either behind the centre of the eye, or at the centre of the eye, or between the eye and the mirror [297–300].

---

7. V 2.VII.32 is a lengthy conclusion that does not repeat the error of V 2.VII.30. Here, Ibn al-Haytham states, correctly, that, for an eye and visible point not at an equal distance from the centre of the mirror, there can be three reflection points on the arc between the two radii, namely two inside the auxiliary circle and one outside it, and one reflection point on the opposite arc. The total number of reflection points can thus be four, but not more than four. Ibn al-Haytham must have had the complete solution in mind, but either his exposition was incomplete, or a proposition may be missing from the extant Arabic manuscripts of the *Optics*.

⟪V 2.VIII.2⟫

Visible points on the perpendicular from the eye to the axis of the cylinder can only be reflected to the eye in the plane parallel to the base of the cylinder, whose intersection is a circle. If the eye is not at the centre, the intersection of the eye with the perpendicular will be reflected to itself in the mirror at one of the endpoints of the diameter through the eye [301–2].

⟪V 2.VIII.3⟫

If the centre of the circle is between the visible point and the eye, the visible point may be reflected to the eye at two (symmetrically located) points on the circle in which the plane through the visible point and eye parallel to the base of the cylinder intersects the cylinder. These two reflection points are not on the diameter through the eye, but the images will be at the centre of the eye [303].

⟪V 2.VIII.4⟫

If the centre of the eye is on the axis of the cylinder, then the point where a perpendicular (to the axis) intersects the eye will be reflected at all points on the circumference of the circle through the eye parallel to the base of the cylinder, and the image is the entire circumference of the circle [304].

⟪V 2.VIII.5⟫ [Diagram V.42]

In a reflection plane that intersects the concave cylinder in an ellipse, it is possible that the visible point may be reflected to the eye at two different reflection points on that ellipse [305].

⟪V 2.VIII.6⟫

In a reflection plane that intersects the cylindrical concave mirror in an ellipse, a visible point cannot be reflected to the eye at more than two reflection points in the same plane [306]. [Thus, if there are more reflection points, they must be in other planes through the eye and visible point.]

⟪V 2.VIII.7⟫ [Diagram V.43]

If the eye and visible point are both on the axis of the right-cylindrical concave mirror, the visible point is reflected to the eye from all points on a circle on the surface of the mirror, and the image of the visible point will be on another circle, one that is twice as large [307–9].

⟪V 2.VIII.8⟫ [Diagram V.44]

Reflection of the visible points in the right-cylindrical concave mirrors when the eye and visible point are not on the same diameter and not on [a line parallel to] the axis of the cylinder. Discussion of the special case where the eye and visible point are in one plane parallel to the base of the cylinder, which intersects the cylinder in a circle. As shown previously, the reflection will in this case only take place at reflection points on the circle, of which there may be one, two, three, or four, but never more than four. The general case of reflection (in planes with elliptical intersection) is then reduced to the special case through the intermediary of the perpendicular projection of one of the points (for example the visible point) onto the plane through the other point (eye) parallel to the base of the cylinder [310–14]. Discussion of the images [315]. For a given position of the eye and visible point, there can be up to four, but no more than four reflection points [316–18]. Summary of results [319].

⟨⟨V 2.VIII.9⟩⟩

Construction of the point or points of reflection on the surface of the right-cylindrical concave mirror, if the visible point and the eye's point are given and not situated on [a line parallel to] the axis of the cylinder, etc. Analogously with Inquiry V 2.VIII.8, the problem is reduced to the previous construction of images for (circular intersections of) the spherical concave mirror [320].

⟨⟨V 2.VIII.10⟩⟩

Binocular perception in a cylindrical concave mirror is reduced to the case of the spherical mirror [321].

⟨⟨V 2: Ninth Inquiry: On the Images of the Conical Concave Mirrors, Consisting of an Introduction and Eight Propositions⟩⟩ [322–45]

⟨⟨Introduction⟩⟩

In the conical concave mirror, the intersections between the surface of the mirror and the reflection planes are all conic sections, except for the plane through the axis, which intersects the mirror in a straight line [as noted in Book IV, Chapter 4]. Images of visible points reflected in the plane through the axis are behind the mirror [322]. [A plane parallel to the circular base of the cone cannot be a reflection plane.]

⟨⟨V 2.IX.1⟩⟩

After reflection in other planes that intersect the mirror in conic sections, the images can be behind the mirror, or between the eye and the mirror, or at the centre of the eye, or behind the eye. This is shown in the same way as for the cylindrical concave mirror [323].

⟨⟨V 2.IX.2⟩⟩

If the eye is between the axis and the conical concave mirror, and a perpendicular is imagined through the eye to the surface of the mirror, the eye will perceive the intersection of its own surface with this perpendicular as reflected by the mirror at the endpoint of the perpendicular [324].

⟨⟨V 2.IX.3⟩⟩

If the eye is on the axis, and a perpendicular is drawn from the eye to any straight line on the conic surface, then it is stated that the intersection between this perpendicular and the surface of the eye will be reflected to the eye from all points on the circumference of a circle, but not from any other point [325].[8]

⟨⟨V 2.IX.4⟩⟩

If the eye and visible point are on the same perpendicular to the surface of the mirror, which perpendicular will necessarily intersect the axis, and if the intersection point with the axis is between the eye and visible point, then the visible point may be reflected to the eye by the surface of the mirror [326].

⟨⟨V 2.IX.5⟩⟩ [Diagram V.45]

If the centre of the eye and the visible point are on the axis of the conical concave mirror, the visible point will be reflected to the eye from the circumference of a circle on the mirror. Construction of the points of reflection and of the images, which are also located on the circumference of a [larger] circle [327–29]. Reference to Book IV, Chapter 5 [327].

⟨⟨V 2.IX.6⟩⟩ [Diagram V.46]

Let the eye and visible point be given, and let at most one of them be on the axis of the cone. Construction of the point or points of reflection in the conical concave mirror. First Ibn al-Haytham deals with the special case where the eye and visible point are in a plane parallel to the circular base of the cone. The plane intersects the cone in a circle and Ibn al-Haytham considers the (up to four) reflection points in that circle, which were constructed in the section on the spherical concave mirror. These reflection points in the circle are not the same as the required reflection points on the conical mirror, but using the (up to four) reflection points in the circle, Ibn al-Haytham is able to construct the actual points of reflection on the conical concave mirror [330–32]. There can be multiple images, as many as the (up to four) reflection points in the circular intersection [333]. The general case [Diagram V.47] is related to the special case by central projection of one of the given points (eye, visible point) from the apex of the cone to the plane through the other given point parallel to the base of the cone [334–38]. Possibility of multiple images [339].

⟨⟨V 2.IX.7⟩⟩ [Diagram V.47]

In the conical concave mirror, reflection of a given visible point to a given eye can take place at no more than four reflection points, and their images are not more than four [340–44].

8. It is apparently assumed that the centre of the eye coincides with the surface of the eye.

⟪V 2.IX.8⟫

Binocular perception in the conical concave mirror is analogous to that in the spherical concave and cylindrical concave mirrors [345].

End of Book V, and date of MS F [346–47].

ARABIC–ENGLISH GLOSSARY
for Books IV and V

a

a.b.d.

abadan always

a.b.r.

ibrah needle
ibrah daqīqah fine needle

a.th.r.

aththara to impress, to make an impression
taʾthīr effect
athar impression
athar maḥsūs sensible effect

a.r.b.

ʿalā-l-taʾrīb slantwise

a.r.j.

urjuwānī purple (IV 76b [142])

a.r.ḍ.

arḍ al-bayt chamber's floor

a.ṣ.l.

aṣl hypothesis

a.f.q.

ufq, ufuq horizon

a.l.f.

maʾlūf familiar

a.m.l.

taʾammala to contemplate, to examine, to gaze at, to observe

a.w.l.

ālah instrument

a.y.ḍ.

ayḍan again, further, furthermore, moreover

b

b.ḥ.th.

baḥatha to inquire
mabḥath (pl. *mabāḥith*) inquiry

b.d.ʾ.

mabdaʾ origin, principle

b.d.l.

mutabādil alternate

b.d.h.

badīhah at a glance
idrāk bi l-badīhah perception at a glance; see also d.r.k.

b.r.h.

burhān geometrical demonstration, geometrical proof (more specific than *bayān*; cf. b.y.n.)

b.z.l.

bāzil bright green (IV 7a [13])

b.s.ṭ.

basaṭa to expand
munbasiṭ expanded
inbisāṭ expansion

b.ṣ.r.

abṣara to see
baṣar eye, eyesight, sight
ibṣār seeing, vision, perception
ibṣār bi l-badīhah perception at a glance
ibṣār bi l-inʿikās perception by reflection
ibṣār bi l-inʿiṭāf perception by refraction
anwāʿ al-ibṣār modes of seeing or of vision
kayfiyyat al-ibṣār manner of vision
mubṣar being seen, visible object

b.ṭ.l.

baṭala to cease to be, to vanish, to be annihilated, to cease to be effective

b.ʿ.d.

baʿuda to recede
buʿd distance, remoteness, radius (of a circle)
buʿd mutafāwit excessive remoteness

ilā abʿad ghāyah however far extended
kharaja ilā abʿad ghāyah produced (extended) as far as possible
tabāʿada ʿan to recede from

b.q.y.

baqiya to remain
fa-yabqā by subtraction

b.y.t.

bayt chamber

b.y.n.

bayyana to make clear, to show, to reveal, to demonstrate, to prove
wa dhālika mā aradnā an nubayyin and that is what we wished to prove
tabayyana to be clear
bayyin clear, manifest, conspicuous, demonstrated
bayān clarity, demonstration, proof; cf. sh.r.ḥ, b.r.h.

t

t.b.ʿ.

tabiʿa to follow, to behave like

t.ḥ.t.

taḥt below (in a geometrical diagram)

t.m.m.

tammama to complete (e.g., a circle)
tamām (= *tatimma*) complement(ary) (V 217a [153])

th

th.b.t.

thabata, *thābit*, *muthbat* fixed

th.q.b.

thaqb aperture, hole
mithqab drill

th.q.l.

thaqīl heavy
al-thaqīl the heavy body
thiqal heaviness; see also *ḥarakat al-thiqal*, *quwwat al-thiqal*

th.l.th.

muthallath triangle
raʾs al-muthallath vertex of the triangle

th.n.y.

thānī second, secondary
thānī thānin second of a secondary (i.e., tertiary)

j

j.dh.b.

jadhaba to pull

j.z.ʾ.

juzʾ (pl. *ajzāʾ*) part, portion, segment (of a line or surface)

j.s.m.

jism body, object
jism ṣaqīl polished body, mirror

j.ṣ.ṣ.

jiṣṣ gypsum

j.ʿ.l.

jaʿala to make, to construct

j.m.ʿ.

jamaʿa to come together, to gather together
ijtamaʿa to converge

j.m.l.

bi l-jumla at all (V 274b)

j.n.b.

ʿalā jānib sideways

j.n.s.

jins kind

j.w.z.

ajāza ʿalā to let pass, to draw (or rather, imagine) (a plane through a point) (V 244b)
jawwaza to presume

ḥ

ḥ.a.ṭ.

aḥāṭa to contain, to surround (encompass)
muḥīṭ circumference

ḥ.j.b.

ḥajaba to conceal, to hide
muḥtajib concealed (IV 6b [11])

ḥ.j.m.

muqtadir al-ḥajm of a fair size

ḥ.d.b.

muḥaddab convex

ḥ.d.th.

aḥdatha to make, to produce (e.g., a line)

ḥ.d.d.

ḥadda to bound, to mark a limit
ḥaddada to determine, to indicate clearly, to specify, to describe precisely
maḥdūd bounded, determinate, determined, specified
taḥdīd determination, qualification
ḥadd edge, limit
zāwiyah ḥāddah acute angle

ḥ.d.s.

ḥadasa to conjecture
ḥads conjecture
idrāk bi l-ḥads perception by conjecture

ḥ.r.r

muḥarrar determined, precise

ḥ.r.k.

taḥarraka to move, to shift
ḥarakah motion, movement; see q.w.y.
ḥarakah sarīʿah fast movement
ḥarakah qawiyyah forceful motion
ḥarakah murakkabah compound motion
ḥarakat al-thiqal motion due to weight
ḥarakah ʿaraḍiyya accidental motion

ḥ.r.y.

taḥarrā to ensure, to secure, to make sure

ḥ.s.b.

bi-ḥasab in accordance with, dependent on

ḥ.s.s.

ḥiss sense
fī-l-ḥiss (relative to) sense
musāwī fī-l-ḥiss sensibly alike, sensibly the same
bi-l-qiyās ilā l-ḥiss as far as appears to sense
al-ḥāss the sentient (faculty); cf. a.th.r.

ḥ.ṣ.l.

ḥaṣala to occur
al-ḥāṣil summation, summing up

ḥ.f.r.

ḥafara to engrave
ḥafr cavity, depression (depressed area)

ḥ.q.q.

ḥaqqaqa to ascertain
ʿalā-l-taḥqīq surely, truly
taḥaqqaqa to identify
muḥaqqaq accurate (V 256a [208]) distinct
ghayr muḥaqqaq indistinct

ḥ.k.m.

ḥakama to make a judgement

ḥ.l.q.

ḥalqah ring, especially the segment of a wooden ring forming the main part of the instrument (*ālah*) for examining the behaviour of light in reflection

ḥ.n.w.

munḥanī curved

ḥ.w.l.

ḥāl condition, situation, state, state of affairs
fī-l-ḥāl immediately
aḥwāl cases
ʿalā jamīʿ al-aḥwāl in all cases
muḥāl impossible
ḥawla around

kh

kh.r.j.

kharaja to proceed, to go out, to extend, to issue (V 308b [297])
kharaja bi-ghayr nihāyah to extend indefinitely
akhraja to draw, to produce (a line or plane), to drop (a perpendicular)

istakhraja to find (e.g., a point)
khārij ʿan (to lie) outside
al-khārij min (line) issuing from, drawn from (a point)

kh.r.ṭ.

makhrūṭ cone
makhrūṭ qāʾim right cone
makhrūṭ mustadīr cone whose base is a circle
makhrūṭ muṣmat solid cone
makhrūṭ qāʾim mutawahham imagined right cone
raʾs al-makhrūṭ apex of cone
sahm al-makhrūṭ axis of cone
saṭḥ al-makhrūṭ surface of cone
inkhirāṭ curvature

kh.ṣ.ṣ

khāṣṣ special
khāṣṣat (al-jism) property (of a body)

kh.ṭ.ṭ.

khaṭṭ (pl. *khuṭūṭ*) line
khaṭṭ mustaqīm straight line

kh.l.f.

mukhtalifān (said of two images) different, unequal

kh.m.r.

khamrī wine (colour) (IV 80a [143])

kh.y.l.

khayāl image
khayālān mutadākhilān two overlapping images
nuqṭat al-khayāl image point

d

d.kh.l.

dākhil inside
mutadākhilān (said of two images) overlapping, coinciding

d.r.k.

adraka to perceive (visually)
idrāk (visual) perception
idrāk bi-l-badīhah perception at a glance
idrāk bi-l-ḥads perception by conjecture
idrāk ʿalā istiqāmah direct or rectilinear perception

idrāk bi-l-inʿikās perception by reflection
idrāk bi-l-inʿiṭāf perception by refraction
idrāk muḥaqqaq perceived distinctly
mudrak perceived
mudrak bi-l-iʿtibār perceived by means of experiment (or experimenting)

d.f.n.

dafana to embed

d.q.q.

daqqa to become thin or subtle
daqīq subtle, thin
fī ghāyat al-diqqah extremely subtle

d.l.l.

dalla to indicate, to prove
istidlāl inference

d.w.r.

dāra to rotate
adāra ʿalā to circumscribe
dāʾirah circle
niṣf dāʾirah semicircle
istidārah circularity

d.w.n.

dūna lower, before (V 279b)

dh

dh.ā.t.

dhātiyy essential

dh.r.ʿ.

dhirāʿ cubit

dh.k.r.

dhakara to mention, to describe

dh.l.k.

ka-dhālik similarly

r

r.ʾ.s.

raʾs vertex
raʾs al-qiṭāʿ vertex of sector

raʾs al-makhrūṭ apex of cone
raʾs al-muthallath vertex of triangle

r.ʾ.y.

mirʾāh (pl. *marāyā*) mirror, 'polished body'
mirʾāh musaṭṭaḥah plane mirror
mirʾāh kuriyyah muḥaddabah spherical convex mirror
mirʾāh kuriyyah muqaʿʿarah spherical concave mirror
mirʾāh usṭuwāniyyah cylindrical mirror
mirʾāh usṭuwāniyyah muḥaddabah cylindrical convex mirror
mirʾāh usṭuwāniyyah muqaʿʿarah cylindrical concave mirror
mirʾāh makhrūṭah (or *makhrūṭiyyah*) (right-)conical mirror
mirʾāh makhrūṭah muḥaddabah conical convex mirror
mirʾāh makhrūṭah muqaʿʿarah conical concave mirror

r.b.b.

rubba-mā maybe, perhaps

r.b.ʿ.

murabbaʿ square

r.s.m.

rasama to draw, to trace out

r.gh.m.

arghama to press (into)

r.f.ʿ.

irtifāʿ elevation

r.q.q

raqīq thin
raqīq delicate

r.k.b.

murakkab ʿalā mounted on

r.k.z.

markaz centre
markaz al-baṣar centre of the eye

r.w.h.

rayḥānī bright green (IV 7a [13])

z

z.r.ʿ.

zarʿī light green (colour) (IV 7a [13])

z.w.y.

zāwiyah angle
zāwiyah ḥāddah acute angle
zāwiyah qāʾimah right angle
zāwiyah munfarijah obtuse angle
zāwiyah khārijah exterior angle
zāwiyah dākhilah interior angle
zāwiyah mushtarakah common angle
ayyatu zāwiyah kānat a random angle (V 312a [305])

z.y.d

zāda to exceed, to add
al-qaṭʿ or *al-qiṭʿ al-zāʾid* hyperbolic section

s

s.ʾ.l.

masʾalah mathematical or geometrical problem

s.t.r.

satara to screen

s.d.d.

sadda to close, to obstruct

s.ṭ.ḥ.

saṭḥ surface
saṭḥ mustawī plane surface
saṭḥ kurī spherical surface
saṭḥ al-makhrūṭ surface of cone
saṭḥ muḥaddab convex surface
saṭḥ muqaʿʿar concave surface
saṭḥ aʿlā-l-ḥalqah top surface of the wooden ring
saṭḥ al-inʿikās reflection plane
saṭḥ muqābil li-l-baṣar surface facing the eye
saṭḥ mumāss tangent plane

s.ṭ.r.

al-masṭarah (holding) post (rectangular block of wood used in an upright or inclined position as a support of a mirror, facing the inner, concave surface of the ring)
al-masṭarah al-ḥāddah sharp-edged ruler

s.f.r.

musfir al-lawn of a light colour

s.q.ṭ.

masqaṭ al-ʿamūd foot of perpendicular; see *ʿamūd*

s.l.k.

salaka to follow

s.m.r.

mismār laṭīf fine nail

s.m.s.

simsimah sesame seed

s.m.k.

sumk thickness (of the ring)

s.h.m.

sahm arrow (V 71a [124]); sagitta (of an arc, technical term for the segment joining the midpoint of an arc to the midpoint of the chord subtending the arc)
sahm axis of cylinder or cone
sahm al-unbūb axis of tube

s.w.d.

sawād al-baṣar see l.w.n

s.w.f.

masāfah distance; cf. b.ʿ.d.

s.w.y.

musāwī equal, identical, the same
mustawī unreversed (in mirror)

s.y.f.

musayyaf sharp

sh

sh.b.h.

shabbaha to assimilate
shabīh similar
mutashābih the same
mutashābih fī-l-ghāyah perfectly similar
mushtabah uncertain
shubhah false apprehension

sh.t.t.

mushattat dispersed

sh.r.ḥ.

sharaḥa to explain, to elucidate
sharḥ wa bayān explanations and demonstrations (V 300b [284])

sh.r.q.

ashraqa to radiate, to shine

sh.r.k.

mushtarak common
faṣl mushtarak intersection

sh.ʿ.r.

shaʿīrah barley-seed

sh.ʿ.ʿ.

shaʿʿa to radiate, to shine
khuṭūṭ al-shuʿāʿ radial lines

sh.f.f.

mushiff transparent
ajsām mushiffah transparent bodies

sh.k.l.

shakl shape
shakl (geometrical) figure, proposition (i.e., geometrical theorem)

sh.m.s.

shams sun

sh.m.ʿ.

shamʿ wax
shamʿ abyaḍ white wax

sh.h.r.

shihr (Persian) lathe

Ṣ

ṣ.b.ʿ.

iṣbaʿ digit, as unit of measurement

ṣ.ḥ.b.

ṣāḥaba to accompany

ṣ.ḥ.ḥ.

ṣaḥḥa to be possible
laysa yaṣiḥḥu it is not possible (V 262b)

ṣ.d.r.

ṣadara (min) to emanate (from)
ṣadr preface

ṣ.d.f.

ṣādafa to encounter

ṣ.r.f.

ʿalā taṣārīf al-aḥwāl always, in all cases

ṣ.r.w.

ṣarw cypress

ṣ.ʿ.w.

ṣaʿwī red (IV 7a [13], IV 79a [142])

ṣ.gh.r.

ṣaghīr small
aṣghar smaller

ṣ.f.ḥ.

ṣafīḥah plate
ṣafīḥah (copper) plate

ṣ.q.l.

ṣiqāl polish
ṣaqīl polished, smooth
jism ṣaqīl polished body (i.e., mirror)
al-ajsām al-ṣaqīlah polished bodies, mirrors

ṣ.n.b.r.

ṣanawbar pine

ṣ.w.r.

ṣūrah form, (sometimes) figure, diagram, special case in a proposition
ṣūrah muʿayyanah definite form
ṣuwar bi-l-quwwah potential forms
ṣūrah mustaqīmah rectilinear form
ṣūrat al-lawn form of colour

ḍ

ḍ.b.ṭ.

ḍābiṭ kullī precision, general rule

ḍ.ʿ.f.

ḍaʿīf weak
ḍiʿf double, twice

ḍ.l.ʿ.

ḍilʿ side
mutasāwī l-aḍlāʿ equilateral

ḍ.m.m.

(khuṭūṭ) mutaḍāmmah compacted (lines)

ḍ.w.ʾ.

ḍawʾ awwal primary light
ḍawʾ thānī secondary light
(ḍawʾ) thānī thānin secondary of secondary [i.e., tertiary] (of light)
ḍawʾ mushriq shining light
ḍawʾ dhātī essential light
ḍawʾ ʿaraḍī accidental light
ḍawʾ ḍaʿīf weak light
aḍwāʾ ḍaʿīfah weak lights
ḍaʿīf al-ḍawʾ poorly lit
ḍawʾ qawī strong light
jism muḍīʾ luminous body, illuminated body (and, hence, luminous body)
jism muḍīʾ min dhātihi self-luminous body
ṣūrat al-ḍawʾ form of light

ṭ

ṭ.b.ʿ.

ṭabaʿa to impress
inṭabaʿa to be impressed
ṭabīʿah nature
ʿilm al-ṭabīʿah science of nature
aṣḥāb al-ʿilm al-ṭabīʿī natural scientists

ṭ.b.q.

ṭabbaqa to apply, to fasten
inṭabaqa to coincide
munṭabiq ʿalā coinciding with

ṭ.r.b.

iḍṭirāb corrugation

ṭ.r.f.

ṭaraf edge, end, extremity, tip

ṭ.r.q.

ṭarīqah method
ṭarīq (al-ʿamal) method (of construction)

ṭ.l.b.

ṭalaba to seek
maṭlūb question

ṭ.w.q.

ṭawwaqa to solder

ṭ.w.l.

ṭūl length

ẓ

ẓ.l.m.

aẓlama to be darkened
muẓlim dark
ẓulmah darkness

ẓ.h.r.

ẓahara to appear
ẓāhir apparent
ka-mā yaẓhar li l-ḥiss as far as appears to the senses

ʿ

ʿ.b.r.

iʿtabara to make or carry out experiments
iʿtibār experiment; experimental examination
iʿtibār muḥarrar precise experimental examination
iʿtibār see also d.r.k.
muʿtabir experimenter, observer

ʿ.r.ḍ.

ʿaraḍa to occur, to arise
ʿaraḍ accident
ʿaraḍī accidental
ḥarakah ʿaraḍiyyah accidental motion
ʿarḍ breadth, width
muʿtariḍ, khaṭṭ muʿtariḍ transverse line (V 156b [44])

ʿ.r.ʿ.r.

ʿarʿar juniper, a soft wood (IV 16a [5])

ʿ.ṭ.f.

ʿaṭafa to bend
inʿaṭafa to be refracted
inʿiṭāf refraction

ʿ.ẓ.m.

ʿiẓam magnitude, size
aʿẓam greater, larger

ʿ.k.s.

ʿaks reverse
inʿakasa to be reflected
inʿikās reflection
bi l-inʿikās in reverse
kayfiyyat al-inʿikās manner of reflection
nuqṭat al-inʿikās reflection point; see also d.r.k.

ʿ.l.q.

yataʿallaqu bi- to be associated with

ʿ.l.l.

ʿillah cause, reason (answering ‘why’) (e.g., V 162a)

ʿ.l.m.

ʿilm knowledge, science
maʿlūm that which is known
taʿallama to mark

ʿ.m.d.

iʿtamada ʿalā to press against
iʿtimād impact, pressure (IV 73a [129])
ʿamūd perpendicular line
masqaṭ al-ʿamūd foot of the perpendicular

ʿ.m.l.

ʿamila to do, to make, to construct (a figure)
istaʿmala to employ, to make use of

ʿ.n.b.

ʿunnāb jujube

ʿ.n.y.

maʿnā (pl. maʿānī) notion, state of affairs, quality, property (as in ‘visible property’; cf. IV 41a [72])

ʿ.w.d.

aʿāda to redraw, to repeat
nuʿīdu we repeat (the same geometrical diagram)
ʿūd mustaqīm straight rod

ʿ.y.n.

ʿayn eye, eyeball
(shayʾ) bi-ʿaynih the very same thing
muʿayyan definite, certain; see also *ṣūrah*

gh

gh.d.r.

mughdir dark (IV 5b [8])

gh.l.ṭ.

ghalaṭ error
aghlāṭ al-baṣar errors of sight

gh.y.r.

lā yataghayyaru it is invariable

f

f.r.j.

munfarij (of angle) obtuse

f.r.d

afrada to set apart
munfarid isolated, separate(d), by itself

f.r.ḍ.

faraḍa to assume
mafrūḍ assumed, given
farḍ assumption

f.r.f.r.

firfirī purpure (IV 7a [13])

f.r.q.

tafarraqa to disperse
tafarruq dispersion, separation
mutafarriqah dispersed

f.s.ḥ.

fasīḥ wide

f.ṣ.l.

faṣala to cut off, to mark off
fuṣila to be cut off
munfaṣil discrete
faṣl chapter, section
faṣl mushtarak intersection

f.ḍ.ḍ.

fiḍḍah silver

f.ḍ.w.

faḍāʾ open space

f.ʿ.l.

infiʿāl affection

f.l.dh.

fūlādh steel

f.n.y.

faniya to expire

f.w.t.

tafāwut maḥsūs appreciable difference
mutafāwit excessively different

f.w.q.

fawq above (in a geometrical diagram)

f.w.h.

fawhah orifice

q

q.b.l.

qabila to admit (a circular segment admitting an angle, as in Euclid III)
qābala to face, to be opposed to
muqābil facing, opposed to
qaṭʿān or *qiṭʿān mutaqābilān* two opposed sections (i.e., the two branches of a hyperbola)

q.d.r.

qadr, *miqdār* magnitude
muqtadir al-ḥajm fairly large

q.d.m.

qaddama to put forward, to advance
muqaddamah (pl. *muqaddamāt*) (geometrical) preliminary, lemma

muqaddimah introduction
quddām in front of

q.r.b.

qaruba to be near
qarīb min close to

q.r.ṭ.s.

qirṭās (piece of) paper

q.s.m.

qasama to divide
qasama ... bi-niṣfayn bisect
qism division, part, segment (of a line or of surface)

q.ṣ.d.

qaṣada to intend
qaṣd, maqṣid (or *maqṣad*) (pl. *maqāṣid*) intention, proposition(s) (as used by Kamāl al-Dīn)

q.ṣ.ṣ.

iqtiṣāṣ recapitulation

q.ṭ.b.

quṭb al-dāʾira pole of a circle on the sphere

q.ṭ.r.

quṭr diameter of a circle or conic section
niṣf al-quṭr radius of a circle

q.ṭ.ʿ.

qaṭaʿa to cut, to cut off, to intersect
qaṭʿ, *qiṭʿ* piece, (conic) section
al-qaṭʿ or *al-qiṭʿ al-zāʾid* hyperbolic section
al-qaṭʿ or *al-qiṭʿ al-nāqiṣ* ellipse
qaṭʿān or *qiṭʿān mutaqābilān* two opposed sections (i.e., the two branches of a hyperbola)
taqāṭuʿ intersection
qiṭāʿ sector
qiṭʿah piece, portion

q.ʿ.d.

qāʿidah base

q.ʿ.r.

muqaʿʿar concave

q.l.b.

maqlūb reversed; cf. s.w.y.

q.l.l.

aqallu-l-qalīl min al-ḍawʾ the least light
qalla-mā rarely

q.w.s.

qaws arc
taqwīs curvature
muqawwas arc-shaped

q.w.l.

maqālah (pl. *maqālāt*) Book(s)

q.w.m.

aqāma to erect, to set up
qāʾim (ʿalā) orthogonal, perpendicular to
qawm group
mustaqīm rectilinear
istiqāmah straightness
kharaja or *imtadda ʿalā istiqāmah* to extend in a straight line, rectilinearly
taqawwama to be constituted (of)

q.w.y.

qawiya to be forceful, strong
quwwah force, strength
quwwah qawiyyah great force
qawiyy forceful, strong
(ḥarakah) qawiyyah forceful (motion)
ghāyat al-quwwah maximum force
quwwat al-thiqal force of weight

q.y.s.

qiyās (pl. *maqāyīs*) reasoning
al-qiyās wa-l-burhān reasoning and geometric proof
bi-l-qiyās ʾilā with respect to

k

k.th.f.

kathīf opaque

k.ḥ.l.

kuḥlī dark blue

k.r.w.

kurah sphere
kurī spherical
kurah muqtadirat al-ʿiẓam fairly large sphere

k.s.f.

kasafah to dim (a light)
inkasafa to be dimmed

k.f.ʾ.

takāfuʾ equivalence

l

l.ʾ.m.

multaʾim coherent

l.b.s.

multabis indistinct, uncertain
multabis bi- confused with

l.z.m.

lazima to follow
lāzim inseparable

l.sh.y

talāshā to vanish (IV 63a [106])

l.ṣ.q.

alṣaqa to affix, to fasten

l.ṭ.f.

laṭīf thin

l.q.y.

laqiya to meet (e.g., a point)
laqiya to encounter
iltiqāʾ meeting
nuqṭat al-iltiqāʾ meeting-point

l.m.m.

alamma bi- to broach

l.w.ḥ.

lāḥa to appear to be
lawḥ board
lawḥ khafīf light panel (of soft wood)

l.w.n.

lawn colour; see ṣ.w.r
lawn sawād al-baṣar dark colour of the eyeball

l.w.y.

iltiwāʾ bending

m

m.th.l.

mithl equal, similar, corresponding to
mithāl example, setting-out (as in Euclid's *Elements*)
tamthīl exemplification

m.d.d.

madda to extend, to produce, to stretch
imtadda to be extended, etc.
imtidād extension

m.r.r.

marra bi- to pass through

m.z.j.

mumtazij mixed

m.s.s.

māssa to touch, to be tangent to
mumāss tangent, touching (line, surface, plane)
mumāssah, tamāss tangency
mawḍiʿ al-tamāss point of tangency

m.s.k.

miskī musk

m.s.m.

masāmm pores

m.ʿ.n.

maʿnā matter
al-maʿnā state of affairs
al-maʿānī qualities, properties
maʿānī laṭīfah subtle features

m.k.k.

mak(k)ūk drinking cup (IV 6b [11])

m.k.n.

makkana to settle, to stabilize

m.l.s.

amlas smooth
malāsah smoothness

m.n.ʿ.

mumānaʿah resistance
mumtaniʿ impossible
imtināʿ total resistance
laysa yamtaniʿ not impossible
laysa bi-mumtaniʿ not impossible

m.y.z.

mayyaza to discern, to single out, to distinguish
tamayyaza to be distinguished
mutamayyiz distinguished, distinct

m.y.l.

māla to shift, to incline (V 306b [293])
mayyala to cause to incline, to tilt
māʾil inclined, tilted
mayl, *tamayyul* inclination
mīl (Greek *mēlē*) a surgeon's probe

n

n.b.b.

unbūb tube

n.t.ʾ.

nutūʾ protrusions

n.ḥ.t.

naḥata carve

n.ḥ.s.

nuḥās copper

n.z.l.

bi-manzilat being (or functioning) as

n.s.b.

nisbah mafrūḍah assumed ratio
nisbah muʾallafah compound ratio (as in Euclid, Book V)
fī nisbat al-musāwāh *ex aequali* (as in Euclid, Book V)

n.sh.b.
intashaba to adhere

n.ṣ.f.
niṣf dāʾirah semicircle

n.ẓ.r.
al-nāẓir the viewer, the observer
naẓīr corresponding
naẓīr similar (V 299b [282])

n.f.dh.
nafadha to carry through, to extend, to emerge

n.q.r.
minqār laṭīf fine chisel

n.q.ṣ.
nāqiṣ less, reduced, deficient
nuqṣān reduction, deficiency
al-qaṭʿ al-nāqiṣ or *al-qiṭʿ al-nāqiṣ* ellipse

n.q.ṭ.
nuqṭah point
nuqṭat al-baṣar eye-point
al-nuqṭah al-mabṣūrah the visible point
nuqṭat al-taqāṭuʿ point of intersection
nuqṭat al-iltiqāʾ meeting point
nuqṭat al-inʿikās reflection point
nuqṭa naẓīrah corresponding point

n.q.l.
naqala to transfer

n.k.s.
mankūs upside-down

n.h.y.
nihāyah edge, end, limit
bi-lā nihāyah indefinite, without end (of a line segment)
kharaja ʿalā (or *bi-ghayr*) *nihāyah* to extend indefinitely

n.w.r.
nār flame
nār layyinah gentle flame

n.w.ʿ.

nawʿ al-ibṣār see *baṣar*

h

h.n.d.

muhandam (adj., of board) neat, fitting, secure in place

h.w.n.

bi-l-huwaynā carefully (IV 3a [2])

h.w.y.

hawāʾ air

h.y.ʾ.

hayʾah figure, shape, manner

w

w.t.r.

watara, yatiru; *wattara, yuwattiru* to subtend
watar chord

w.j.b.

awjaba to entail

w.j.d.

wajada to find

w.j.h.

jihah direction, side, way
fī jihat ... on the side of, towards, in the direction of
muwājih facing, opposed to

w.r.d.

warada min to come from

w.r.y.

warāʾ, *min warāʾ* behind

w.z.y.

muwāzī parallel to
mutawāzī parallel to one another
mutawāzī al-aḍlāʿ parallelogram

w.ṣ.f.

waṣafa to describe
ṣifah condition, description (V 205a [292]), state, manner

w.ṣ.l.

waṣala to join (two points)
ittaṣala bi- to adjoin
muttaṣil continuous
muttaṣil bi- continuous with
muttaṣil ʿalā istiqāmah extended rectilinearly
ittiṣāl continuity
ittiṣāl compactness (of parts)

w.ḍ.ʿ.

waḍʿ position
mutashābih al-waḍʿ similarly situated (V 332b [345])
mawḍiʿ location
mawḍiʿ place
mawḍiʿ maḥdūd determinate location
mawḍiʿ al-tamāss point of tangency

w.f.q.

kayfa-mā ittafaqa at random

w.q.t.

waqt time
fī-l-waqt al-wāḥid at one and the same time, simultaneously

w.q.ʿ.

waqaʿa to fall
waqaʿa to occur (V 298b [281])
wāqiʿ min (of perpendicular) drawn from; see *ʿamūd*

w.l.d.

wallada to produce, to engender
tawlīd, tawallud production, engendering

w.l.y.

waliya to be (placed) next
(al-zāwiyah) al-tāliyah adjacent (angle)
yalī it is (placed) next

w.h.m.

tawahhama to imagine
mutawahham imagined

CONCORDANCE
between the Chapters and Paragraphs in the Arabic and Latin Editions of the *Optics* by Ibn al-Haytham

Jan P. Hogendijk

Tanq subdivisions in the revision of the *Optics* (*Tanqīḥ al-Manāẓir*)

Arabic paragraphs in Sabra's Arabic edition and English translation

Latin paragraphs in Smith's Latin edition and English translation

BOOK IV

Tanq	*Arabic*	*Latin*
Ch. 1	1–2	1.1–1.2
Ch. 2		
IV 2.1	1	2.1
	2	2.2
	3	2.3
	4–5	2.4
IV 2.2	6–8	2.5
	8–9	2.6
	10	2.7
	11–13	2.8
	14	2.9
IV 2.3	15	2.9–2.10
	16	2.11
IV 2.4	17	2.12
IV 2.5	18	2.13
	19	2.14–2.15
	20	2.16
IV 2.6	21	2.17
	22	2.18–2.19
	23	2.20
	24	2.21

BOOK IV

Tanq	*Arabic*	*Latin*
Ch. 3, introduction	1	3.1
IV 3.1	2	3.2–3.3
	3–4	3.4–3.8
	5–6	3.9–3.11
	7–8	3.12–3.13
	9–10	3.14–3.15
	11	3.16
	12	3.17
	13–14	3.18–3.19
	15	3.20
	16	3.21
	17	3.22
	18–19	3.23
	20	3.24
	21	3.25
	22–23	3.26
	24–25	3.27
	26	–
	27	3.28
	28–29	3.29–3.30
	30–31	3.31
	32	3.32
	33	3.33
	34	3.34
	35	3.35
	36	3.36
	37	3.37
	38	3.38
	39	3.39–3.40

BOOK IV

Tanq	*Arabic*	*Latin*
	40	3.41
	41–46	3.42–3.45
	47–48	–
	49–50	3.46–3.47
	51–52	3.48
	53	3.49
	54	3.50
	55–56	3.51
	57	3.52
	58	3.53–3.56
	59–60	3.57
	–	3.58
	61	3.59
	62–65	3.60–3.61
	66–72	3.62–3.64
IV 3.2	73	3.64
	74	3.65–3.67
	75–77	3.68–3.74
	78–80	3.75–3.78
	81–83	3.79
	84	3.80
	85	3.81–3.82
	86	3.83
	87	3.84
	88–89	3.85
	90–91	3.86
IV 3.3	92–96	3.87
	97	–
	98	3.88

BOOK IV

Tanq	*Arabic*	*Latin*
	99	3.89–3.90
IV 3.4	100–1	3.91–3.92
IV 3.5	102	3.93
IV 3.6	103	3.94
IV 3.7	104–5	3.95–3.96
	106	3.97
	107	3.98
	108	–
IV 3.8	109–12	–
	113–19	3.99–3.100
	120	3.101
	121–27	3.102
	128–30	3.103
	131–33	3.104–3.106
IV 3.9	134–41	3.107
	142–44	3.108
	145	–
Ch. 4, introduction	1	4.1
IV 4.1	2–3	4.2–4.3
	4–7	4.4
	7–9	4.5
	10–11	4.6
Ch. 5, introduction	1–2	5.1
	3–4	5.2
IV 5.1	5	5.2
	6–9	5.3
IV 5.2	10	5.4
IV 5.3	11	5.5
IV 5.4	12–13	5.6

BOOK IV

Tanq	*Arabic*	*Latin*
	14	5.7
	15–17	5.8
IV 5.5	18	5.9
	19	5.10
IV 5.6	20	5.11
	21–22	5.12
	22–23	5.13
	24	5.14
IV 5.7	25	5.15–5.16
	26	5.17
	27–28	5.18
	29–31	5.19–5.20
	32–34	5.21–5.22
	35	5.23
	36–37	5.24
	38–39	5.25
	40–41	5.26
IV 5.8	42	5.27–5.29
	43	5.30
	44	5.31
	45	5.32–5.33
	46–47	5.34–5.35
	48–50	5.36–5.38
	51–53	5.39
	54	5.40
	55–56	5.41–5.42
	57	5.43–5.44
	58	5.45
IV 5.9	59	5.46

BOOK IV

Tanq	*Arabic*	*Latin*
	60	5.47
	61	5.48–5.49
	62	5.50
IV 5.10	63	5.51
	64	5.52–5.53
	65	5.54–5.56
IV 5.11	66	5.57
	67–68	5.58
	69	5.59
	70–72	5.60
	73–76	5.61
	77–81	5.62

BOOK V

Tanq	*Arabic*	*Latin*
Ch. 1		
V 1	1	1.1
Ch. 2	1	2.1
V 2:1.1	2–8	2.2–2.3
	9–11	2.4
	12–16	2.5
V 2:1.2	17	2.6
	18–20	2.7
	21–24	2.8–2.9
V 2:1.3–4	25	2.10
	26	2.11
	27–28	2.12
	29	2.13
	30	2.14

BOOK V

Tanq	*Arabic*	*Latin*
	31–33	2.15
	34	2.16
	35–36	2.17
	37	2.18
	38	2.19
	39	2.20
V 2:I.5	40	2.21
	41–43	2.22–2.23
	44	2.24
	44–46	2.25
	46	2.26
V 2:I.6–7	47	2.27
	48	2.28
	49	2.29
	50	2.30
	51–53	2.31
	54–55	2.32
V 2:II.1	56–57	2.33–2.34
V 2:II.2	58–62	2.35–2.36
	63–64	2.37–2.38
	65–66	2.39
	67	2.40
	68	–
	69	2.41
V 2:III.1	70	2.42
	71	2.43
	72	2.44
	73	2.45
	74	2.46

BOOK V

Tanq	*Arabic*	*Latin*
V 2:III.2	75	2.47–2.48
V 2:III.3	76	2.49–2.50
V 2:III.4	77	2.51–2.53
	78–80	2.54–2.57
	81	2.58
V 2:III.5	82	2.59–2.61
	83	2.62
	84–85	–
	86	2.63
	87	2.64–2.66
	88	2.67
	89	–
V 2:IV.1–4	90–94	2.68–2.76
V 2:IV.5	95	2.77–2.80
V 2:IV.6	96	2.81–2.84
	97	2.85–2.87
	98	2.88–2.91
	99–101	2.92–2.94
V 2:IV.7–8	102–3	2.95
V 2:IV.9–10	104–5	2.96–2.102
V 2:IV.11	106	2.103–2.106
V 2:IV.12	107	2.107–2.110
	108	2.111–2.114
	109–10	2.115–2.119
	111–16	2.120–2.124
V 2:IV.13	117–18	2.125–2.129
	119	2.130
	120	2.131
V 2:IV.14	121	2.132

BOOK V

Tanq	*Arabic*	*Latin*
V 2:IV.15	122	–
V 2:IV.16	123	2.133–2.136
V 2:IV.17	124	2.137–2.139
	125	2.140
	126	2.141–2.145
	127–29	2.146–2.148
	130	2.149–2.152
	131	2.153–2.155
	132	2.156–2.157
	133–37	2.158–2.166
	138	2.167–2.173
	139–44	2.174–2.185
	145–47	2.186–2.192
	148–49	2.193–2.198
	150–52	2.199–2.208
	153	2.209–2.214
	154	2.215
V 2:IV.18	155	2.216–2.218
	156	2.219
	157–60	2.220–2.221
	161–62	–
V 2:V	163	2.222–2.225
V 2:V.1	164	2.226–2.229
V 2:V.2	165	–
V 2:V.3	166	–
V 2:V.4	167	2.230
V 2:V.5	168	2.231–2.232
	169	2.233–2.235
	170	2.236

BOOK V

Tanq	*Arabic*	*Latin*
	171	2.237
	172–73	–
V 2:V.6	174	2.238
	175	2.239
	176	2.240
	177–81	2.241–2.247
	182	–
V 2:V.7	183	2.248–2.249
V 2:VI	184	2.250–2.251
	185–86	2.252
V 2:VI.1	187–88	2.253–2.260
	189	2.261–2.262
	190	2.263–2.268
V 2:VI.2	191	2.269–2.278
	192	2.279–2.283
	193–94	2.284–2.288
	195	2.289–2.292
	196	2.293
	197–98	2.294–2.297
V 2:VI.3	199	2.298
	200	2.299
V 2:VII	201	2.300
V 2:VII.1	202	2.301–2.305
	203	2.306–2.310
	204	2.311
	205	–
V 2:VII.2–3	206–7	2.312
V 2:VII.4–7	208–12	2.313

BOOK V

Tanq	*Arabic*	*Latin*
V 2:VII.8	213–14	2.314
	215–17	2.315
V 2:VII.9	218–20	2.316–2.323
V 2:VII.10	221	2.324–2.325
V 2:VII.11	222–23	2.326–2.330
	224–27	2.331–2.333
	228–29	–
V 2:VII.12	230	2.334–2.337
	231	2.338
V 2:VII.13	232	2.339–2.341
	233	2.342
	234–36	2.343–2.347
V 2:VII.14	237	2.348
	238	–
	239	2.349–2.350
	240	–
V 2:VII.15	241–44	2.351–2.358
	245	–
V 2:VII.16	246	2.359–2.363
	247	2.364–2.365
	2.248	2.366
V 2:VII.17	249	2.367–2.369
	250–51	2.370–2.371
	252	2.372–2.373
V 2:VII.18	253–57	2.374–2.383
V 2:VII.19	258	2.384–2.387
	259	2.388
V 2:VII.20	260	2.389–2.390

BOOK V

Tanq	*Arabic*	*Latin*
V 2:VII.21	261	2.391–2.395
V 2:VII.22	262	2.396
V 2:VII.23	263–64	–
V 2:VII.24	265–67	2.397–2.405
V 2:VII.25	268	2.406–2.409
V 2:VII.26	269	2.410–2.411
	270–71	2.412–2.418
	272–75	2.419–2.424
V 2:VII.27	275	–
V 2:VII.28	276	2.425–2.438
V 2:VII.29	277–80	2.439–2.454
	281	2.457–2.461
V 2:VII.30	282	2.462–2.464
V 2:VII.31	283	2.465–2.468
V 2:VII.32	284	2.469
	285	2.470
	286	2.471–2.475
	287	2.476–2.477
	288	2.478
	289	2.479–2.485
	290	2.486
	291	–
	292–93	2.487–2.488
	294	2.489
V 2:VII.33	295	2.490
V 2:VIII.1	296–98	2.491–2.493
	299	2.494
	300	–

BOOK V

Tanq	*Arabic*	*Latin*
V 2:VIII.2	301	2.495–2.496
	302	2.497
V 2:VIII.3	303	–
V 2:VIII.4	304	2.498
V 2:VIII.5–6	305–6	2.499
V 2:VIII.7	307–8	2.500–2.502
	309	2.503
V 2:VIII.8	310–11	2.504–2.511
	312	2.512–2.515
	313–19	2.516–2.518
V 2:VIII.9–10	320–21	2.519
V 2:IX.1	322–23	2.520
V 2:IX.2	324	2.521
V 2:IX.3–4	325–26	2.522–2.523
V 2:IX.5	327	2.524–2.526
	328	2.527
V 2:IX.6	330–31	2.528–2.534
	332–39	2.536–2.545
V 2:IX.7–8	340–45	2.546–2.547
	346–47	–

SOURCES
for the Geometrical Diagrams in Book V

Jan P. Hogendijk

Book V of Ibn al-Haytham's *Optics* included nearly 50 diagrams, but some are lost, as the Arabic manuscript tradition is defective. MSS F (Istanbul, Fatih, no. 3215) and S (Istanbul, Ayasofya, no. 2448) have empty spaces without diagrams. MS K (Istanbul, Köprülü, no. 952) has diagrams, but is unfortunately incomplete, and what remains is disorganized, as can be seen in the concordance between MSS K and F in Sabra's Arabic edition (Kuwait, 2002, I, pp. 17–20). Thus, 23 diagrams from the original have not come down to us in Arabic manuscripts of Book V of the *Optics*. Sabra reconstructed these missing diagrams (V.19–V.39 and V.44–V.45) on the basis of the *Revision of the* Optics *(Tanqīḥ al-Manāẓir)* by Kamāl al-Dīn al-Fārisī (c.1300 AD), the same work that he used to restore problematic passages in the text. Sabra was well aware that Kamāl al-Dīn paraphrased Ibn al-Haytham's text, with the result that his diagrams may not agree with Ibn al-Haytham's originals in all details. He added extra diagrams for further clarification of the geometric proofs; these are often simplifications of very complicated diagrams by Ibn al-Haytham or Kamāl al-Dīn.

The following table has been provided for readers who want to study Sabra's diagrams in connection with the sources. The columns contain the label of the diagram in Sabra's edition and the corresponding paragraphs in Sabra's translation of Chapter 2 of Book V. This is followed by the folio (if any) where the corresponding diagram is found in MS K of Book V of the *Optics*, and the folio in the autograph MS A of the *Tanqīḥ al-Manāẓir*, in which all the diagrams are found. MS K is available online at archive.org (Fazilahmedps952). MS A, which belonged to the Adilnor Collection in Stockholm, is now in Qatar and hopefully will be online soon. I have added references to corresponding figures in two important publications.[9] Hyd. is volume I of the Hyderabad edition of the *Tanqīḥ al-Manāẓir* by Kamāl al-Dīn, where the figures are found on separate pages between pp. 436 and 573. S is Smith's edition and translation of the medieval Latin version of Books IV and V of the *Optics*; the figures from the Latin manuscript tradition are in volume II, pp. 563–613, and the reconstructed figures are in volume I, pp. 220–88. The figures are numbered the same way in both volumes: thus, Fig. 5.2.31 in the following table refers to the group of figures numbered 5.2.31 and 5.2.31a–e in II, pp. 587–89, and I, pp. 246–49. Sometimes the reader will find in the table the same group of figures (S5.2.31) corresponding to two of Sabra's diagrams (V.22 and V.23). It should be noted that the table is approximate, in the sense that the exact correspondences between Arabic and Latin traditions of the *Optics* of Ibn al-Haytham are subject to clarification by future research.

9. For full details, see the List of Works Cited, p. 267.

Sabra (edition)	Sabra (translation)	MS K	MS A	Hyd.	S
V.1	75–76	K106b	A121b	H15	S5.2.1
V.2	77–81	K74a	A122a	H16	S5.2.2
V.3	82–83	K75a	A122a	H17	S5.2.3
V.4	84–89	K75b	A123a	H18–20	S5.2.4
V.5	90–95	K76b	A123b	H21	S5.2.5
V.6	96–103	K77b	A124b	H23	S5.2.9
V.7	104–5	K78b	A125b	H24	S5.2.10
V.8	106	K79a	A126a	H25	S5.2.11
V.9	107–16	K80a	A127a	H26	S5.2.12–13
V.10	117–23	K81b	A128b	H27	S5.2.16
V.11	126	K89b	A129a	H28	S5.2.19
V.12	127–32	K107b	A130b	H29–35	S5.2.19
V.13	133–37	K90a	A131a	H36	S5.2.20
V.14	138	K90b	A133a	H37–40	S5.2.21
V.15	139–44	K91a	A134a	H41–42	S5.2.22
V.16	145–47	K91b	A134b	H43–44	S5.2.23
V.17	148–49	K92a	A135a	H45	S5.2.24
V.18	150–54	K93a	A136b	H46–47	S5.2.25
V.19	164–66	K–	A138b	H48	S5.2.26
V.20	168–72	K–	A139b	H49	S5.2.27
V.20+	174–83	K–	A141b	H50	S5.2.28
V.21	187–90	K–	A143a	H51	S5.2.30
V.22	191–95	K–	A144b	H52	S5.2.31
V.23	197–98	K–	A145a	H53	S5.2.31

Sabra (edition)	Sabra (translation)	MS K	MS A	Hyd.	S
V.24	202–5	K–	A146a	H54	S5.2.32
V.25	218–19	K–	A148a	H55–59	S5.2.33
V.26	221–23	K–	A148b	H60	S5.2.34
V.27	224–25	K–	A149a	H61	S5.2.34C
V.28	230–31	K–	A150a	H62	S5.2.35
V.29	232–35	K–	A152a	H66–67	S5.2.36
V.30	241–44	K–	A153a	H68	S5.2.37
V.31	246–47	K–	A153b	H69	S5.2.37
V.32	249–51	K–	A154b	H70	S5.2.37
V.33	253–57	K–	A155b	H71–72	S5.2.38
V.34	258–59	K–	A156a	H73	S5.2.39
V.35	260	K–	A156a	H74	S5.2.40
V.36	261–67	K–	A157a	H75	S5.2.41–42
V.37	268	K–	A157a	H76	S5.2.43
V.38	268–75	K–	A157b	H77	S5.2.44
V.39	276	K–	A159a	H78	S5.2.45
V.40	277–81	K30a	A160b	H79–81	S5.2.46
V.41	283–84	K30b	A161b	H82	S5.2.48
V.42	296–305	K97a	A165b	H85	S5.2.50
V.43	307–9	K98a	A166a	H86	S5.2.51
V.44	310–20	K–	A168a	H87	S5.2.52
V.45	327–29	K–	A169a	H88	S5.2.53
V.46	330–33	K134a	A169b	H89	S5.2.54
V.47	334–43	K134b	A170b	H90	S5.2.54

ERRATA
to the Arabic Edition of Books IV and V

Jan P. Hogendijk

There follows a list of errata[1] to the Arabic edition *Kitāb al-Manāẓir (the* Optics*) of al-Ḥasan ibn al-Haytham. Books IV–V: On Reflection and Images Seen by Reflection*, prepared by Abdelhamid I. Sabra and published in 2002 in two volumes.[2] The large majority of these errata were found by Sabra himself when preparing his English translation of Books IV and V, and they have been verified by checking the translation against the Arabic edition, the apparatus, and the Arabic manuscripts. Some of the errata are obvious printer's errors. A few further errata relate to the second part of Book V, where the mathematical sense sometimes requires minimal changes in one or more letters denoting points on a geometrical figure, in cases where confusion between the letters is likely (for example ﺟ and ح, or ر and ز, or ك and ﺪ, and so on). Below, I have listed these minimal changes without detailed references to manuscripts or apparatus.

A notation here such as 170:8 V 150a [30] refers to a correction to page 170, line 8, of volume I of the Arabic edition; the folio number in the part of MS F containing Book V is 150a (150 recto), and the correction occurs within paragraph [30] of the relevant chapter (Chapter 2 of Book V). Similarly, 8:2 IV 5b [7] is a correction to page 8, line 2, of volume I of the Arabic edition; the folio number in the part of MS F containing Book IV is 5b (5 verso), and the correction occurs within paragraph [7] in the relevant chapter (Chapter 2 of Book IV). Using the folio number, it is easy to consult the Arabic critical apparatus in volume II, where (for example) fol. V 150b is keyed as فه ١٥٠ ظ and fol. IV 5a as فد ٥ و. Hyd. denotes references to volume I of the Hyderabad edition of the *Revision of the* Optics.[3]

Book IV, Chapter 2

7:2 IV 4b [5]	Change to	عند ميل عند حضور الميل
7:2 IV 4b [5], at end of line	Change to	وضعها ووضعها
7:2–4 IV 4b [5]	Delete[4]	‹كنسبة› ‹من› ‹ظهر على الجسم المقرب›
8:2 IV 5b [7]	After insert[5]	الانعكاس «قرّت المعتبر إلى المرأة جسمًا أبيض ثالث من غير جهة الانعكاش»

1. With warmest thanks to Uwe Vagelpohl for his assistance with the typography of this section.
2. For full details, see the List of Works Cited, p. 267.
3. For full details, see ibid.
4. These expressions were added by Sabra.
5. See Hyd., p. 333. Compare the experiment in paragraph [23], and also the Latin [2.5] 'third white body'.

Book V, Chapter 2

170:8 V 150a [30]	Change to[6]	فم فهو عمود على الخط المماسّ للدائرة التي هي الفصل المشترك بين سطح المرأة وبين سطح أعلى الحلقة إذا توهم
170:11 V 150a [30]	Change to	هذ هذا
177:11 V 157a [45]	Change to[7]	مركز البصر مركز الكرة
186:13 V 165a [64], between في and تنطبق	Insert[8]	سطح المرأة الكريّة المحدّبة على هذه الصفة صورة شبيهة بالصورة التي حصلت في ذلك الجزء لكانت تلك الصورة
216:15 V 189a [104]	Change to	خط ز د خط ز ه
255:7 V 215a [151]	Delete[9]	ونجعل زاوي ل د ح مثل زاويه ا جـ ب
256:23 V 216b [152]	Change to[10]	زاوية ب د ل زاوية ب د ه
257:11 V 216b [153]	Change to[11]	ونصل ا د ب ب جـ ا جـ ونصل ا د ب د جـ د
258:15 V 217b [153]	Change to	مثلث د ش ز مثلث د ش ر
258:16 V 217b [153]	Change to	إلى ك ز إلى ك ر
281:2 V 237a [179]	Change to	ونقطتا ز جـ ونقطتا ز ح

6. A section (found in MS F, 150a:7–8, and MS K, 105b:12–13) was omitted by a printer's error.
7. The changed text is the reading of MS K; cf. the apparatus in II, p. 71, to V 157a, no. 1.
8. This is found in MS F, 165a:7–9, and MS K, 104b:16–17.
9. This was a printer's error; the beginning of the line should read ح جـ ا د ح ك. Cf. MS F, 215a:7, and MS K, 92b:10.
10. Cf. the apparatus in II, p. 103, to V 216b, no. 1.
11. This is closer to the readings of MSS K and F; cf. the apparatus in II, p. 103, to V 216b, no. 4.

Book V, Chapter 2 (continued)

284:12 V 239b [183]	Change to	ونصل جز ا ز ونصل ح ز ا ز
291:1 V 244b [190]	Change to	ونصل خطوط خ و ر و ث و ص س ت ر ت ط ونصل خطوط خ و ر و ث و ض ش ت ي ت ط ت ظ
291:1 V 244b [190]	Delete	《أقول: وكذا ي ط》
292:21–22 V 246a [191]	Change to	وليكن خط فليكن خط
296:13 V 249a [194]	Change to	خطي ح ه ه ي خطي ح ه ه ق
303:15 V 254b [203]	After insert	ونقطة ق تمتدّ ⟨صورتها⟩
311:11 V 260b [220]	Delete[12]	[[في الصورة الخامسة]]
321:22 V 269a [234]	Delete[13]	وتكون نسبة ه ج إلى ج ح كنسبة ه ك إلى ك ج التي هي نسبة ه ك إلى ك د
322:3 V 269a [234]	Change to	كنسبة ه ج فنسبة ه د إلى د ح كنسبة ه ج
322:3 V 269a [234], at the end of the line	Change to	كنسبة ه ج فنسبة ه د إلى د ح كنسبة ه ج
322:16 V 269a [234]	Change to	نقطة د نقطة ك
322:18 V 269b [234]	Change to	نقطتي ز ج نقطتي ب ج
322:21 V 269b [235]	Change to	خط ه ح خط ه ج

12. This is not found in the manuscripts.

13. This passage occurs twice in MS F (V 269a:5–6 and V 269b:8–9), apparently by scribal error, and also twice in the Arabic edition (321:22 and 322:1–2). The second occurrence is correct in the mathematical argument, in agreement with Hyd., p. 521:11–13.

Book V, Chapter 2 (continued)

323:1 V 269b – V 270a [235], at the folio break, after ح	Insert[14]	إلى تلك النقطة أصغر من خط ح ج وإن كانت النقطة
323:2 V 270a [235]	Change to	نقطة ج نقطة ه
323:6 V 270a [235]	Change to	خطي ج د د ح خطي ه د د ح
323:8 V 270a [235]	Change to[15]	يقسم ا ج الزاوية التي يحيط بها ذانك الخطان بنصفين فليس تنعكس صورة نقطة ح إلى بصر ه من سطح مرأة ا ب ج
325:14–15 V 272a [239]	Change to	زاوية ا ح ج زاوية ا د ج
325:15 V 272b [239]	Change to	ج ح وتر ج ب وتر
325:18 V 272b [239]	Change to	دائرة ج ح ب دائرة ج د ب
325:18 V 272b [239]	Change to	نقطة ج نقطة ح
328:11 V 274b [244]	Change to	قوسي ا ك ز ح قوسي ا ك ⟪ب ج⟫
329:12 V 275a–b [246], before the folio break	Insert[16]	من قوس ا ب فلتتعكس صورتها
329:13 V 275b [246]	Delete	[[اوّلًا]]

14. A brief passage is missing from MS F and the edition, so this is a conjectural reconstruction of the text. The second manuscript, K, has a lacuna, so cannot be consulted. Kamāl al-Dīn (Hyd., p. 522) deleted the section corresponding to 322:15–323:9.

15. A passage in MS F (V 270a:7–8) was omitted from the edition by a printer's error.

16. This text is at the end of MS F (V 275a:13), but was omitted by a typing error.

Book V, Chapter 2 (continued)

333:7 V 278a [251], between 334:8 and 334:9	Insert[17]	⟪حاصل الأشكال الثلثة⟫
333:9 V 278b [252]	Change to[18]	⟨داخل⟩ ⟪داخل⟫
333:11 V 278a–b [252]	After insert[19]	بها الخط ⟨الواصل بين البصر والمرأة والخط⟩
333:12 V 278b [252]	Change to[20]	⟨مركز⟩ ⟪مركز⟫
335:21 V 280b [254]	Change to	د ط د ا
336:7 V 280b [254]	Change to	فخط ل ش فخط ع ش
336:8 V 280b [254]	Change to	من ش ل من ش ع
336:9 V 280b [254]	Change to	ق ف ف ل ق ص ص ل
337:5 V 281a [256]	Change to	فنسبة ق ص فنسبة ف ص
337:7 V 281a [256]	Change to	ونسبة فنسبة
337:12 V 281b [256]	Change to	كنسبة و ي كنسبة ت ي
339:15 V 283a [258]	Change to	نقطتي ف م نقطتي ن م

17. Cf. Hyd., p. 532, where the detailed discussion in 333:8–334:7 is omitted.
18. Cf. Kamāl al-Dīn (Hyd., p. 531:21).
19. The addition is for mathematical sense: text is missing at the page break.
20. Cf. Hyd., p. 532:2.

Book V, Chapter 2 (continued)

340:3 V 283b	Change	نقطة ح
[259]	to	نقطة ⟨ز⟩
340:12 V 283b	Change	إلى نقطة
[259]	to	⟪من ز⟫ إلى نقطة
342:13 V 285b	After	نقطة ب
[261]	insert[21]	⟪وهو إمّا مساوٍ لـ ب ا أو أعظم أو أصغر فإن كان أعظم⟫
342:15 V 285b	Change	و النقطة
[261]	to[22]	⟪وإن كان مساويا⟫ فالنقطة
349:1 V 290a	Change	خط ق ج
[271]	to	خط ف ج
350:12 V 291b	Change	نقطتي ط ن
[274]	to	نقطتي ط ب
354:4 V 294a	Change	ن ط إلى ط ق
[276]	to	ن ي إلى ن ق
356:8 V 296a	Change	زاوية د ق س
[277]	to	زاوية د ف س
356:14 V 296a	Change	نصف ط ت ع
[277]	to	نصف ت ط ع
363:8 V 301a	After	كل واحد من الخطين
[285]	insert[23]	⟨أو خطّ واحدٌ من الخطين⟩
373:3 V 309b	Change	أكبر
[298]	to[24]	أكثر
374:1 V 310b	Change	بصر ط
[299]	to[25]	بصر ن

21. Cf. Hyd., p. 537:10.
22. Cf. Hyd., p. 537:11.
23. According to Hyd., p. 553:12. The words are missing from MS F (V 301a:11), and K has a lacuna here.
24. According to Hyd., p. 560:11, and in agreement with MS K (96b:29) and MS F (V 309b:7).
25. Cf. MS F, V 310b:2, and MS K, 97a:8. This was a printer's error.

Book V, Chapter 2 (continued)

374:1 V 310b [299], between فيما and التي هي	Insert[26]	مركز البصر. فإن كان مركز البصر عند نقطة ط فإن صورة نقطة م تتعكس إلى بصر ط من نقطة ج
378:11 V 314a [309]	Change[27] to[28]	على كل فصل مشترك يقطع محيط تلك الدائرة 《على جميع الخطوط التي في سطح الأسطوانة》
394:10 V 326b [332]	Change to	خط ز م خط ك م

26. This is from MS K (97a:9); it is missing from both MS F (V 310b:2) and the Arabic edition.
27. In MS F (V 314a:1–2) and MS K (98a:6–7), but incorrect.
28. Cf. Hyd., p. 563:12.

www.ingramcontent.com/pod-product-compliance
Ingram Content Group UK Ltd.
Pitfield, Milton Keynes, MK11 3LW, UK
UKHW060959120926
8556IPUK00033B/906

9 781908 590589